AF552475

PLANT NUTRITION FOR FOOD SECURITY

A Guide for Integrated Nutrition Management

PLANT NUTRITION FOR FOOD SECURITY

A Guide for Integrated Nutrition Management

By

R.N. Roy
Land and Water Development Division
FAO, Rome, Italy

A. Finck
University of Kiel
Kiel, Germany

G.J. Blair
University of New England
Armidale, Australia

H.L.S. Tandon
Fertiliser Development and Consultation Organisation
New Delhi
India

DISCOVERY PUBLISHING HOUSE
NEW DELHI-110002

"Published by arrangement with the Food and Agriculture Organization of the United Nations by Discovery Publishing House FOR SALE AND DISTRIBUTION IN INDIA ONLY

"This Book was Originally Published by the Food and Agriculture Organization of the United Nations (FAO).

First Indian Edition–2007

ISBN 978-81-8356-191-4

Published by

DISCOVERY PUBLISHING HOUSE
4831/24, Ansari Road, Prahlad Street, Darya Ganj, New Delhi-110002 (India)
Phone: 23279245 • Fax: 91-11-23253475
E-mail: dphbooks@rediffmail.com, dphtemp@indiatimes.com

Printed at:
Sachin Printers, Delhi

Contents

List of figures

List of tables

List of plates

Preface

An expanding world population and the urgency of eradicating hunger and malnutrition call for determined policies and effective actions to ensure sustainable growth in agricultural productivity and production. Assured access to nutritionally adequate and safe food is essential for individual welfare and for national, social and economic development. Unless extraordinary efforts are made, an unacceptably large portion of the world's population, particularly in developing countries, could still be chronically undernourished in the coming years, with additional suffering caused by acute periodic shortages of food.

For biomass synthesis, which serves as the food resource for humans and animals, nutrient supply to plants is a prerequisite. Therefore, an adequate and appropriate supply of plant nutrients, is a vital component of a crop production system. Agricultural intensification, one of the basic strategies for enhanced food production, is dependent on increased flows of plant nutrients to the crops for securing high yields. Unless supported by adequate nutrient augmentation, the process of agricultural intensification would lead to land degradation and threaten the sustainability of agriculture.

In the past two decades, it has been increasingly recognized that plant nutrient needs in many countries can best be provided through an integrated use of diverse plant nutrient resources. An integrated plant nutrition system (IPNS) or integrated nutrient management (INM) enables the adaptation of the plant nutrition and soil fertility management in farming systems to site characteristics, taking advantage of the combined and harmonious use of organic, mineral and biofertilizer nutrient resources to serve the concurrent needs of food production and economic, environmental and social viability.

FAO has been engaged actively in the development of INM in the last two decades. Through its field projects, expert consultations and publications, the FAO has focused global attention on the need for large-scale adoption of INM. Propagation of the INM concept and methodology application at the farm level requires that the scientific community, extension workers, decision-makers, and other stakeholders concerned with agricultural development have a clear understanding of the subject.

This guide on integrated plant nutrient management, dealing with various aspects of plant nutrition, is an attempt to provide support to the ongoing efforts directed at enhanced and sustainable agricultural production. It seeks to bridge the scientific knowledge gap, and it presents updated information on plant nutrition with emphasis on INM. In helping stakeholders to improve their ability to identify and resolve constraints relating to plant nutrition – be they of a technical, economic, social or policy nature – and to demonstrate on the field practical ways of increasing production through efficient plant nutrition, the guide should assist in achieving the goal of food security.

Acknowledgements

The contribution of R.N. Roy to the conceptualization, initiation, technical guidance, inputs, reviewing and editing of this publication is duly acknowledged. Special thanks are due to A. Finck, G.J. Blair and H.L.S. Tandon for their contributions in progressively shaping this document. The contribution of R.V. Misra in reviewing and providing constructive suggestions is most gratefully acknowledged.

Thanks also go to J. Plummer for editing this publication and to L. Chalk for its preparation.

List of abbreviations and acronyms

^{40}K	Potassium-40
ADP	Adinosine diphosphate
AEC	Anion exchange capacity
AFS	Apparent free space
AISA	Adequate-input sustainable agriculture
Al	Aluminium
AN	Ammonium nitrate
ANP	Ammonium nitrate phosphate
APS	Ammonium phosphate sulphate
AS	Ammonium sulphate
As	Arsenic
ATP	Adinosine triphosphate
B	Boron
BCR	Benefit–cost ratio
BGA	Blue green algae
BNF	Biological nitrogen fixation
C	Carbon
Ca	Calcium
CAN	Calcium ammonium nitrate
Cd	Cadmium
CDU	Crotonylidene urea
CEC	Cation exchange capacity
CFS	Committee on World Food Security
Cl	Chlorine
CL	Critical level, critical limit
Co	Cobalt
CO_2	Carbon dioxide
CPM	Carbonation press mud
Cr	Chromium
CRH	Critical relative humidity
Cu	Copper
DAP	Di-ammonium phosphate
DRIS	Diagnosis and recommendation integrated system
DTPA	Diethylenetriamine pentaacetic acid
EAAI	Essential amino acid index
EDDHA	Ethylenediamine (o-hydroxyphenyl) acetic acid
EDTA	Ethylenediamine tetraacetic acid
ESP	Exchangeable sodium percentage

ET	Evapotranspiration
EU	European Union
F	Fluorine
Fe	Iron
FYM	Farmyard manure
GA	Gibberellic acid
GPS	Geographical Positioning System
H	Hydrogen
H_2S	Hydrogen sulphide
H_3BO_3	Boric acid
HCN	Hydrocyanic acid
Hg	Mercury
HYV	High-yielding variety
I	Iodine
IAA	Indole acetic acid
IBDU	Isobutylidene diurea
IFA	International Fertilizer Industry Association
IFOAM	International Federation of Organic Agriculture Movements
IKS	Indigenous potassium supply
INM	Integrated nutrient management
INS	Indigenous nitrogen supply
IPNS	Integrated plant nutrition system
IPS	Indigenous phosphorus supply
IR	Irrigation water requirement
Iw	Irrigation water
K	Potassium
KCl	Potassium chloride
LCC	Leaf colour chart
LIFDC	Low-income food-deficit country
LISA	Low-input sustainable agriculture
MAP	Mono-ammonium phosphate
Mg	Magnesium
Mn	Manganese
Mo	Molybdenum
MOP	Muriate of potash (potassium chloride)
MPP	Mono-potassium phosphate
N	Nitrogen
N_2	Dinitrogen
N_2O	Nitrogen dioxide
Na	Sodium
NENA	Near East and North Africa
NH_3	Ammonia
NH_4^+	Ammonium ion
Ni	Nickel

NO	Nitrous oxide
NUE	Nutrient-use efficiency
O_3	Ozone
P	Phosphorus
PAPR	Partially acidulated phosphate rock
Pb	Lead
PGPR	Plant-growth-promoting rhizobacteria
pH	Potential hydrogen (negative log of H^+ concentration)
PMD	Profit-maximizing dose
PR	Phosphate rock
PSB	Phosphate-solubilizing bacteria
PSM	Phosphate-solubilizing micro-organism
PSR	Pore space ratio
RNA	Ribonucleic acid
Rw	Rainfall
S	Sulphur
Se	Selenium
Si	Silicon
SO_2	Sulphur dioxide
SOM	Soil organic matter
SOP	Sulphate of potash, or potassium sulphate
SPFS	Special Programme for Food Security
SPM	Sulphitation press mud
Sr	Strontium
SSA	Sub-Saharan Africa
SSNM	Site-specific nutrient management
SSP	Single superphosphate
Sw	Water stored in soil profile
Th	Thorium
TSP	Triple superphosphate
U	Uranium
UAP	Urea ammonium phosphate
UN	United Nations
UNESCO	United Nations Educational and Scientific Cooperation Organization
USDA	United States Department of Agriculture
USG	Urea supergranule
V	Vanadium
VAM	Vesicular-arbuscular mycorrhizae
VCR	Value–cost ratio
WHC	Waterholding capacity
WHO	World Health Organization
WR	Water requirement
WUE	Water-use efficiency

Y	Economic crop yield
YMD	Yield-maximizing dose
Zn	Zinc

Chapter 1
Introduction

Of the essential material needs of humankind, the basic requirement is for an adequate supply of air, water and food. People have free access to the air they breathe. However, access to drinking-water and food, while easily obtained for some, is difficult for many. In addition to being physically available, these materials should also be of acceptable quality and continuously so.

Hunger and diseases have affected humankind since the dawn of history. Throughout time, there have been periods of famine leading to suffering and starvation, making the fight against hunger and the diseases caused by malnutrition a permanent challenge. For many centuries until about 1800, the average grain yield was about 800 kg/ha, providing food only for a few people. The main problems were the low fertility of most soils (mainly caused by the depletion of nutrients) and the great yield losses from crop diseases and pests.

Efforts to achieve freedom from hunger became successful only after the discovery of the nutritional needs of crops in the mid-nineteenth century. In order to supplement plant nutrients of low fertility soils or poor soils, the value of manures was stressed and mineral fertilizers were developed. Mineral fertilization started about 1880, became a common practice in the 1920s and adopted on larger scale since 1950. In Europe, cereal yields have increased at an annual rate of 1.5–2.5 percent for many decades, from an average of 2 tonnes/ha in 1900 to 7.5 tonnes/ha in 2000. The impact of fertilizers on wheat yields is best demonstrated by results from the Broadbalk Experiment, which was started in 1844 at the Rothamsted Experimental Station, the United Kingdom, and is still continuing (Figure 1).

FIGURE 1

The effect of fertilizer on wheat grain yields in the Broadbalk Experiment, Rothamsted Experimental Station, the United Kingdom

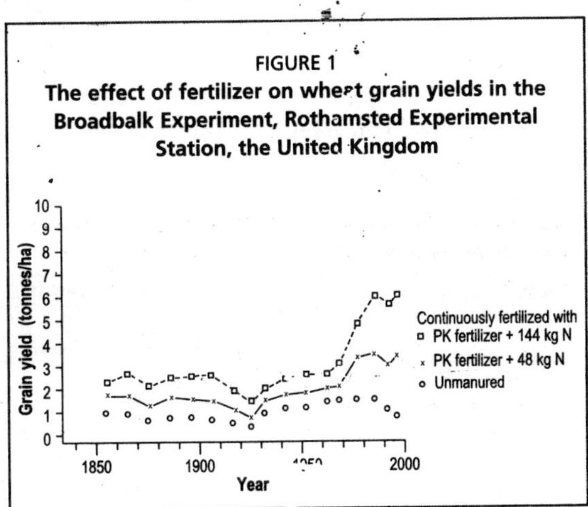

Source: Johnston, 1997.

Even with restrictions on the present land area utilized for farming, a modern ecotechnological-oriented system of agriculture has the potential for large production increases. In comparison, a strictly environment-oriented agriculture without mineral fertilizers and other manufactured inputs, would be 2–3 times less productive, and incapable of sustaining even the present world population on

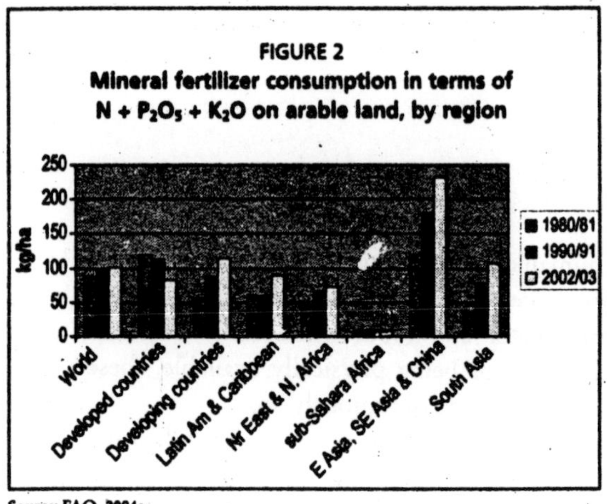

FIGURE 2
Mineral fertilizer consumption in terms of $N + P_2O_5 + K_2O$ on arable land, by region

Source: FAO, 2004a.

the already utilized land (IFPRI, 1995a).

PRESENT AND FUTURE DEMANDS FOR PLANT NUTRIENTS IN DEVELOPING REGIONS

As most of the additional food required must come from already cultivated land, intensification of agriculture with high (optimal but not excessive) and balanced use of nutrient inputs will be required. Even with a high degree of nutrient recycling through organics, mineral fertilizers will continue to be of central importance for meeting future food demands. Figure 2 shows the present level of mineral fertilizer application in developing regions. About 50 percent of all mineral fertilizer nutrients are used for the production of cereals (wheat, rice and maize), and 50 percent of all mineral fertilizer nutrients are consumed by China, the United States of America, and India.

Worldwide mineral fertilizer nutrient use is expected to increase from 142 million tonnes in 2002/03 to 165 million tonnes in 2009/2010, to 175 million tonnes in 2015 and to 199 million tonnes in 2030 (FAO, 2000a, 2005). The projections of mineral fertilizer demand differ considerably among the regions (Table 1). The largest share of mineral fertilizers will be used by East Asia, followed by South Asia. These two regions together will account for about half of world mineral fertilizer use by 2030. The growth rate in mineral fertilizer use is predicted to be highest in sub-Saharan Africa (SSA) and the Near East and North Africa (NENA).

Although the obstacles to higher food production seem almost insurmountable in problem areas, available land and inputs need not be limiting factors. However, production increases on low fertility soils will require special expertise, large investment in nutrients and major initiatives on a sustained basis.

TABLE 1
Mineral fertilizer use and projected nutrient demand to 2030 in developing regions

Region	$N + P_2O_5 + K_2O$	
	2002/03	2009/2010
	(million tonnes)	
Sub-Saharan Africa (including South Africa)	2.3	
Near East and North Africa	7.9	
East Asia	50.6	59.5
South Asia	20.9	25.7
Latin America	13.2	18.3
World	141.6	165.0

Source: FAO, 2000a, 2005.

Steps that promote optimal and efficient plant nutrition are required on a large scale in order to achieve food security. The aim should be to develop and adopt production systems that are productive, sustainable and least burdensome on the environment. Organic sources and recycling do not on their own suffice to meet increased demands for food on a fixed land area. On the other hand, because of possible environmental concerns and economic constraints, crop nutrient requirements often cannot be met solely through mineral fertilizers. Hence, a judicious combination of mineral fertilizers with organic and biological sources of nutrients is being promoted. Such integrated applications are not only complementary but also synergistic as organic inputs have beneficial effects beyond their nutrient content.

Therefore, the nutrient needs of such production systems can best be met through integrated nutrient management (INM). The concept of INM aims to increase the efficiency of use of all nutrient sources, be they soil resources, mineral fertilizers, organic manures, recyclable wastes or biofertilizers. Extension staff who are to translate research data into practical recommendations will need to take stock of both farmers' expertise and the applicability of research results. Available knowledge will need to be summarized competently and evaluated economically in order to provide practical guidelines for the adoption of INM by farmers having a range of investment capacities for achieving food security on a sustained basis. At the same time, plant nutrition research must continue to develop new techniques while refining existing ones based on feedback from the field.

Chapter 2

Food security and agricultural production

STRIVING FOR FOOD SECURITY

Past and present efforts

Nobody would have forecast 100 years ago that world agriculture could produce sufficient food, feed and other agricultural commodities for almost four times as many people as existed in 1900 (1 600 million in 1900 compared with 6 000 million in 2000). This apparently unattainable goal has been achieved through a combination of many factors, the combined impact of which triggered the so-called green revolution. Here, a combination of irrigation, fertilization and high-yielding varieties (HYVs) of crops resulted in the greatest progress ever made in food production. While it is difficult to envisage a repetition on this scale, further progress is certainly possible and urgently required.

National food self-sufficiency has been achieved in many countries through the combined efforts of farmers, industry, farm advisers and scientists. In the countries of Western Europe, in the United States of America and in other developed areas, there is a surplus of food production, and food is cheap. In the past, average workers with a family of four persons spent 50 percent of their income on food. This has now dropped to 15 percent, enabling them to purchase a wide range of other goods and services, the result being a higher standard of living.

Nonetheless, in large regions, consisting mainly of developing countries, hunger and malnutrition still exist. However, current food shortages are only partly caused by production problems. Disturbances to food production resulting from poor economic conditions, widespread poverty, civil war, inappropriate food pricing policies and logistical constraints contribute significantly to the problem. According to Borlaug (1993): "The dilemma is feeding a fertile population from infertile soils in a fragile world."

Recent international efforts towards food security

In 1974, the World Food Conference proclaimed that every person has the inalienable right to be free from hunger and malnutrition. As this goal was not achieved after more than two decades (there being more than 800 million people, mainly in developing countries, without sufficient food), a new attempt was made at the World Food Summit in Rome in 1996 to renew the commitment at the highest political level to eliminate hunger and malnutrition, and to achieve sustainable food security for all people. According to the summit:

- Food security exists where all people, at all times, have physical and economical access to sufficient, safe and nutritious food to meet their dietary needs and food preferences for an active and healthy life.
- World food security is the concern of members of the international community because of its increasing interdependence with respect to political stability and peace, poverty eradication, prevention of, and reaction to, crisis and disasters, environmental degradation, trade, global threats to the sustainability of food security, growing world population, transborder population movements, and technology, research, investment, and financial cooperation.

The summit adopted the "*Rome Declaration on World Food Security*" and seven commitments as a "*Plan of Action*". The preliminary aim was to halve the number of undernourished people by no later than 2015. In addition, world food production should increase by more than 75 percent in the next 30 years to feed about 8 000 million people by 2025. To meet the target of halving malnutrition in developing countries by 2015, this number needs to be cut by at least 20 million/year, more than twice as fast as the current reduction of about 8 million/year. With a growing world population, this situation will worsen unless very determined and well-targeted actions are taken to improve food security.

It was against the above-mentioned background that the Special Programme for Food Security (SPFS), launched by FAO in 1994, was further strengthened, expanded and its implementation accelerated after the 1996 World Food Summit. The main objective of the SPFS is to help developing countries, in particular the low-income food-deficit countries (LIFDCs), to improve food security at household and national level through rapid increases in food production and productivity. It aims to achieve this by reducing year-to-year variability in food production on an economically and environmentally sustainable basis and by improving people's access to food. The programme is currently operational in about 75 countries. The FAO Committee on World Food Security (CFS) was made responsible for monitoring, evaluating and consulting on the international food security situation.

The underlying assumption is that viable and sustainable means of increasing food availability exist in most of the 83 LIFDCs but that they are not being realized because of a range of constraints that prevent farmers from responding to needs and opportunities. By working with farmers and other stakeholders to identify and resolve such constraints – be they of a technical, economic, social, institutional or policy nature – and to demonstrate in the field practical ways of increasing production, the SPFS should open the way for improved productivity and broader food access.

To achieve the target, the focus of action is at the country level. This means that food security is largely a national task. This is not easy for poor countries, and international organizations should give both advice and financial assistance.

For many well-fed people, food security refers less to food shortage and more to secure food (i.e. nutritious and safe food, free of toxic substances). According to the present-day demands of urban consumers, food should be abundant, diverse, tasty, nutritious, safe and cheap. Chapter 10 examines some of these aspects in detail.

Food production vs environment preservation

The discussion of potential food supply somewhat overshadows another aspect, namely the tolerance or capacity of the earth to support an ever-increasing number of people, including domestic animals. The production and consumption of essential goods such as food and industrial goods through intensive production systems is connected inevitably with some negative side-effects on the environment. Chapter 11 explores environmental issues in relation to plant nutrition.

Long before the maximum food production capacity of the world's agriculture is reached, retarding effects caused by environmental damage will become increasingly apparent. Global warming is one of its indicators. The damaging effects are caused partly by agriculture. Today, a common view is that agriculture places a heavy burden on the environment. However, this is so because people demand abundant and cheap food. The vital question is not only how many people this planet can feed and clothe but how many people it can support at an environmentally sustainable level.

FOOD SECURITY FOR A GROWING WORLD POPULATION

World and regional population until 2020

The world population doubled within 40 years after 1960. Despite some efforts to slow the growth rate, the global population will be about 7 500 million in 2020 according to a forecast by the United Nations (UN) using a medium-fertility model. In the more distant future, there may be 9 000 million people by 2050, and the number may stabilize at slightly more than 10 000 million after 2100 (IFPRI, 1997, 1999).

The population increase during the next two decades will occur almost entirely in 93 developing countries. With a growth rate of 1.5 percent/year, there will be 1 500 million more people by 2020, half of them urban and mostly young. This increase is comparable with the entire population of the developed countries (Figure 3).

FIGURE 3
Population of developing regions and cereal yields

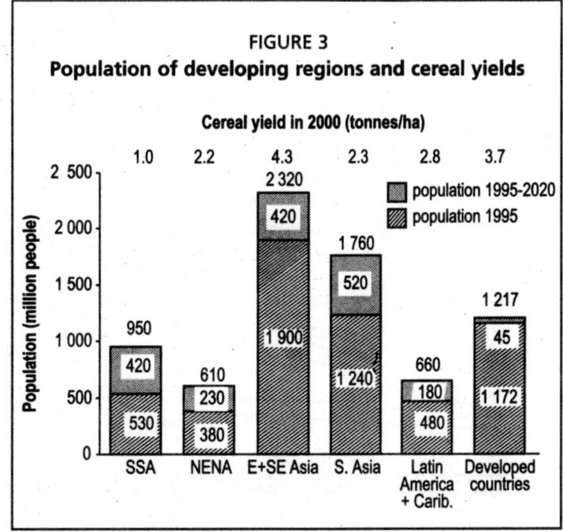

Note: SSA = sub-Saharan Africa, NENA = Near East + North Africa, E + SE Asia = East + Southeast Asia.
Source: IFPRI, 1999; Finck, 2001.

The highest population growth (80 percent) will be in SSA, a region that already has the most critical food supply situation. For NENA, the growth is predicted to be 40 percent, but the food supply is slightly better. East and Southeast Asia plus China have the smallest predicted population increase (20 percent), but the greatest existing population. The absolute increase is greatest in South Asia, with about 500 million people (40 percent). Latin America and the Caribbean are predicted to have strong population growth, but good food prospects as well.

In most regions, present food grain yields range from 2.2 to 2.8 tonnes/ha, but crop yields are only 1 tonne/ha in SSA. These yields are insufficient to feed the growing population. The task for the near future is to feed 700 million more people, and about 1 500 million more people in 2020. Thus, it is clear that:

- The increase in the global population will be entirely in the developing countries.
- An additional 1 500 million people will have to be fed by 2020, mainly in areas with present food shortages.
- SSA is the most critical region for future food supply.
- The bulk of the population (4 000 million) will have to be fed in Asia (East, Southeast and South Asia).
- Additional food must come mainly from higher production on existing agricultural land.

The necessity to feed so many more people in regions with "critical" food supply is an enormous challenge for food production and requires great efforts. One such effort will be to provide adequate crop nutrition so that the required amount of food and other crop products can be produced on a sustained basis.

Food production capacity of the world

An estimation of the biophysical limits of food production reveals that a much greater number of people than the expected equilibrium population (of about 10 000 million) could be supplied with sufficient food. According to FAO (2000a): "For the world as a whole there is enough or more than enough food production potential to meet the growth of effective demand." Intensive agriculture, while observing ecological requirements, can feed an ever-growing world population. Figure 4 highlights the impact that soil fertility, mineral fertilizers and animal manure have had on cereal production.

FIGURE 4
Estimated global trends in population, cereal yields and source of plant nutrients

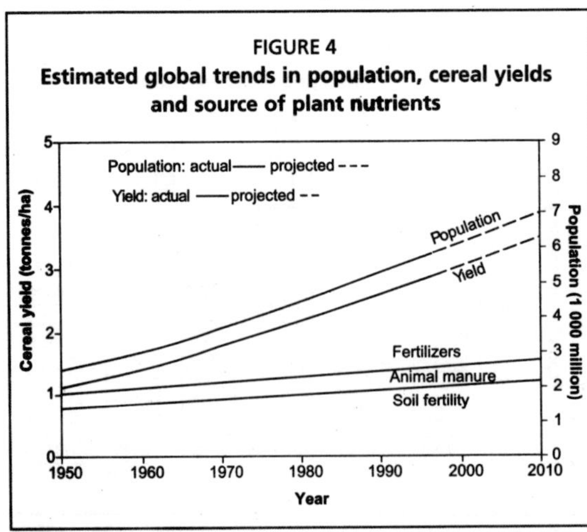

Note: The distance between the lines indicates the contribution from different sources.
Source: Kaarstad, 1997.

While enormous gains have been made in increasing cereal yields worldwide, there are very large differences between the progress made in the different regions, particularly when compared with population growth in those regions. Figure 5 presents data for six key areas plotted by Evans (2003) using data from FAO production yearbooks. The population–yield relation is most favourable in North America and Europe while it is least favourable in Africa.

FIGURE 5
Relationship between human population and average cereal yields in six regions

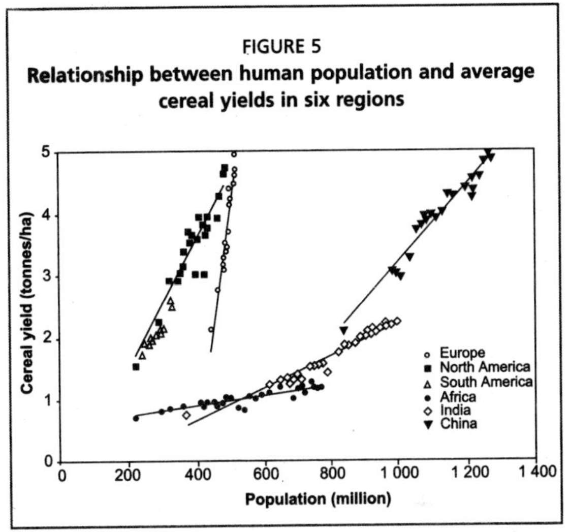

Source: Evans, 2003.

Food demand in developing countries

Compared with global food prospects, the challenge for the developing countries is much greater. Based on a projection from 1995 data, the global annual demand for cereals will increase by about 40 percent until 2020. Out of the globally required additional 700 million tonnes, developing countries will need about 600 million tonnes more cereals in 2020 (40 percent for China and India). About 80 percent of this additional food supply will have to come from already cultivated areas, as newly cropped land is likely to supply less than 20 percent of the increase.

The large increase in cereal demand will not only result from population growth but also from an increasing demand for meat, which will almost double to 30 kg/capita/year by 2020. As a consequence, the cereal demand for livestock feed will double, and the area of maize grown for animal feed is likely to exceed that of rice and wheat grown for human consumption. The cereal demand for 6 300 million people including both food and feed has been estimated at about 1 700 million tonnes, which amounts to 270 kg/capita/year or 0.75 kg supplying 2 800 kcal/day (IFPRI, 1999). Table 2 shows the regional food supply situation in 2000.

Developing countries have an average food and feed supply of about 250 kg/capita, which is considered satisfactory. In order to maintain this level in 2020, the average yield of 2.8 tonnes/ha in 2000 will need to increase to 3.5 tonnes/ha, but correspondingly less if the present cropping area is expanded. The above goal seems to be within reach, especially for Latin America and the Caribbean.

The situation in both NENA and South Asia is less satisfactory. With actual yields of about 2.2 tonnes/ha, both regions require a substantial yield increase in order to meet future demands (about 70 percent for NENA, and about 50 percent for South Asia). East and Southeast Asia consist of two rather different blocks. China has high yields and rather good food supply prospects, whereas the other countries of this region are in a position similar to South Asia. SSA is in the least

TABLE 2
Cereal production, supply and demand in developing regions

Data	Unit	Developing countries	SSA	NENA[1]	E + SE Asia	South Asia	Latin America + Caribbean
Population (2000)	million	4 800	590	400	1 860	1 320	510
Area, harvested	million ha	442	75	40	147	131	47
Production, total	million tonnes	1 227	75	88	633	305	133
Yield, average	tonnes/ha	2.8	1.0	2.2	4.3	2.3	2.8
Supply, total/capita	kg/year	256	127	220	340	230	260
Supply for human consumption/capita	kg/year	170	114	213	201	158	129
Demand 2020[2]:							
Additional; same level	million tonnes	384	45	46	156	101	39
Additional; yields required	tonnes/ha	3.7	1.6	3.4	5.4	3.1	3.6
Total; average demand[3]	million tonnes	1 575	238	153	x	440	165
Total; yields required[4]	tonnes/ha	3.5	3.1	3.8	x	3.4	3.5

Notes:
[1] Data for NENA estimated from FAO (1993a).
[2] Additional demand 2020 on basis of supply level of 2000.
[3] Total demand based on average supply of developing countries 2000 (250 kg/capita/year).
[4] Yields required: on cereal area in 2000.
x No average data because of great differences on the two blocks.

favourable position with a yield level of only 1 tonne/ha, which needs to rise by 50 percent just to maintain the supply level of 2000 in 2020. However, compared with Asia, there are greater prospects in Africa for using more fallow land.

Food quantity and quality, and malnutrition

About 800 million people in developing countries (20 percent of the population) are undernourished. The percentage of malnourished children is estimated to be 35 percent in SSA and 70 percent in South Asia. The term malnutrition refers mainly to suboptimal food energy intake, the required daily supply being 2 600–3 000 kcal (2 500 kcal/day corresponds to 0.7 kg of cereals per day or 250 kg/year). However, malnutrition in a complete sense also includes shortages of protein (essential amino acids), vitamins and essential mineral nutrients (e.g. phosphate and micronutrients).

Even with a satisfactory average supply, the problem of food shortage and malnutrition will persist in 2020, albeit at a reduced scale in most regions. However, in SSA, 15 percent of the people will probably still be undernourished in 2030 (FAO, 2000a). Sufficient food energy is only the first goal, and sufficient nutritious food the final one. In developing countries, protein deficiency (less than 50 g/day for an adult weighing 60 kg or shortages of some essential amino acids such as lysine) and a deficiency in vitamin A and iron (Fe) are common, particularly among women and children. A lack of Fe is associated with anaemia.

In order to prevent diseases resulting from nutritional deficiencies, the production of high-quality food is essential. Equally important is the knowledge of maintaining food quality through the selection and the preservation of its quality components during food processing and preparation. The neglect of food quality is widespread and by no means restricted to hungry people. Apparently well-fed people may also suffer from avoidable diseases induced by a deficiency in essential nutrients. Sufficient healthy food not only alleviates hunger but also prevents many diseases resulting from malnutrition. Chapter 10 examines the importance of adequate food of high quality and the role of plant nutrition in producing it.

FOOD PRODUCTION PROSPECTS IN DEVELOPING COUNTRIES

The food production prospects of developing regions with 6 700 million people in 2030 are of global concern. The challenge is to feed almost 2 000 million more people on the available land base. Data from a detailed study (FAO, 2000a) indicate that, for developing countries as a whole, food production will increase in the next 15 years by 2.1 percent/year, food demand by 2.2 percent/year and population growth by 1.4 percent/year. However, there are great regional differences. For example, in SSA, production may grow by 2.6 percent/year, demand by 2.8 percent/year, and population by 2.4 percent/year. The future food production in different regions will depend largely on land resources, inputs and the efforts to use them.

Land resources

Important indicators of available land resources are: total suitable land area for cropping, land suitability for different production systems, land actually put into production, cropping intensity, potential for expansion in area, and amount of irrigated land. Table 3 summarizes some basic data on these indicators for various developing regions.

The comparison of total suitable land with actual arable land shows that there is large potential for increasing cropped area only in SSA and Latin America. NENA and South Asia have very little potential for area expansion. The estimated expansion in arable land by 2030 will be highest in SSA (25 percent) and lowest in South Asia (4 percent). Total harvested land is expected to show the highest increase (about 40 percent) in SSA and the lowest (14 percent) in South Asia in the next three decades. In terms of the proportion of harvested land that is

TABLE 3

Crop production base in developing regions

Developing region	Land suitable for cropping	Arable land used, 1997–2030	Harvested land 1997–2030	Very good + good land as % of suitable land	% of harvested land irrigated (1997)
		(million ha)		(%)	
SSA	1 031	231–288	146–205	75	3
NENA	99	87–94	71–86	26	37
East Asia	366	232–278	301–327	72	37
South Asia	220	207–216	230–262	88	43
Latin America	1 066	202–243	128–173	80	12
Total	2 780	960–1 079	877–1 053	76	29

Source: FAO, 2000a.

FIGURE 6
World population, arable area and cereal yields from 1800 to 1999

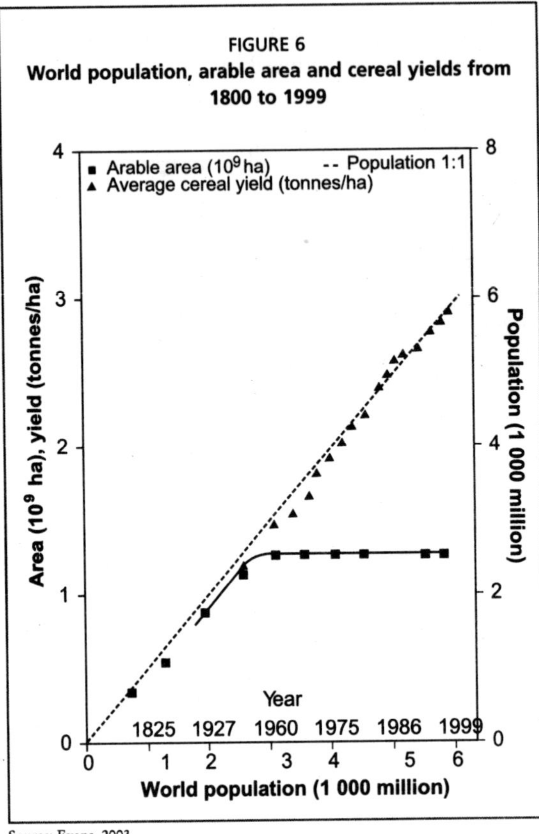

Source: Evans, 2003.

irrigated, there are large differences between the regions, the figure ranging from 3 percent in SSA to 12 percent in Latin America and about 40 percent in NENA and East and South Asia.

The suitability of land for cropping can be estimated from the percentage of very good and good land compared with total suitable land, the remainder being moderately suitable or unsuitable. The proportion of good land is very high in South Asia and Latin America (more than 80 percent), somewhat less in SSA and East Asia (about 73 percent), but only 26 percent in NENA.

Combining the prospects for land expansion and cropping intensification indicates that there is still considerable potential for higher food production in all regions. Figure 6 provides an indication of the considerable progress made in intensification. Whereas the arable area has remained constant since 1960, the average cereal yield per hectare has continued to increase linearly. In the future, SSA will face the greatest problems in this respect. In South Asia, India is a good example of the progress that has been made through intensive cropping.

There are already serious problems in large arid areas as a result of a shortage of irrigation water. This is caused by overutilization by agriculture and conflicts of interest between irrigation, drinking-water and industrial supplies. However, the problem of a shortage of freshwater may be reduced if an economical and environmentally acceptable method of desalinization can be developed.

Plant nutrients

In time, the shortage of the essential plant nutrient phosphate may also seriously limit crop production. The major plant nutrients are nitrogen (N), phosphorus (P) and potassium (K). Of these, N is abundant in the air, and deposits of K are ample, but the phosphate reserves will become scarce. This may lead to conflicts for a share of phosphate fertilizers long before the phosphate rock (PR) deposits are exhausted. Only strict rules for recycling and efficient use could postpone this first serious shortage of an essential plant nutrient.

FIGURE 7
Yield gap at various levels in relation to production factors

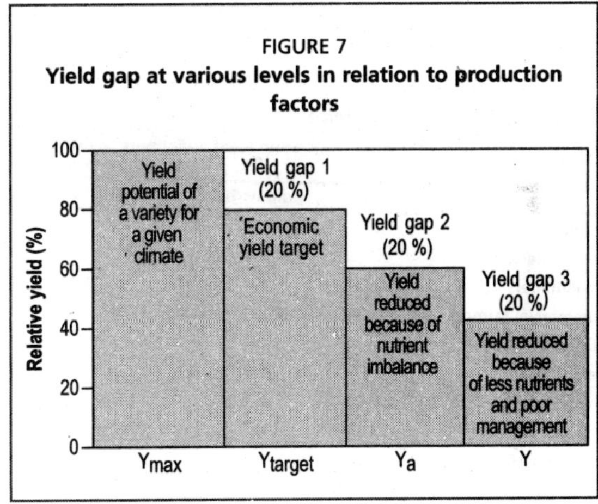

Source: Fairhurst and Witt, 2002.

Yield levels

Worldwide, and also in many developing countries, the annual growth in cereal yields is still increasing although the global rate of increase has dropped from slightly over 2 percent to about 1.5 percent. For developing countries as a whole, the trend is from an increase of about 1.5 percent/year in the last few decades to less than 1 percent/year in the future (IFPRI, 1999). However, several countries in problem areas such as SSA (Ethiopia, Nigeria, etc.) show a stagnation in cereal yields (at a low level) or even a declining trend, e.g. Zambia and Zimbabwe (FAO, 1999).

In many developing countries, there is still a very large gap between the economically achievable yield and average yield obtained. Many rice farmers in Asia achieve less than 60 percent of the potential yields (Figure 7). In Figure 7, Y_{max} is the maximum yield potential, Y_{target} is the highest yield that can be obtained through optimal and efficient use of inputs, Y_a is the yield with optimal water and crop management but with the farmer's current nutrient management practices, and Y is the actual yield in the farmer's field. Such a yield gap analysis gives rise to the following considerations (Fairhurst and Witt, 2002):

- Yield gap 1: It is usually uneconomic to attempt to close this yield gap because of the large amounts of inputs required and the high risk of crop failure caused by pests, infection and lodging.
- Yield gap 2: To close this yield gap, it is essential to manage N, based on seasonal plant needs, and follow long-term strategies for other nutrients including P and K.
- Yield gap 3: The greatest benefit from improving nutrient management is found on farms with good crop management and few pest problems. Farmers need to know what factors can be changed to increase productivity (knowledge-based management) and should know that larger yield increases result where several constraints (e.g. pest and disease problems and inappropriate nutrient management) are overcome simultaneously.

Many factors contribute to stagnating or declining yields in spite of farmers' efforts to achieve higher output. Production under adverse conditions faces many natural obstacles, e.g. insufficient and unreliable rainfall, poor or eroded soils, low soil fertility, shortage of irrigation water, crop-damaging and soil-eroding typhoons in humid regions or dust storms in arid regions, and rapidly spreading pests and

TABLE 4
Yields of sorghum and maize on smallholder and commercial farms in Zimbabwe

Crop and farm type	Area	Grain yield	Farmers' yields as % of record yield
	(ha)	(tonnes/ha)	(%)
Sorghum			
Record yield		3.6	
Smallholders	160 000	0.44	12
Commercial farmers	9 000	2.3	64
Maize			
Record yield		5.0	
Smallholder	1 000 000	0.9	18
Commercial farmers	200 000	3.7	74

Source: FAO, 1999.

plant diseases. In addition, there are often economic issues such as high prices for inputs like fertilizers, low produce prices, and poor infrastructure. A combination of some of these factors diminishes the possibility of and incentive for higher yields and production beyond subsistence level.

There are great differences in cereal yield even on similar soils in similar climates. This indicates the significant gaps between usually obtained yields and those obtainable. One example of the impact of expertise and management on yield levels can be seen from the data for Zimbabwe from 1980 to 1996 (Table 4). The yields differed considerably whether obtained in smallholder areas or on commercial farms. There are wide gaps between the average yield and record yield, especially under climate conditions of frequent drought. Smallholders obtained less than 20 percent of the sorghum or maize yields obtained in record years. The better performance of commercial farmers is the result of their greater expertise and better access to inputs. However, even for this group, the long-term average yield is only about 70 percent of that in a record year.

PROBLEMS AND POSSIBILITIES

Two different cases of the problems and possibilities are cited here, one pertaining to SSA and the other to India.

Example of sub-Saharan Africa

With a population of about 500 million, SSA will pose the greatest challenge to food production because of its high population growth rate. This is occurring on top of a decline in available food per capita in recent decades (FAO, 2000a, 2001a, 2001b).

Shortage of productive land

Including dry areas, an estimated 0.4 ha/person was available in 1995. Production increases will have to come mainly from the already cultivated land. In areas receiving satisfactory rainfall, where most people live, the cultivated area was only 0.25 ha/person or less. There is a possibility of a substantial area becoming available (2 ha/person) for cultivation from fallowed land or land under shifting cultivation. However, this will require a massive recapitalization of plant nutrients.

Soil degradation

Soil degradation, particularly that of soil fertility, is a major cause of stagnating or even decreasing yields in some countries. Apart from widespread soil erosion, the major causes are: loss of organic matter resulting in reduced biological activity;

nutrient depletion as a result of erosion, mining or inactivation of nutrient (e.g. sorption of phosphate); and reduced nutrient retention. High levels of soil acidity and aluminium (Al) toxicity are a problem in 30 percent of the area.

The estimated average nutrient depletion in 2000 was about 50 kg of nutrients ($N + P_2O_5 + K_2O$) per year. Without at least a medium level of plant nutrient input, many countries will not be able to meet their food needs, and some may not do so even with high inputs.

Low crop yields

Cereal yields are low at 1 tonne/ha. This is partly the result of soil degradation, a harsh climate, low levels of external nutrient application, and frequent droughts, and partly the result of a lack of economic incentives. Average fertilizer use is only 10 kg/ha of total nutrients (ranging from 0 to 50 kg/ha). Although some areas have shown a distinct yield increase in the last decade, sorghum yields have been stagnant in Burundi, Ethiopia, Ghana, Kenya and Nigeria, and maize yields have been stagnant in Zambia and Zimbabwe. Cassava yields have fallen sharply in Angola and Malawi (FAO, 1999, 2000a).

Regional differences

In the mainly dry semi-arid area, with 250–700 mm rainfall, water supply is the critical factor, as in the Sahel region. Maximum use must be made of the limited rainfall by all kinds of water harvesting techniques. Soils are mainly sandy and of low fertility. The input of minimum nutrients and irrigation of suitable land is often limited by water shortages. Maize and sorghum grain yields range from 0.2 to 1.5 tonnes/ha but much higher yields could be achieved if more water were available.

About 80 percent of the population live in the humid and subhumid agro-ecological zones (700–1 500 mm of rain). In these areas, the main soil problems (besides erosion) are low organic matter and poor biological activity, structural deterioration and nutrient deficiencies. Improvements in plant nutrient supply may start with locally available PR applied to legumes such as *Sesbania*, and adoption of INM. Grain yields of 1–2 tonnes/ha of maize are far lower than they should be. Under these favourable rainfall conditions, grains yields of 3–4 tonnes/ha are possible.

Recent attempts to improve soil fertility have been successful (FAO, 2000a), e.g.:

- Uganda: Soil improvement by farmers association with mulches, manure and fertilizers.
- The United Republic of Tanzania: Water and soil conservation by agroforestry.
- Zambia: Sustainable cropping by replacing grass fallow with legumes plus fertilizer application.
- Burkina Faso: Production increase by use of indigenous PR and more legumes.

To summarize the situation in SSA, there are reasonable prospects of food production, but as indicated in FAO (2000a): "It is necessary to recognize and build upon many indigenous farming systems and soil and management practices that have maintained and sustained agriculture for generations."

Example of India

India is the largest country in South Asia and contains 70 percent of the total regional population. In spite of a rapidly growing population (nine times the growth in area under grains since 1950), it has made significant progress in food production and achieved cereal self-sufficiency with even a sizeable surplus. India, with a population of 1 000 million people, produced 220 kg of cereals per person from an area of 100 million ha in 2000. Such a level of progress has been achieved through intensification and the use of modern production inputs.

Since the green revolution in the 1960s, enormous progress with modern HYVs, irrigation and fertilizer application has been made. Fertilizer consumption rose from almost zero in 1950 to 17 million tonnes of N + P_2O_5 + K_2O in 2000 (Tandon, 2004). This corresponds to an average nutrient application rate of 92 kg/ha, of which 65 percent is N, 25 percent P_2O_5 and 10 percent K_2O. A significant amount of sulphur (S) and zinc (Zn) is also applied.

Average cereal yields are now 2.2 tonnes/ha (2.6 tonnes/ha for rice) and the cropping intensity is 130 percent. Because of the scarcity of land, cereal yields of 3.8 tonnes/ha will be required in order to feed the future population and 500 million domestic animais. Careful use of all kinds of organic nutrient sources would be very desirable given the very large nutrient requirements of Indian agriculture and the persisting gap of 8–10 million tonnes of N + P_2O_5 + K_2O between nutrient additions and removals. It is estimated that 25 percent of the total NPK need could be supplied by organic resources including rural, urban and industrial wastes.

However, the key component will be proper nutrient management with more mineral fertilizers and more balanced nutrient use. This would entail less N and relatively more P and K, which should be supplemented by yield-limiting macronutrients and micronutrients. Even with the present progress, there is still a wide yield gap to be narrowed within safe input limits. Only by more intensive farming backed with INM can 300 million more people be fed by 2020. The alternative of low-input extensive farming would threaten the food security of about 400 million people (FAO/IFPRI, 1998).

These two contrasting scenarios concerning food security can be summarized as follows:

- SSA is the region offering maximum challenges because of rapid population growth and very low cereal yields. The non-utilized yield potential and the substantial fallow land available offers prospects for progress.
- India is a good example of successful past yield increases (and probably future ones) through intensification of agriculture in spite of relatively small additional suitable land reserves.

DEMANDS ON AGRICULTURE FOR PROVIDING FOOD SECURITY

Need for productive and sustainable agriculture

In the foreseeable future, the majority of affordable food must be produced by soil-based agriculture. In order to maintain increased food production, modern agriculture must be very productive and yet sustainable.

There are many definitions of sustainability. The concept of sustainable agriculture set out by FAO (1989) is quite relevant to many countries. It states: "The goal of sustainable agriculture should be to maintain production at levels necessary to meet the increasing aspirations of an expanding world population without degrading the environment." Moreover, "Sustainable agriculture should involve the successful management of resources for agriculture to satisfy changing human need while maintaining or enhancing the quality of the environment and conserving natural resources. No single resource is more important in achieving a sustainable agriculture than the soil which contains essential nutrients, stores the water for plant growth and provides the medium in which plants grow." (FAO, 1989).

According to FAO (1995): "Sustainable agricultural development is the management of the natural resource base in such a manner as to ensure the attainment and continued satisfaction of material human needs for present and future generations. It conserves or increases land capacity to produce agricultural goods, water availability, plant genetic resources, is environmentally non-degrading, technically appropriate, economically viable and socially acceptable."

High-yielding crop production at a sustainable level is based on five factors, which must be integrated efficiently. These factors are:

- productive crops with high-yield potential that are managed properly from seed to harvest;
- fertile soils as the basis for high and sustainable production;
- adequate water supply by rainfall or irrigation;
- adequate nutrient supply for crops, and efficient use of applied nutrients;
- protection of crops against weeds, diseases and pests on the field and post-harvest care in storage.

In principle, sustainable cropping can be carried out at low, medium or high yield levels. The appropriate level is the one that meets the needs and aspirations of the population. Sustainability at a low yield level, termed low input sustainable agriculture (LISA), means a lot of work for small results – a system that many farmers may have no other choice but to use. According to Borlaug (1997): "Most farmers loath to adopt low-input, low-output cropping, because it tends to perpetuate human drudgery and the risk of hunger." The preferred goal is sustainable production at a high level of productivity using adequate inputs. Here, adequate may mean high or medium input depending on the production conditions and targets.

The minimum goal should be sustainable production at medium yield levels. It is likely that most farmers would prefer highly productive sustainable agriculture, a system that makes the use of all inputs and capital worthwhile and results in abundant products and an economic profit. Agriculture should not plunder the soil resource by "exhaustion cropping" for short-term profit, but rather maintain or even improve it for the benefit of future generations. Using banking terminology, agriculture is supposed to live off the interest, not off the inherited capital.

With the results of new research and the extension of new technologies to farmers' fields, considerable progress can be made. However, the optimal utilization of any improved factor requires its integration into the whole production system through a "holistic approach". Individual production factors should not only be improved and applied, but the whole combination of factors must be optimized. This is not a simple task. It requires considerable investment and much expertise.

The five factors listed above are equally important and indispensable for supporting modern agriculture. The yield potential or resistance of crops to diseases may be greatly increased in future, but better crop nutrition with a high nutrient efficiency will remain a central component for productive and sustainable agriculture, and thus for future food security.

Food production adjusted to consumer demands and environmental issues

In most societies, farmers produce food and other agricultural products for a market. Therefore, they must accept market rules and the corresponding economic system, which involves them in a web of special conditions and regulations. However, market demands may be partly contradictory to the demands of the society. Several less desirable developments in modern agriculture are not just the result of modern technology as such but of conflicting demands of urban consumers, mostly in the developed countries, who are politically dominant and increasingly determine the basic rules for farmers. Three examples of this are:

- Urban consumers want food to be cheap but many of them dislike the consequences of "mass production" of so-called "industrial" agriculture. For example, in order to produce cheap meat, farmers are forced to keep large numbers of pigs in sheds where the wastes are collected as slurry instead of straw containing farmyard manure (FYM). Slurry was practically unknown in Europe 50 years ago, but is now the dominant form of animal manure and probably the most important source of plant nutrient losses from agriculture to the environment. This represents a "consumer-driven" undesirable development in modern agriculture.
- Many urban consumers, largely for supposed health reasons, prefer so-called "natural" food, supposedly produced by low-input production, but also want much land left to natural vegetation in order to preserve biodiversity. So-called "organic" farming, being connected with nostalgic reminiscences, seems to guarantee healthy and uncontaminated food from crops growing without "chemicals" and from "happy" farm animals on "natural" green pastures. However, low-input production, without actually producing better food, is not only more costly, but requires more than twice as much land for cropping. Therefore, this demand comes into conflict with the demand of urban populations for large recreational areas with natural parks, etc.

- Urban consumers return their partly contaminated waste materials to agriculture, but tend to criticize farmers for selling "contaminated" food to urban markets. However, while the enormous amounts of waste materials need to be recycled as cheaply available nutrient sources, many of these products are contaminated by inorganic and/or organic toxic substances, which may damage soil fertility or food quality. This problem needs to be solved at the expense of urban populations who are causing this problem, otherwise farmers will be reluctant or even unwilling to use such urban wastes.

From these examples, it seems that urban consumers, most of them lacking a basic understanding of agricultural production, can put a great strain on agriculture with contradictory demands. Farmers have to react to conflicting requests, and in any case, should not be held responsible for the consequences of recycling contaminated urban waste products.

For the goal of food security, farmers should not be made responsible for the results of conflicting demands of urban consumers who set the principles and laws with little regard for the unique rules of basic agriculture production. Much work is needed in laying down the ground rules for the on-farm recycling of wastes. Steps should be taken to ensure that urban wastes processed for recycling meet appropriate quality standards so that their use on farmland does not harm the land, produce, waterbodies, people or the environment.

NUTRIENTS IN PRODUCTION AND CONSUMPTION CYCLES AND NUTRIENT TRANSFERS

All harvested crops remove plant nutrients from the soil. Whether used for food, feed or as industrial raw materials, the various crop products are often consumed far away from the production sites, some times thousands of kilometres away in another country. When crop products are moved, the nutrients contained in them are also transported. This implies a loss of nutrients for the production area and a gain for the area where these are finally utilized. Although soils gain as well as lose nutrients, agricultural production and food security is threatened whenever the nutrients removed or lost are not replenished adequately. In the end, it is the balance between the amounts gained and the amounts lost that determines whether the soil nutrient status is being depleted, maintained or improved, and this in turn determines the productivity level that a soil can sustain.

Whether at the farm level or across national boundaries, nutrient cycling takes place to varying extents. Quite often, where nutrients are circulating in small or large cycles, taken up and partly transformed by plants, microbes, animals or humans, they reappear in waste materials, which can again serve as nutrient sources (Figure 8).

Such cycles operate continuously in soils at various levels. Nature, which operates these cycles, does not discriminate between organic or mineral forms of nutrient and allows both forms to enter and leave the same cycle. However, intersite nutrient transfers bring about different types of changes in the nutrient balance than do normal nutrient cycles.

FIGURE 8
Plant nutrients in production and consumption cycles

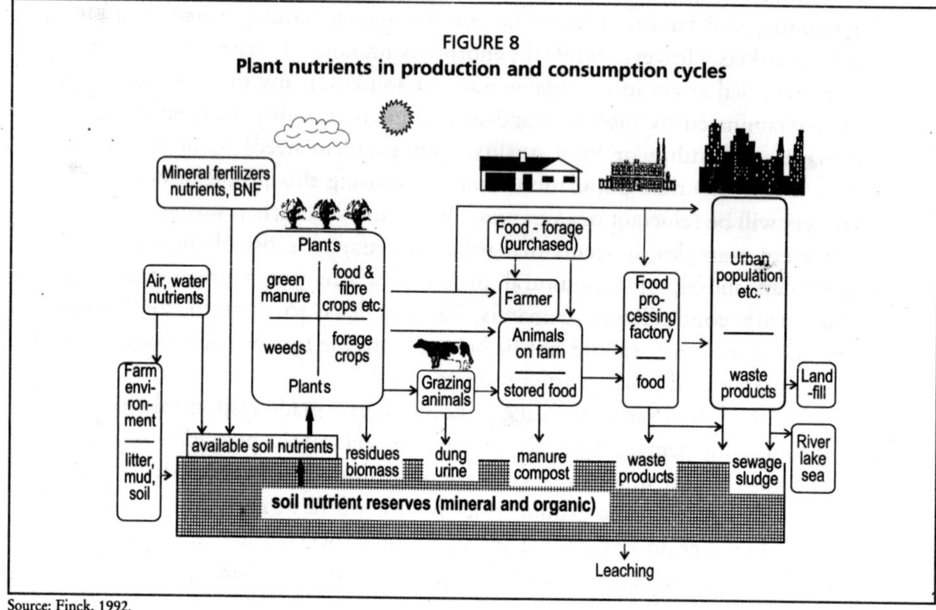

Source: Finck, 1992.

Natural nutrient transfers

A steady flow of nutrients occurs naturally with surface water or groundwater movement in hilly and mountainous areas as part of the natural erosion process, even under natural vegetation cover. The annual losses of nutrients in the soil solution from hilly parts of the landscape are often relatively small as are the gains for the low-lying land. However, the amount of nutrients transferred in solid form by soil erosion can be considerable. Over geological time, this transfer has produced impoverished hilly areas and many fertile alluvial soils in river basins that now represent the best agricultural lands in the world. Where this process is accelerated by human-induced soil erosion, it can lead to serious declines in soil fertility in hilly areas and to excessive losses of nutrients into water.

The problem of natural nutrient transfers can be considerable in the plains as well. This is caused primarily by the movement of nutrient-rich surface soils through wind and water erosion. In India, about 5 300 million tonnes of soil are estimated to be displaced annually through water erosion alone, resulting in a movement of 8 million tonnes of N + P_2O_5 + K_2O (Prasad and Biswas, 2000). These cannot be considered real losses because a significant proportion of this tonnage is intersite transfers. Such a large transfer of plant nutrients is close to one-third of the nutrients removed by harvested crops and nearly half of the amounts added through fertilizers (Tandon, 2004).

On-farm nutrient cycles

The nutrient cycles in the field or on the farm are not closed. Some nutrients are removed (exported) from the field or farm with food, feed and raw materials, others are just lost. At the same time, a field also gains nutrients through biological nitrogen fixation (BNF) and the addition of fertilizers and manures. Some losses are inherent to crop production because some production factors cannot be controlled. However, many losses can be avoided by more efficient management and recycling. In order to remain sustainable, nutrient cycles need some input from soil reserves and/or from external nutrient sources. An indicator of nutrient status is the input/output balance at the field or farm level. With higher productivity, the amount of circulating nutrients usually increases as a result of increased nutrient input as well as output in order to sustain the process at a high level of productivity. In areas where groundwaters are pumped for irrigation, some of the leached nutrient can return and re-enter the cycle as an input.

Regional nutrient transfer

Many nutrients leave the farm or the village and are transferred to urban areas. These transfers could be a few kilometres away to the nearest town or even several hundred kilometres away from a food-surplus to a food-deficit region within the same country. Ideally, they should be completely recycled to agricultural soils. However, in most cities, large amounts of nutrients are deposited into landfills or into the sea, which is a wasteful procedure, especially for nutrients in limited supply, e.g. phosphate. The transfer of phosphate to cities is used as an example to demonstrate the magnitude of this problem. Humans need 1.0–1.5 g of P per day, which translates into a supply of about 1.7 g of P per day. Therefore, a city of 1 million people requires 1.7 tonnes of P per day or 620 tonnes of P per year (1 400 tonnes of P_2O_5). As phosphate is used in human metabolism, but not destroyed, a large proportion of the P intake amount appears in solid or liquid wastes. Ideally, these should be recycled.

However, P recovery from city wastes varies from 10 to 80 percent depending on the sophistication of the recycling systems. Some urban areas have exemplary P-recovery systems with precipitation of Fe or Al phosphate from wastewaters, and agricultural use of these mineral phosphates as nutrient sources. The rate of recycling usually decreases with increasing size of the settlement. Many cities are proud of their sewage disposal system, which often disposes of biologically treated sewage water into rivers or into the sea. This action of just disposing of waste materials is not the best solution. It means an enormous loss of plant nutrients with secondary effects of pollution, health hazards and eutrophication.

The main obstacles to complete recycling of plant nutrients from urban areas are the unwanted side-effects that urban waste products can have on farmers fields, even where they are composted. Therefore, knowledgeable farmers are increasingly reluctant to apply composted sewage or garbage as nutrient sources, even if offered free of charge, because of the problems of toxic heavy metals and possibly toxic organic substances. With environmental laws in some areas becoming more severe, farmers suspect that the critical limits for soil contamination might be decreased, thus putting otherwise fertile land out of production.

In addition, farmers dislike being accused by urban people of "poisoning" the soil and so decreasing food quality while at the same time using or rather misusing their fields to dispose of urban wastes. Recycling of urban wastes in many developing countries does not exist beyond dumping. Any recycling is rather casual because of a lack of quality standards and adequate information for producers and consumers. Farmers near urban areas are sometimes known to willingly use urban wastes, sewage sludge, etc. for vegetable production meant for sale but do not use these wastes on the small patch of land reserved for growing crops for home consumption.

In the future, ever-increasing urbanization will result in an enormous nutrient transfer into the cities. Hence, steps must be taken to enhance the recycling of plant nutrients. This can be achieved through the composting of urban wastes and utilization of sewage as well as slaughterhouse waste for manuring. At the same time, quality standards must be established and enforced, supplemented by proper education at all levels along the recycling chain, on a continuous basis.

International nutrient transfer

The export of food and feed results in considerable amounts of nutrients being transferred to other countries, or even other continents, without being recycled. Some developed countries import enormous amounts of plant nutrients with feed for animals. Global nutrient transfer partly results in a paradoxical situation where plant nutrients are mined from poor soils in developing countries and added to already fertile soils in developed areas. The reverse is the case where food is imported by developing countries to meet shortages created by low local production.

Nutrient exports

The export of agricultural products results in an unnoticed export of plant nutrients and, thus, a loss from the national nutrient balance. These nutrient exports to other countries can reach substantial amounts (Table 5).

About 15 kg of N, 5–6 kg of P_2O_5 and 5–6 kg of K_2O are exported from the farm with every tonne of cereal. Thailand and Viet Nam together have a net export through cereals of about 150 000 tonnes of N and 60 000 tonnes each of

TABLE 5
Examples of plant nutrients exported and imported through cereals, 1999

Movement	Country	Commodity	Nutrients		
			N	P_2O_5	K_2O
			(1 000 tonnes)		
Export	Developing countries	Cereals	740	300	300
	Thailand	Cereals	90	36	36
	Viet Nam	Cereals	57	23	23
	Zimbabwe	Cereals	3.25	1.3	1.3
Import	Netherlands	Cereals[1]	100	40	40
	Germany	Cereals[1]	45	18	18

[1] Imported mainly for feed.

phosphate and potash. The developing countries as a whole have a loss of about 1.3 million tonnes of nutrients, mainly through cereal exports. The amounts are shown in terms of NPK only as an example. In reality, all nutrients present in the exported produce are also moved across national boundaries. Such exports cause a considerable loss of nutrients, which are largely obtained by nutrient mining of often already poor soils and are not compensated for by imports.

On the other hand, as long as this transfer of nutrients with agricultural products is, or can be, compensated for by re-imports of mineral fertilizers, the nutrient loss from the developing countries will not be a serious problem. However, it is necessary to consider the overall economic and environmental aspects of importing fertilizer nutrients and at what level of efficiency these will be used for crop production. In any case, nutrients exported through crops represent net removals, while 2–4 units of fertilizer nutrients are needed for every unit of nutrient contained in the crops exported.

As trade barriers for the export of agricultural produce from developing countries are removed, the issue of international nutrient transfers will need re-examining. In any case, nutrients exported through crop products cannot be equated with nutrients imported through fertilizers on a 1:1 basis. This is because a fraction of the fertilizer nutrients ends up in the exported product. In addition, it cannot be assumed that when fertilizers are imported by a country, these are used in the areas that produced the exportable surplus. This is one reason why macrolevel nutrient balances fail to provide insights into nutrient balances at the microlevel.

Import of nutrients

The Netherlands and Germany import about 150 000 tonnes of N and 60 000 tonnes each of phosphate and potash in grains imported for animal feed. After consumption, the animal wastes are used as nutrient sources for manuring the fields. Often, animal slurry is added to soils that are already well supplied with available P and K in these countries. This could be because the farmland is easily accessible for the disposal of slurry.

Many developed countries with high animal production produce sufficient feedstuff from their own agriculture but import substantial amounts of feedstuff because of cheaper prices. For the Netherlands, these imports are outstandingly high, and in the cases of P and K have been estimated to represent about two-thirds of total fertilizer imports (Cooke, 1982). Food-deficit developing countries also import plant nutrient whenever they import food grains or other farm produce (grain legumes, oilseeds and sugar) whether from developed countries or from other developing countries.

In such countries, it makes sound agro-economical and ecological sense to import fertilizers and develop their agricultural production capability rather than import food grains or other "finished" crop products. By putting the plant nutrient to work, they can make value-added products out of their abundant supplies of sunlight, air, carbon dioxide (CO_2)and human labour. International nutrient transfer is a subject that will become increasingly relevant and also provide a basis for developing the most effective strategies of international trade of inputs as well as output. Towards this end, the optimization of plant nutrients has a role to play because by maximizing the efficiency of production inputs, unit-product cost can be reduced and farm produce made more competitive. At the same time, national farm policies may be needed that ensure that the highly productive agricultural soils are replenished with adequate nutrients in order to sustain their productivity. These are also the areas where crop production skills have reached a satisfactory level and where efficient use of applied nutrients can be expected. All these factors will contribute towards increasing agricultural production and ensuring food security.

Chapter 3

Plant nutrients and basics of plant nutrition

Plants convert light energy into biomass through photosynthesis and produce various products of economic value (grain, fibre, tubers, fruits, vegetables and fodder) among others. To do this, plants need sufficient light, suitable temperature, substances such as water, CO_2, oxygen, and a number of nutrients. The survival and well-being of humans and animals depends on plant production, which in turn depends heavily on the availability of mineral and other nutrients. This is why plants and animals (including humans) have several essential nutrients in common.

Like all organisms, higher green plants need nutrients for their growth and development. Nutrients are indispensable as plant constituents, for biochemical reactions, and for the production of organic materials referred to as photosynthates (carbohydrates, proteins, fats, vitamins, etc.) by photosynthesis. In agriculture (including horticulture), optimal crop nutrition is an important prerequisite for obtaining high yields and good-quality produce. The nutrients required are obtained by plants both from soil reserves and external nutrient sources (fertilizers, organic manures, the atmosphere, etc). Almost all of the 90 natural elements can be found in green plants although most of them have no function (e.g. the heavy metal gold).

PLANT NUTRIENTS

Essential plant nutrients

A total of only 16 elements are essential for the growth and full development of higher green plants according to the criteria laid down by Arnon and Stout (1939). These criteria are:

- A deficiency of an essential nutrient makes it impossible for the plant to complete the vegetative or reproductive stage of its life cycle.
- Such deficiency is specific to the element in question and can be prevented or corrected only by supplying this element.
- The element is involved directly in the nutrition of the plant quite apart from its possible effects in correcting some unfavourable microbiological or chemical condition of the soil or other culture medium.

The essentiality of most micronutrients for higher plants was established between 1922 and 1954. The essentiality of nickel (Ni) was established in 1987 by Brown et al, although there is no unanimity among the scientists as to whether Ni is essential or beneficial. However, this list may not be considered as final and it is probable that more elements may prove to be essential in future.

TABLE 6
Essential plant nutrients, forms taken up and their typical concentration in plants

Nutrient (symbol)	Essentiality established by	Forms absorbed	Typical concentration in plant dry matter
Macronutrients			
Nitrogen (N)	de Saussure (1804)	NH_4^+, NO_3	1.5%
Phosphorus (P, P_2O_5[1])	Sprengel (1839)	$H_2PO_4^-$, HPO_4^{2-}	0.1–0.4%
Potassium (K, K_2O[1])	Sprengel (1839)	K^+	1–5%
Sulphur (S)	Salm-Horstmann (1851)	SO_4^{2-}	0.1–0.4%
Calcium (Ca)	Sprengel (1839)	Ca^{2+}	0.2–1.0%
Magnesium (Mg)	Sprengel (1839)	Mg^{2+}	0.1–0.4%
Micronutrients			
Boron (B)	Warington (1923)	H_3BO_3, $H_2BO_3^-$	6–60 µg/g (ppm[2])
Iron (Fe)	Gris (1943)	Fe^{2+}	50–250 µg/g (ppm)
Manganese (Mn)	McHargue (1922)	Mn^{2+}	20–500 µg/g (ppm)
Copper (Cu)	Sommer, Lipman (1931)	Cu^+, Cu^{2+}	5–20 µg/g (ppm)
Zinc (Zn)	Sommer, Lipman (1931)	Zn^{2+}	21–150 µg/g (ppm)
Molybdenum (Mo)	Arnon & Stout (1939)	MoO_4^{2-}	below 1 µg/g (ppm)
Chlorine (Cl)	Broyer *et al.*, (1954)	Cl^-	0.2–2 percent

Notes:
[1] Oxide forms are used in extension and trade.
[2] ppm = parts per million = mg/kg = µg/g; 10 000 ppm = 1 percent.

Out of these 16 elements, carbon (C) and oxygen are obtained from the gas CO_2, and hydrogen (H) is obtained from water (H_2O). These three elements are required in large quantities for the production of plant constituents such as cellulose or starch. The other 13 elements are called mineral nutrients because they are taken up in mineral (inorganic) forms. They are traditionally divided into two groups, macronutrients and micronutrients, according to the amounts required. Regardless of the amount required, physiologically, all of them are equally important. The 13 mineral elements are taken up by plants in specific chemical forms (Table 6) regardless of their source.

Oxygen, C and H make up 95 percent of plant biomass, and the remaining 5 percent is made up by all other elements. The difference in plant concentration between macronutrients and micronutrients is enormous. The relative contents of N and molybdenum (Mo) in plants is in the ratio of 10 000:1. Plants need about 40 times more magnesium (Mg) than Fe. These examples indicate the significant difference between macronutrients and micronutrients. Chapter 6 provides more detailed on nutrient concentration in crops and crop products.

Beneficial nutrients

Several elements other than the essential nutrients have beneficial functions in plants. Although not essential (as the plant can live without them), beneficial nutrients can improve the growth of some crops in some respects. Some of these nutrients can be of great practical importance and may require external addition:

- Nickel (Ni): a part of enzyme urease for breaking urea in the soil, imparts useful role in disease resistance and seed development.
- Sodium (Na): for beets, partly able to replace K (uptake as Na^+).

- Cobalt (Co): for N fixation in legumes and for other plants (uptake as Co^{2+}).
- Silicon (Si): for stalk stability of cereals particularly rice (uptake as silicate anion).
- Aluminium (Al): for tea plants (uptake as Al^{3+} or similar forms).

Other important nutrients

As humans and domestic animals require several nutrients in addition to those required by plants, these additional nutrients should also be considered in food or feed production, and their deficiencies corrected by appropriate inputs. In addition to plant nutrients, the elements essential for humans and domestic animals are: Cobalt (Co), selenium (Se), chromium (Cr) and iodine (I).

NUTRIENTS – THEIR FUNCTIONS, MOBILITY IN PLANTS AND DEFICIENCY/ TOXICITY SYMPTOMS

Some knowledge of the properties and functions of plant nutrients is helpful for their efficient management and, thus, for good plant growth and high yields. Available nutrients in the soil solution can be taken up by the roots, transported to the leaves and used according to their functions in plant metabolism.

Nutrient ions are of extremely small size, i.e. like atoms. For example, there are more than 100 000 million K^+ cations within a single leaf cell and more than 1 000 000 molybdate anions, the micronutrient required in the smallest amount. In general, N and K make up about 80 percent of the total mineral nutrients in plants; P, S, Ca and Mg together constitute 19 percent, while all the micronutrients together constitute less than 1 percent.

Most plant nutrients are taken up as positively or negatively charged ions (cations and anions, respectively) from the soil solution. However, some nutrients may be taken up as entire molecules, e.g. boric acid and amino acids, or organic complexes such as metal chelates and to a very small extent urea. Whether the original sources of nutrient ions in the soil solution are from organic substances or inorganic fertilizers, ultimately, the plants absorb them only in mineral forms.

Plants exhibit many shades of greenness but a medium to dark green colour is usually considered a sign of good health and active growth. Chlorosis or yellowing of leaf colour can be a sign of a marginal deficiency and is often associated with retarded growth. Chlorosis is a light green or rather yellowish discoloration of the whole or parts of the leaf caused by a lower content of chlorophyll. Because the cells remain largely intact, the chlorotic symptoms are reversible, i.e. leaves can become green again after the missing nutrient (responsible for chlorophyll formation) is added. A severe deficiency results in death of the tissue (necrosis). Necrosis is a brownish discoloration caused by decaying tissue, which is destroyed irreversibly. Necrotic leaves cannot be recovered by addition of the missing nutrient, but the plant may survive by forming new leaves.

Deficiency symptoms can serve as a guide for diagnosing limiting nutrients and the need for corrective measures. However, chlorotic and necrotic leaves might also result from the toxic effects of nutrients, pollution and also from disease and insect attacks. Therefore, confirmation of the cause is important before corrective measures are taken.

Nitrogen

N is the most abundant mineral nutrient in plants. It constitutes 2–4 percent of plant dry matter. Apart from the process of N fixation that occurs in legumes, plants absorb N either as the nitrate ion (NO_3^-) or the ammonium ion (NH_4^+). N is a part of the chlorophyll (the green pigment in leaves) and is an essential constituent of all proteins. It is responsible for the dark green colour of stem and leaves, vigorous growth, branching/tillering, leaf production, size enlargement, and yield formation.

Absorbed N is transported through the xylem (in stem) to the leaf canopy as nitrate ions, or it may be reduced in the root region and transported in an organic form, such as amino acids or amides. N is mobile in the phloem (the plant tissue through which the sap containing dissolved food materials passes downwards to the stem, roots, etc.); as such, it can be re-translocated from older to younger leaves under N deficiency and translocated from leaves to the developing seed or fruit. The principal organic forms of N in phloem sap are amides, amino acids and ureides. Nitrate and ammonium ions are not present in this sap.

N deficiency in plants results in a marked reduction in growth rate. N-deficient plants have a short and spindly appearance. Tillering is poor, and leaf area is small. As N is a constituent of chlorophyll, its deficiency appears as a yellowing or chlorosis of the leaves. This yellowness usually appears first on the lower leaves while upper leaves remain green as they receive some N from older leaves. In a case of severe deficiency, leaves turn brown and die. As a result, crop yield and protein content are reduced (percent N in seed × 6.25 = percent protein content).

The effects of N toxicity are less evident than those of its deficiency. They include prolonged growing (vegetative) period and delayed crop maturity. High NH_4^+ in solution can be toxic to plant growth, particularly where the solution is alkaline. The toxicity results from ammonia (NH_3), which is able to diffuse through plant membranes and interfere with plant metabolism. The potential hydrogen (pH – negative log of H^+ concentration) determines the balance between NH_3 and NH_4^+.

Phosphorus

P is much less abundant in plants (as compared with N and K) having a concentration of about one-fifth to one-tenth that of N in plant dry matter. P is absorbed as the orthophosphate ion (either as $H_2PO_4^-$ or HPO_4^{2-}) depending on soil pH. As the soil pH increases, the relative proportion of $H_2PO_4^-$ decreases and that of HPO_4^{2-} increases. P is essential for growth, cell division, root lengthening, seed and fruit development, and early ripening. It is a part of several compounds including oils and amino acids. The P compounds adenosine diphosphate (ADP) and adenosine triphosphate (ATP) act as energy carriers within the plants.

P is readily mobile within the plant (unlike in the soil) both in the xylem and phloem tissues. When the plant faces P shortage (stress), P from the old leaves is readily translocated to young tissue. With such a mobile element, the pattern of redistribution seems to be determined by the properties of the source (old leaves, and stems) and the sink (shoot tip, root tip, expanding leaves and later into the developing seed).

Plant growth is markedly restricted under P deficiency, which retards growth, tillering and root development and delays ripening. The deficiency symptoms usually start on older leaves. A bluish-green to reddish colour develops, which can lead to bronze tints and red colour. A shortage of inorganic phosphate in the chloroplast reduces photosynthesis. Because ribonucleic acid (RNA) synthesis is reduced, protein synthesis is also reduced. A decreased shoot/root ratio is a feature of P deficiency, as is the overall lower growth of tops.

Extremely high levels of P can result in toxicity symptoms. These generally manifest as a watery edge on the leaf tissue, which subsequently becomes necrotic. In very severe cases, P toxicity can result in the death of the plant.

Potassium

K is the second most abundant mineral nutrient in plants after N. It is 4–6 times more abundant than the macronutrients P, Ca, Mg and S. K is absorbed as the monovalent cation K^+ and it is mobile in the phloem tissue of the plants. K is involved in the working of more than 60 enzymes, in photosynthesis and the movement of its products (photosynthates) to storage organs (seeds, tubers, roots and fruits), water economy and providing resistance against a number of pests, diseases and stresses (frost and drought). It plays a role in regulating stomatal opening and, therefore, in the internal water relations of plants.

The general symptom of K deficiency is chlorosis along the leaf boundary followed by scorching and browning of tips of older leaves. The affected area moves inwards as the severity of deficiency increases. K-deficiency symptoms show on the older tissues because of the mobility of K. Affected plants are generally stunted and have shortened internodes. Such plants have: slow and stunted growth; weak stalks and susceptibility to lodging; greater incidence of pests and diseases; low yield; shrivelled grains; and, in general, poor crop quality. Slow plant growth can be accompanied by a higher rate of respiration, which means a wasteful consumption of water per unit of dry matter produced. K-deficient plants may lose control over the rate of transpiration and suffer from internal drought.

Calcium

Calcium (Ca) ranks with Mg, P and S in the group of least abundant macronutrients in plants. It is absorbed by plant roots as the divalent cation Ca^{2+}. Ca is a part of the architecture of cell walls and membranes. It is involved in cell division, growth, root lengthening and activation or inhibition of enzymes. Ca is immobile in the phloem.

Ca deficiency is seen first on growing tips and the youngest leaves. This is the case with all nutrients that are not very mobile in the plants. The Ca-deficiency problems are often related to the inability of Ca to be transported in the phloem. The problems occur in organs that do not transpire readily, i.e. large, fleshy developing fruits. Ca-deficient leaves become small, distorted, cup-shaped, crinkled and dark green. They cease growing, become disorganized, twisted and, under severe deficiency, die. Although all growing points are sensitive to Ca deficiency, those of the roots are affected more severely. Groundnut shells may be hollow or poorly filled as a result of incomplete kernel development.

Magnesium

Mg ranks with Ca, P and S in the group of least abundant macronutrients in plants. Plants take up Mg in the form of Mg^{2+}. Mg occupies the centre-spot in the chlorophyll molecule and, thus, is vital for photosynthesis. It is associated with the activation of enzymes, energy transfer, maintenance of electrical balance, production of proteins, metabolism of carbohydrates, etc. Mg is mobile within the plants.

As Mg is readily translocated from older to younger plant parts, its deficiency symptoms first appear in the older parts of the plant. A typical symptom of Mg deficiency is the interveinal chlorosis of older leaves in which the veins remain green but the area between them turns yellow. As the deficiency becomes more severe, the leaf tissue becomes uniformly pale, then brown and necrotic. Leaves are small and break easily (brittle). Twigs become weak and leaves drop early. However, the variety of symptoms in different plant species is so great that their generalized description is more difficult in case of Mg than for other nutrients.

Sulphur

S is required by crops in amounts comparable with P. The normal total S concentration in vegetative tissue is 0.12–0.35 percent and the total N/total S ratio is about 15. Plant roots absorb S primarily as the sulphate ion (SO_4^{2-}). However, it is possible for plants to absorb sulphur dioxide (SO_2) gas from the atmosphere at low concentrations.

S is a part of amino acids cysteine, cystine and methionine. Hence, it is essential for protein production. S is involved in the formation of chlorophyll and in the activation of enzymes. It is a part of the vitamins biotin and thiamine (B_1), and it is needed for the formation of mustard oils, and the sulphydryl linkages that are the source of pungency in onion, oils, etc.

S moves upwards in the plant as inorganic sulphate anion (SO_4^{2-}). Under low S conditions. mobility is low as the S in structural compounds cannot be translocated. As the S status of the plant rises, so does its mobility. This pattern of mobility means that in plants with adequate S, sulphate is preferentially translocated to young, actively growing leaves. As the supply of S becomes more limiting, young leaves lack S and, hence, show deficiency symptoms.

In many ways, S deficiency resembles that of N. It starts with the appearance of pale yellow or light-green leaves. Unlike N deficiency, S-deficiency symptoms in most cases appear first on the younger leaves, and are present even after N application. Plants deficient in S are small and spindly with short and slender stalks. Their growth is retarded, and maturity in cereals is delayed. Nodulation in legumes is poor and N fixation is reduced. Fruits often do not mature fully and remain light green in colour. Oilseed crops deficient in S produce a low yield and the seeds have less oil in them.

S toxicity can occur under highly reduced conditions, possibly as a result of sulphide (H_2S) injury. Most plants are susceptible to high levels of atmospheric SO_2. Normal SO_2 concentrations range from 0.1 to 0.2 mg SO_2/m^3, and toxicity symptoms are observed when these exceed 0.6 mg SO_2/m^3. S-toxicity symptoms appear as necrotic spots on leaves, which then spread over the whole leaf.

Boron

Boron (B) is probably taken up by plants as the undissociated boric acid (H_3BO_3). It appears that much of the B uptake mainly follows water flow through roots. B in a plant is like the mortar in a brick wall, the bricks being the cells of growing parts such as tips (meristems). Key roles of B relate to: (i) membrane integrity and cell-wall development, which affect permeability, cell division and extension; and (ii) pollen tube growth, which affects seed/fruit set and, hence, yield. B is relatively immobile in plants and, frequently, the B content increases from the lower to the upper parts of plants.

B deficiency usually appears on the growing points of roots, shoots and youngest leaves. Young leaves are deformed and arranged in the form of a rosette. There may be cracking and cork formation in the stalks, stem and fruits; thickening of stem and leaves; shortened internodes, withering or dying of growing points and reduced bud, flower and seed production. Other symptoms are: premature seed drop or fruit drop; crown and heart rot in sugar beet; hen- and chicken-type bunches in grapes; barren cobs in maize; hollow heart in groundnut; unsatisfactory pollination; and poor translocation of assimilates. Death of the growing tip leads to sprouting of auxiliary meristem and a bushy broom-type growth. Roots become thick, slimy and have brownish necrotic spots.

B toxicity can arise under excessive B application, in arid or semi-arid areas, and where irrigation water is rich in B content (more than 1–2 ppm B). B-toxicity symptoms are yellowing of the leaf tip followed by gradual necrosis of the tip and leaf margins, which spreads towards the midrib (central vein). Leaves become scorched and may drop early.

Chlorine

Chlorine (Cl) is absorbed as the chloride anion (Cl^-). It is thought to be involved in the production of oxygen during photosynthesis, in raising cell osmotic pressure and in maintaining tissue hydration. Some workers consider it essential only for palm and kiwi fruit. Deficiency of Cl leads to chlorosis in younger leaves and overall wilting as a consequence of the possible effect on transpiration. Cl-toxicity symptoms are: burning of the leaf tips or margins; bronzing; premature yellowing; leaf fall; and poor burning quality of tobacco.

Copper

Copper (Cu) is taken up as Cu^{2+}. Its uptake appears to be a metabolically mediated process. However, Cu uptake is largely independent of competitive effects and relates primarily to the levels of available Cu in the soil. Cu is involved in chlorophyll formation and is a part of several enzymes such as cytochrome oxidase. As much as 70 percent of the Cu in plants may be present in the chlorophyll, largely bound to chloroplasts. It participates in lignin formation, protein and carbohydrate metabolism, and is possibly required for symbiotic N fixation. Cu is a part of plastocyanin, which forms a link in the electron transport chain involved in photosynthesis. Cu is not readily mobile in the plant and its movement is strongly dependent on the Cu status of the plant.

Cu-deficiency symptoms are first visible in the form of narrow, twisted leaves and pale white shoot tips. At maturity, panicles/ears are poorly filled and even empty where the deficiency is severe. In fruit trees, dieback of the terminal growth can occur. In maize, yellowing between leaf veins takes place, while in citrus the leaves appear mottled and there is dieback of new twigs.

Cu-toxicity symptoms are more variable with species and less established than its deficiency symptoms. Excess Cu induces Fe deficiency and, therefore, chlorosis is a common symptom.

Iron

Fe is absorbed by plant roots as Fe^{2+}, and to a lesser extent as Fe chelates. For efficient utilization of chelated Fe, separation between Fe and the organic ligand has to take place at the root surface, after the reduction of Fe^{3+} to Fe^{2+}. Absorbed Fe is immobile in the phloem. Fe is generally the most abundant of the micronutrients with a dry-matter concentration of about 100 µg/g (ppm). It plays a role in the synthesis of chlorophyll, carbohydrate production, cell respiration, chemical reduction of nitrate and sulphate, and in N assimilation.

Fe deficiency begins to appear on younger leaves first. Otherwise, its deficiency symptoms are somewhat similar to those of Mn, as both Fe and Mn lead to failure in chlorophyll production. Yellowing of the interveinal areas of leaves (commonly referred to as iron chlorosis) occurs. In severe deficiency, leaves become almost pale white because of the loss of chlorophyll. In cereals, alternate yellow and green stripes along the length of the leaf blade may be observed. Complete leaf fall can occur and shoots can die.

Fe toxicity of rice is known as bronzing. In this disorder, the leaves are first covered by tiny brown spots that develop into a uniform brown colour. It can be a problem in highly reduced rice soils as flooding may increase the levels of soluble Fe from 0.1 to 50–100 µg/g Fe within a few weeks. It can also be a problem in highly weathered, lowland acid soils.

Manganese

Manganese (Mn) is taken up by plants as the divalent ion Mn^{2+}. It is known to activate several enzymes and functions as an auto-catalyst. It is essential for splitting the water molecule during photosynthesis. It has certain properties similar to Mg. It is also important in N metabolism and in CO_2 assimilation. Like Fe, it is generally immobile in the phloem.

Mn-deficiency symptoms resemble those of Fe and Mg deficiency where interveinal chlorosis occurs in the leaves. However, Mn-deficiency symptoms are first visible on the younger leaves whereas in Mg deficiency, the older leaves are affected first. Mn deficiency in oats is characterized by "grey-speck" where the leaf blade develops grey lesions but the tip remains green, the base dies and the panicle may be empty. In dicots (e.g. legumes), younger leaves develop chlorotic patches between the veins (somewhat resembling Mg deficiency).

Mn-toxicity symptoms lead to the development of brown spots, mainly on older leaves and uneven green colour. Some disorders caused by Mn toxicity are: crinkle leaf spot in cotton; stem streak; necrosis of potato; and internal bark necrosis of apple trees.

Molybdenum

Mo is absorbed as the molybdate anion MoO_4^{2-} and its uptake is controlled metabolically. Mo is involved in several enzyme systems, particularly nitrate reductase, which is needed for the reduction of nitrate, and nitrogenase, which is involved in BNF. Thus, it is involved directly in protein synthesis and N fixation by legumes. Mo appears to be moderately mobile in the plant. This is suggested by the relatively high levels of Mo in seeds, and because deficiency symptoms appear in the middle and older leaves.

Mo deficiency in legumes can resemble N deficiency because of its role in N fixation. Mo deficiency can cause marginal scorching and rolling or cupping of leaves and yellowing and stunting in plants. Yellow spot disease in citrus and whip tail in cauliflower are commonly associated with Mo deficiency.

Fodders containing more than 5 µg/g Mo in the dry matter are suspected to contain toxic levels of Mo for grazing animals (associated with the disease molybdenosis).

Zinc

Zn is taken up as the divalent cation Zn^{2+}. Early work suggested that Zn uptake was passive, but more recent work indicates that it is active (energy-dependent). Zn is required directly or indirectly by several enzymes systems, auxins and in protein synthesis, seed production and rate of maturity. Zn is believed to promote RNA synthesis, which in turn is needed for protein production. The mobility of Zn is low. The rate of Zn mobility to younger tissue is particularly depressed in Zn-deficient plants.

Common symptoms of Zn deficiency are: stunted plant growth; poor tillering; development of light green, yellowish, bleached spots; chlorotic bands on either side of the midrib in monocots (particularly maize); brown rusty spots on leaves in some crops, which in acute Zn deficiency as in rice may cover the lower leaves; and in fruit trees the shoots may fail to extend and the small leaves may bunch together at the tip in a rosette-type cluster. Little-leaf condition is also a common symptom. Internodes are short. Flowering, fruiting and maturity can be delayed. Shoots may die off and leaves can fall prematurely. Deficiency symptoms are not the same in all plants.

Zn toxicity can result in reduction in root growth and leaf expansion followed by chlorosis. It is generally associated with tissue concentrations greater than 200 µg/g Zn.

BENEFICIAL ELEMENTS

Nickel

Ni is a part of the enzyme urease, which breaks down urea in the soil. It also plays a role in imparting disease resistance and is considered essential for seed development. Information on various aspects of Ni as a micronutrient is gradually becoming available.

Silicon

Si is taken up as the undissociated $Si(OH)_4$ monosilicic acid. The prevalent form of Si in plants is silica gel in the form of hydrated amorphous silica (SiO_2 in H_2O), or polymerized silicic acid, which is immobile in the plant.

The beneficial effects of Si on plants include increases in yield that can result from increasing leaf erectness, decreasing susceptibility to lodging, decreasing incidence to fungal infections, and prevention of Mn and/or Fe toxicity. Thus, Si is able to counteract the effects of high N, which tend to increase lodging.

In lowland or wetland rice that is low in Si, vegetative growth and grain production is reduced severely and deficiency symptoms such as necrosis of the mature leaves and wilting can occur. Similarly, sugar cane suffers growth reduction under conditions of low Si availability.

Cobalt

Co is taken up as the divalent cation Co^{2+}. It is essential for N-fixing micro-organisms, irrespective of whether they are free-living or symbiotic. Co is the metal component of vitamin B_{12}. Thus, Co deficiency inhibits the formation of leghaemoglobin and, hence, N_2 fixation. The Co content of the shoots can be used as an indicator of Co deficiency in legumes, where the critical levels are between 20 and 40 ppb of shoot dry weight.

BASICS OF PLANT NUTRITION

Plant nutrition is governed by some basic facts and principles concerning nutrient supply, their absorption, transport and production efficiency. These should be understood and applied during practical nutrient management, which is covered in detail in Chapters 6, 7 and 8.

Nutrient demand and supply

Plants require nutrients in balanced amounts depending on their stage of development and yield levels. For optimal nutrition of crops, a sufficient concentration of the individual nutrients should be present in the plant leaves at any time. An optimal nutrient supply requires:

- sufficient available nutrients in the rootzone of the soil;
- rapid transport of nutrients in the soil solution towards the root surface;
- satisfactory root growth to access available nutrients;
- unimpeded nutrient uptake, especially with sufficient oxygen present;
- satisfactory mobility and activity of nutrients within the plant.

The nutrient concentrations required in plants, or rather in the active tissues, are usually indicated on a dry-matter basis, as this is more reliable than on a fresh-matter basis with its varying water content. Leaves usually have higher nutrient concentrations than do roots. These are usually stated as a percentage for macronutrients and in micrograms per gram (parts per million) for micronutrients.

The law of the minimum and its implications

In plant nutrition, there is a law known as Liebig's law of the minimum. It is named after its author, Justus von Liebig, who said that the growth of a plant is limited by the nutrient that is in shortest supply (in relation to plant need). Once its supply is improved, the next limiting nutrient controls plant growth. This concept has been depicted in many ways. One is to imagine a barrel with staves of different heights (Figure 9). Such a barrel can only hold water up to the height of its shortest stave. The barrel can be full only when all its staves are of the same size. A plant can also produce to its full potential when all nutrients (production factors in an enlarged sense) are at an optimal level, i.e. without any deficiencies or excesses.

FIGURE 9
Demonstration of the law of the minimum using a barrel with staves of different heights

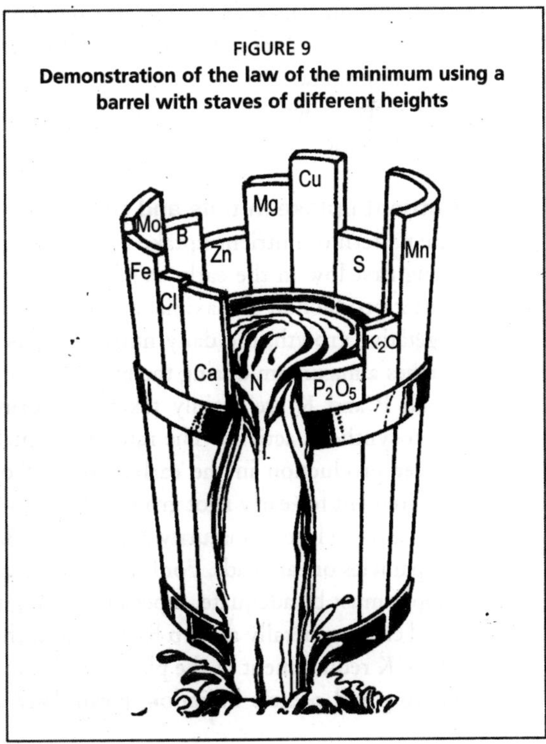

In order to produce high yields, plant nutrition requires a continuous effort to eliminate minimum factors and provide balanced nutrition in the optimal range (Figure 10). Even if the law of the minimum is only a guiding rule, it serves as a useful basis for nutrient management. In a broader

FIGURE 10
Example of yield-limiting minimum factors

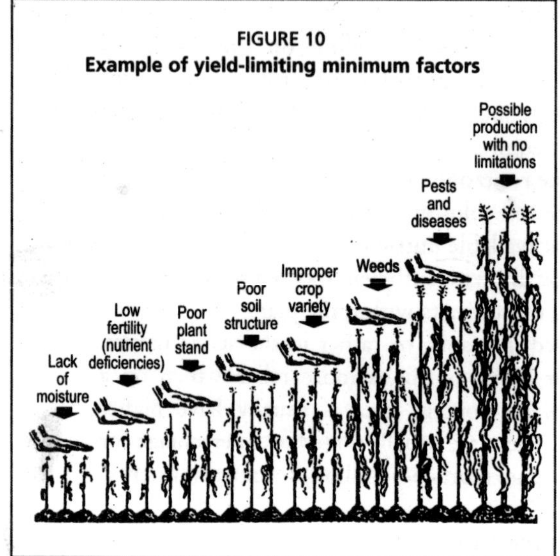

sense, the law of the minimum can be extended to include all production inputs, not only nutrients.

Important aspects of the influence of nutrient supply on plant growth are:

- Plants need certain concentrations of nutrient in their tissue for active growth.
- Nutrient requirement comes somewhat in advance of plant growth.
- Insufficient nutrient uptake results in slight to severe deficiencies.
- Slight deficiencies are not visible and denote "hidden hunger".
- Deficiency symptoms indicate a severe shortage of the nutrient in question.
- High yields are only obtained where all nutrients are in the optimal supply range.
- The nutrient with the lowest (minimum) supply determines the yield level.
- Many mistakes in fertilization can be attributed to disregarding the law of the minimum.
- It is easier to correct nutrient deficiencies than to eliminate nutrient toxicities.

Nutrient uptake in time and contents

The pattern of nutrient uptake follows a sigmoid (S-shaped) curve in most cases, being first low in the early stages of crop growth, increasing rapidly when dry-matter production is maximal and then declining towards crop maturity. During vegetative growth, the daily nutrient uptake increases as growth progresses and reaches a maximum during the main growing period.

N, P and K are mainly taken up during active vegetative growth for high photosynthetic activity. The rate of N uptake generally exceeds the rate of dry-matter production in the early stages. Phosphate has an additional small peak requirement for early root growth. Modern high-yielding grain varieties continue to absorb P close to maturity and, like N, 70–80 percent of absorbed P ends up in the panicles or ear heads. For fast-growing crops and high yields, the daily nutrient supply must be adequate, especially during the period of maximum requirement. Field crops generally absorb K faster then they absorb N and P. In rice, 75 percent of the K requirement of the plant may be absorbed up to boot leaf stage. Between tillering and panicle initiation, mean daily absorption rates can approach 2.5 kg

K_2O/ha/day. Unlike N and P, only 20–25 percent of absorbed K is transferred to the grain, the rest remaining in the straw.

During the final stages of growth as the plant approaches its reproductive phase before maturity, nutrient uptake decreases. Perennial plants retrieve most of the nutrients from the leaves before leaf fall and relocate these for future use. In certain plants, such as jute, a considerable proportion of the absorbed nutrients is returned to the soil through leaf shedding before the crop matures.

While the total amount of a nutrient within the plant steadily increases, the concentration (percentage) of the nutrient generally decreases, even with a good supply. The highest concentrations of nutrients are found in leaves at early growth stages, and the lowest in leaves near harvest. This decrease in nutrient concentration over time is because of the transfer to other organs and also what is called the dilution effect, which results from a larger increase in dry matter than in nutrient content. For example, young plants with 50 kg K in 1 500 kg of dry matter contain 3.3 percent K but plants approaching flowering with 100 kg K in 5 000 kg of dry matter contain 2.0 percent K.

The dilution effect makes the interpretation of plant analysis results difficult, but it can be taken into account by relating plant data to a certain stage of growth.

Nutrient mobility and its effect on deficiency symptoms

While nutrients are transported easily from roots to shoots, their redistribution from the original place of deposition is more difficult for the so-called immobile nutrients. In the event of nutrient deficiency, a partial re-activation is required in order to supply newly formed leaves from the reserves of older ones. The relative mobility of a nutrient within the plant is helpful in understanding the reasons for the differential appearance of nutrient deficiency symptoms as discussed above. For example:

- The appearance of deficiency symptoms on older leaves indicates the shortage of a mobile nutrient because the plant can transport some nutrient quantities from old to new leaves.
- The appearance of deficiency symptoms on younger leaves indicates the shortage of an immobile nutrient because of lack of supply from older to younger leaves.

The range of nutrient supply from deficiency to toxicity

The nutrient status of a plant can range from acute deficiency to acute toxicity. A broad division of nutrient status into three groups namely deficient, optimal and excess may be useful for general purposes. For a more accurate assessment of the nutritional status of plants, detailed categorization is required in which six different ranges can be distinguished (Figure 11):

- Acute deficiency: It is associated with definite visible symptoms and poor growth. The addition of the deficient nutrient usually results in increased

FIGURE 11
Plant growth and yield dependence on nutrient supply

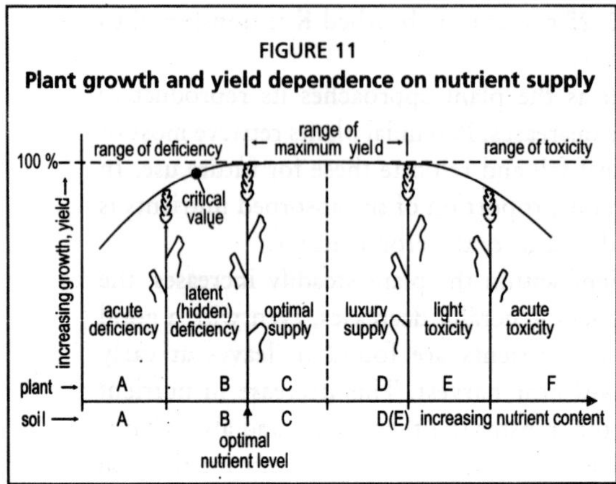

Source: Finck, 1992.

growth and yields. This range should be avoided as its occurrence is a sign of low nutrient supply or poor nutrient management and poor crop performance.

- Marginal or latent deficiency (hidden hunger): It is a small range with or without visible deficiency symptoms. However, growth and yield are reduced. Addition of the yield-limiting nutrient results in higher yields but this may not be visible. Optimal nutrient supply prevents hidden hunger.
- Optimal supply: This is the range to aim for. Here all nutrients are at the most desired level. In this range, healthy green plants, good growth and high yields with good quality can be expected. This range is generally wide for most nutrients. The optimal supply is reached above the critical concentration, which is generally associated with 90 percent of maximum yield. The critical concentration serves as a diagnostic index for nutrient supply through plant analysis (Chapter 4).
- Luxury supply: Although there is no definite borderline between optimal and luxury supply, it is useful to identify this range of unnecessarily high nutrient supply. Even if there may not be any negative effects on plant growth or yield, nutrient input is wasted and product quality as well as disease resistance may be reduced especially in the case of excess N. Therefore, luxury consumption of a nutrient should be avoided. In other words, optimal supply should be maintained and not exceeded except in special cases, such as the need for protein enrichment in grain for quality considerations (Chapter 10).
- Marginal or light (hidden) toxicity: Here the nutrient concentration is moving towards toxicity. Above the critical toxic concentration, crop growth and yield start to decrease because of the harmful effects of a nutrient surplus, or of toxic substances on biochemical processes and imbalances. No symptoms may be evident, as in the case of hidden hunger.
- Acute toxicity: This is the other extreme of excessive supply or poor nutrient management. Plants are damaged by toxic levels resulting in toxicity symptoms, poor or no growth, poor yield, low quality and damage to soil and plant health. The disease resistance of plants may also be lowered and the plant may even die. This range should definitely be avoided for any nutrient.

Nutrient interactions

It is not easy to provide plants with exactly adequate amounts of all nutrients, and the task is made more difficult by numerous interactions between nutrients. On the one hand, nutrients have their individual specific functions as described above. On the other hand, there are also some common functions as well as interactions. These can be positive or negative. Where a nutrient interaction is synergistic (positive), their combined impact on plant production is greater than the sum of their individual effects where used singly. In an antagonistic (negative) interaction, their combined impact on plant production or concentration in tissues is lower than the sum of their individual effects:

➤ synergistic (positive) interaction:
- effect of nutrient A on yield = 100,
- effect of nutrient B on yield = 50,
- effect of combined use of A and B on yield = greater than 150;

➤ antagonistic (negative) interaction:
- effect of nutrient A on yield = 100,
- effect of nutrient B on yield = 50,
- effect of combined use of A and B on yield = lower than 150;

➤ additive effect (no interaction):
- effect of nutrient A on yield = 100,
- effect of nutrient B on yield = 50,
- effect of combined use of A and B on yield = 150.

Where they occur, antagonistic interactions are caused mainly by imbalanced nutrient supply and suboptimal nutrient ratios required for satisfactory growth and development. Therefore, from a practical point of view, many unwanted antagonistic (negative) interactions can be avoided by maintaining a balanced nutrient supply.

The soundness of a nutrient management programme can be judged from the extent to which it is able to harness the benefits that accrue from positive interactions between nutrients and other production inputs. Some available results on the contribution of positive interactions for several pairs of nutrients and other inputs are summarized in Table 7. The synergistic advantage would have been lost and nutrient-use efficiency (NUE) would have been reduced if only one of the two nutrients had been used and the other had been neglected.

Positive interactions have a very high pay-off for farmers, and research must make available all the possible positive interactions for the use of farmers and also tell them how the negative ones can be kept at a safe distance from their fields. The need to harness positive interactions will be felt increasingly

TABLE 7

Some examples of synergistic interactions between nutrients and other inputs

Interacting inputs	Crop	Response attributes to positive interaction (% of total response)
Nitrogen × phosphorus	Wheat	30
Nitrogen × phosphorus	Maize	26
Nitrogen × phosphorus	Sorghum	50
Nitrogen × potassium	Pineapple	46
Nitrogen × potassium	Rice	38
Nitrogen × sulphur	Rapeseed	25
Potassium × boron	Black gram	41
Nitrogen × water	Rice	34
Nitrogen × weed control	Wheat	33
Phosphorus × population	Pigeon pea	26
Phosphorus × weed control	Chickpea	26

Source: Tandon, 2004.

as agriculture becomes more intensive and investments in inputs increase. Cooke (1982) states: "In a highly developed agriculture, large increases in yield potential will mostly come from interaction effects. Farmers must be ready to test all new advances that may raise yield potentials of their crops and be prepared to try combinations of two or more practices."

ROOT GROWTH AND NUTRIENT UPTAKE

As plants absorb nutrient primarily through their roots, regardless of the type of plant, good growth and proliferation of roots is essential for efficient nutrient uptake. Root growth can be favoured or retarded by soil physical and chemical factors. Even small roots must be able to permeate the rooting volume of the soil in both lateral and vertical directions. The major portion of nutrients is taken up by the root hairs, which are about 1–2 mm long and 0.02 mm wide. These are extensions of the epidermal root cells. Root hairs vastly expand the root surface area.

Many plants develop several million of these hairs with a total length of more than 10 km. Because very close contact with the soil is required, the amount of fine roots is critical and the number and efficiency of the root hairs is also important. Many root hairs last only a few days, but this is sufficient to extract the available nutrients from the adjacent soil volume. As the main roots grow, new root hairs are formed and, thus, there is a continuous exploration of the soil volume to access available nutrients. Anything that affects root growth and its activity affects nutrient uptake.

FIGURE 12
Uptake of nutrients from the soil by a root hair, using Ca as an example

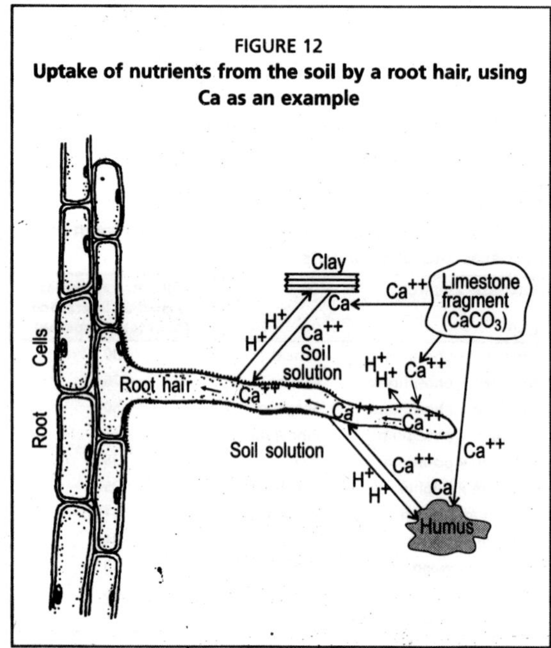

Uptake of nutrients from the soil solution

The available nutrient forms in the soil (Table 6) are free to move in the soil solution by mass flow or diffusion or up and down the soil profile with water movement. Figure 12 illustrates the processes in the vicinity of a root hair. The acquisition of nutrients depends on the size and fineness of the root system, the number of root hairs, the cation exchange capacity (CEC) of the apparent free space (AFS) or the apoplast, etc. A higher CEC results in greater uptake of divalent cations, especially Ca^{2+} (as with legumes). A lower CEC results in greater uptake of monovalent cations such as K^+.

The first step in uptake is the entry of the nutrient ion and its passing the outer layer. Nutrients can enter the cell wall without hindrance. Because of their extremely small size (hydrated K ions have a diameter of about 0.001 µm), they are able to penetrate the cell wall tissue of the root hairs. This tissue seems to be a free space and is, therefore, called AFS or the apoplast, a place different from the cytoplast (the real cell substance).

The second step in nutrient uptake involves movement of the nutrient ion into the cytoplast by crossing the membrane. The nutrients must be actively taken up into the interior of the cell. The energy required for this uptake is delivered by root respiration, a process that needs oxygen from the soil air and special uptake mechanisms. Thus, nutrient uptake by roots can be active or passive:

- Nutrients can flow passively through the cell wall (AFS) of the root hairs along with the water.
- The free flow ends at the membrane surrounding the active cell substance (cytoplast).
- Nutrients are actively transported through the membrane by special ion carriers (ionophores).
- Active uptake needs energy from root respiration, which requires sugar and oxygen (O_2).
- Cations are taken up in exchange for H^+ and anions for bicarbonate ions (HCO_3^-) on the root surface.
- Plants can preferentially select nutrients and attempt to exclude unwanted substances.

The fact that nutrient uptake is an active process explains some of its peculiarities. Plants not only accumulate nutrients against a concentration gradient, but they are also able to select from the nutrients at the root surface according to their requirements (preferential uptake). In addition, owing to their selection capacity, they can exclude unwanted or even toxic substances, but this exclusion capacity is limited. After uptake into the cytoplast, the nutrients are transported to the next cells and finally arrive at the xylem, which is the tissue through which water and dissolved minerals move upward from the roots to the stem and leaves. They move to the leaves in these water-transporting vessels where they are used for photosynthesis and other processes.

Nutrient uptake by leaves

Apart from gaseous forms of nutrients (CO_2, SO_2, etc.), leaves are able to take up nutrient ions (Fe^{2+}) or even molecules (urea). Although the outer layer of the leaf cuticle closely protects the plant against water loss, nutrients enter the leaves either via the stomata, which serve for gas exchange, or mainly via small micropores of the cuticle and into the apoplast. Foliar application of nutrients is carried out through dilute solutions in order not to damage the leaf cells by osmotic effects (Chapters 6 and 7).

EFFICIENT USE OF NUTRIENTS

Most nutrient sources added to the soil involve a monetary expense and, thus, should be utilized, as far as possible, during the vegetative growth period in order to obtain a quick return. Some residual effect during the following season should be acceptable, but losses should be kept low. The magnitude and duration of the residual effect depends on the nutrient, soil properties and cropping intensity. Balanced and adequate supply of plant nutrients is important in order to achieve a high degree of nutrient utilization by crops, which also results in lower losses.

In a wider sense, efficient use of nutrients can only be achieved by considering the whole production system. The nutrition of the plant must be integrated into all aspects of crop management. This requires INM in order to become fully effective (Chapter 6).

Chapter 4
Soil fertility and crop production

Soil fertility is a complex quality of soils that is closest to plant nutrient management. It is the component of overall soil productivity that deals with its available nutrient status, and its ability to provide nutrients out of its own reserves and through external applications for crop production. It combines several soil properties (biological, chemical and physical), all of which affect directly or indirectly nutrient dynamics and availability. Soil fertility is a manageable soil property and its management is of utmost importance for optimizing crop nutrition on both a short-term and a long-term basis to achieve sustainable crop production.

Soil productivity is the ability of a soil to support crop production determined by the entire spectrum of its physical, chemical and biological attributes. Soil fertility is only one aspect of soil productivity but it is a very important one. For example, a soil may be very fertile, but produce only little vegetation because of a lack of water or unfavourable temperature. Even under suitable climate conditions, soils vary in their capacity to create a suitable environment for plant roots. For the farmer, the decisive property of soils is their chemical fertility and physical condition, which determines their potential to produce crops.

Good natural or improved soil fertility is essential for successful cropping. It is the foundation on which all input-based high-production systems can be built.

SOILS AS A BASIS FOR CROP PRODUCTION

Soil, the natural medium for plant growth

Crop production is based largely on soils. For large-scale and low-cost crop production, there is no substitute for natural soils as a substrate for crops in the foreseeable future. Soils are the uppermost part of the earth's crust, formed mainly by the weathering of rocks, formation of humus and by material transfer. Soils vary a great deal in terms of origin, appearance, characteristics and production capacity. Well-developed soils generally show a distinct profile with different layers. The uppermost layer, called topsoil or A horizon, is richest in organic matter, nutrients and various soil organisms. Plants mainly use the topsoil as rooting volume to obtain water and nutrients, but they can also use the subsoil (partly corresponding to B horizon) or even lower layers up to 1 m or even deeper (Figure 13).

Major types of soils are formed from rocks by weathering processes over long periods extending to more than 1 000 years. During weathering, physical disintegration of rocks and minerals occurs, and chemical and/or biochemical soil forming processes lead to their decomposition. The result is the synthesis of new products, e.g. clay minerals and humic substances. Mineral or organic

FIGURE 13
A vertical cross-section of a typical soil profile showing soil horizons (A + B = solum)

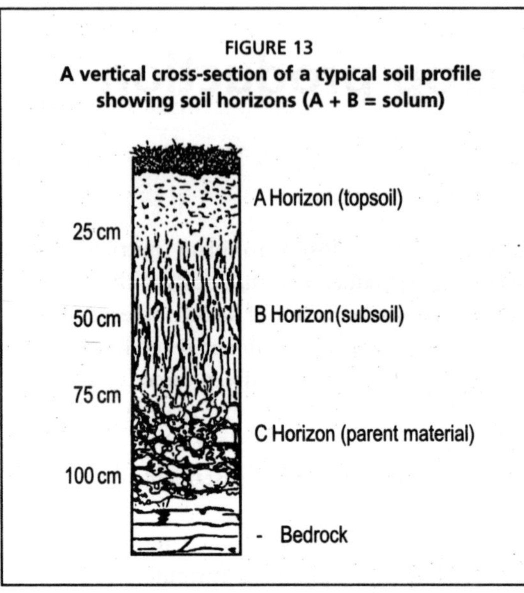

substances can be moved downwards or upwards within the profile, but they may also be lost by transportation to other places by water and wind erosion. Some of the most productive soils are the result of distant long-term geological soil erosion.

Soils vary largely with respect to their natural fertility and productivity resulting in plant growth ranging from practically zero (no growth at all on extreme problem soils) to abundant luxuriant growth of natural vegetation. However, only a small proportion of world's soils has a very good level of fertility. Most soils have only good to medium fertility and some have very low fertility, and are often referred to as marginal soils. Such areas should not generally be used for cropping but only for grazing in a controlled manner. However, under natural vegetation in a suitable climate, even soils of poor fertility may produce luxuriant vegetation where the nutrient cycle is closed, e.g. the Amazon forests.

Well-known fertile soils are deep alluvial soils formed from river mud, organic-matter-rich soils on loess material, nutrient rich Vertisols and volcanic soils. In most countries with large food demand, cropping cannot be restricted to the most fertile soils because of the large population and general shortage of usable land. However, soils with medium fertility can be improved considerably as has been demonstrated in many countries. Naturally poor or degraded soils can also be restored to a satisfactory fertility level. Under poor management, soil fertility can be seriously depleted and soils may become useless for agriculture.

Classification of soils

Soil scientists classify soils by different classification or taxonomic systems. Formerly, the classifications at national level were based on easily recognizable features and relevant soil properties for cropping. Soil-type names were generally well understood by farmers. Even on a higher classification level, the division into zonal soils (mainly formed by climate), intrazonal soils (mainly formed by parent material or water) and azonal soils (young alluvial soils) was easy to understand.

Modern and global-scale classification systems are based on developmental (pedogenic) aspects and resulting special soil properties. A common one is the system of soil types developed by FAO and the United Nations Educational and Scientific Cooperation Organization (UNESCO) used for the World Soil

FIGURE 14
The World Soil Map

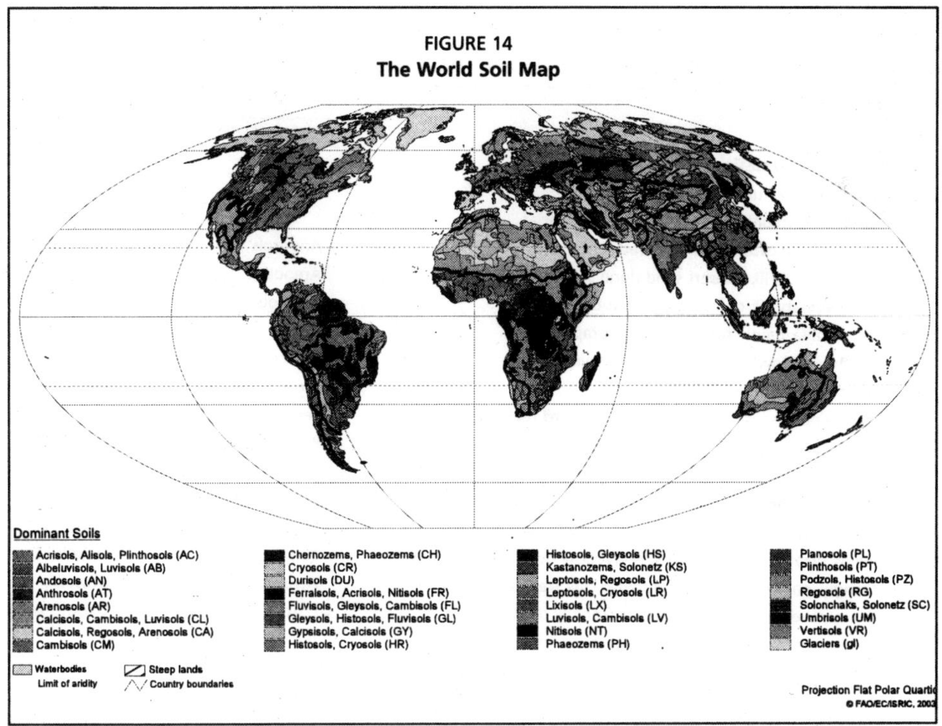

Source: FAO, 1991.

Map (Figure 14) or the international classification based on the soil taxonomy developed by the United States Department of Agriculture (USDA). The major soil units depicted are listed in Table 8, which contains the modern FAO/ UNESCO classification (28 major soil groupings, composed in 9 sets) and USDA equivalents.

The total land surface of the world is covered by the following major soils:

- soils of humid tropics, e.g. Ferralsols (Oxisols), etc.: 20 percent;
- soils of arid regions, e.g. Calcisols (Calcid), etc.: 18 percent;
- mountainous soils, Leptisols (Umbrept): 15 percent;
- soils of steppe region, e.g. Chernozems (Udolls): 7 percent;
- Podzols (Spodosols) and similar soils: 7 percent;
- clay soils of subtropics, Vertisols (Vertisols): 3 percent.

As agriculture develops, the natural properties of soils, especially of the topsoils, become more and more similar and adapted to crop requirements. This means that most cropped soils tend to become Arthrosols.

SOIL CONSTITUENTS

A soil consists of mineral matter, organic matter and pore space, which is shared by air, water and life forms. In addition to the above constituents, the soil also contains a large and varied population of micro-organism and macro-organisms

TABLE 8
Major soil groups of the FAO world soils map and USDA equivalents

No.	Type of soil	Soil order under	
		FAO/UNESCO	USDA
1.	Organic soils, consisting largely of organic matter	Histosols	Histosols
2.	Soils formed by human influence	Arthrosols	Plaggept
3.	Mineral soils, mainly formed by parent material		
	- Volcanic soils, often dark coloured	Andosols	Andisols
	- Shifting sands, like desert sand dune soils	Arenosols	Psamments
	- Dark swell–shrink clay soils of the subtropics	Vertisols	Vertisols
4.	Mineral soils, mainly formed by topography		
	- Soils in level lowlands, like young alluvial soils	Fluvisols	Fluvents
	- Waterlogged soils in level lowlands	Gleysols	Aquept
	- Mountainous soils, like shallow soil on hard rock	Leptosols	Umbrept
	- Poorly developed deeper soils in elevated regions	Regosols	Inceptisols
5.	Recently developed mineral soils, from the tropics to the polar regions	Cambisols	Tropept
6.	Mineral soils formed by climate/vegetation in humid tropics		
	- Strongly weathered soils, irreversibly hardened, laterite	Plinthosols	Udox
	- Deeply weathered acid loams, red/yellow due to iron oxides	Ferralsols	Oxisols
	- Strongly leached very acid soils with clay transfer	Acrisols	Ultisols
	- Strongly weathered soils of the seasonally dry tropics	Lixisols, Nitisols	Oxisols, Alfisols
7.	Mineral soils formed by climate/vegetation in arid regions		
	- Saline soils, with high content of soluble salts	Solonchaks	Salid
	- Sodic soils, with a high percentage of adsorbed Na	Solonetz	Argid
	- Soils with gypsum accumulation	Gypsisols	Gypsid
	- Soils with carbonate accumulation	Calcisols	Calcid
8.	Mineral soils formed by climate/vegetation in steppe region		
	- Brown chestnut soils, in the driest steppe areas	Kastanozems	Ustoll
	- Black earths, with deep dark topsoil	Chernozems	Udoll
	- Soils of prairie regions, e.g. degraded chernozems	Phaeozems	Boroll
	- Soils with high humus content, e.g. grey forest soils	Greyzem	Mollisols
9.	Mineral soils formed by climate/vegetation in subhumid regions		
	- Brown soils, base rich, clay transfer; similar to Podzoluvisols	Luvisols	Alfisols
	- Poorly drained, low-lying soils	Planosols	Alboll
	- Acid soils with ash-grey layer above iron oxide horizon	Podzol	Spodosols
	- Brown soils, base rich, clay transfer; similar to Podzoluvisols	Luvisols	Alfisols
	- Poorly drained, low-lying soils	Planosols	Alboll
	- Acid soils with ash-grey layer above iron oxide horizon	Podzol	Spodosols

Source: Abridged from Driessen and Dudal, 1991.

that play an important role in plant nutrition. Figure 15 indicates the relative proportions of each of these constituents in an "average soil" on a volume basis.

About 45–50 percent of the volume of a normal soil consists of mineral matter, 1–5 percent is organic matter and the remaining 50 percent consists of open pore spaces that are shared by air and water. In a very dry soil, most of these pores are full of air, while in a saturated soil, they are filled with water. Ideally, air and water occupy about equal space, the air residing in the larger pores and water in the smaller ones. Both are needed for the soil to serve as a medium for plant growth. The organic matter and the pores also house a variety of plant and animal life ranging from microscopic bacteria to earthworms and rodents.

Of the various soil components, the mineral matter changes little during a farmer's lifetime. The organic matter can be increased, maintained or depleted depending upon the amounts of organic manures used and the rate at which these are decomposed. The air–water status can change on a day-to-day or even hourly basis.

FIGURE 15
The average proportion of various constituents in a common soil on volume basis

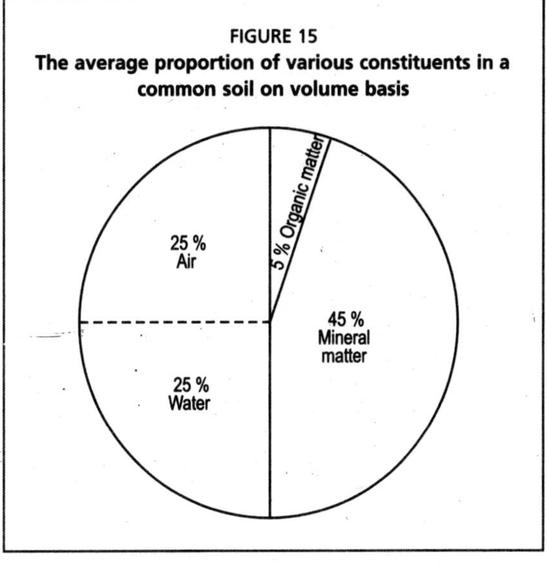

Soil mineral matter

The soil mineral matter ranges from large pieces of gravel, pebbles and nodules to small grains of sand, silt and clay particles. In addition, there are various oxides, sulphates, silicates and carbonates. The mineral matter on the earth's surface is made up largely of oxygen (47 percent) and Si (28 percent). Only eight elements are present in amounts greater than 1 percent. Among the plant nutrients, the most abundant are not N or P but Fe, Ca, K and Mg.

The difference in size between a coarse sand particle (2.0–0.2 mm diameter) and fine clay (less than 0.002 mm) is a thousandfold. The small size of clay particles gives a very large reactive surface area. While all the mineral fractions determine the texture and waterholding capacity (WHC) of a soil, the sand fraction does not do much. The silt acts as a storehouse of certain nutrients, such as K, while the clay fraction is the most active and a class by itself. As described blow, clay particles along with organic matter give the soil its CEC – a property of great importance from a nutrient management point of view.

Some common minerals formed in soils are listed below on the basis on their degree of solubility (Bolt and Bruggenwert, 1976):

- minerals of high solubility:
  - nitrates,
  - chlorides,
  - bicarbonates,
  - sulphates except calcium sulphate;
- minerals of intermediate solubility:
  - gypsum (calcium sulphate),
  - calcite (calcium carbonate),
  - pyrite (under reduced conditions);
- minerals of low solubility:
  - apatite (tricalcium phosphates),
  - oxides and hydroxides of Si, Al and Fe,

- silicates of Ca, Ma and K.

Saturated solutions of salts with a high solubility inhibit plant growth because of high osmotic pressure. These salts in solid phase are only present in significant quantities under exceptional circumstances (e.g. deserts and in saline or sodic soils). Minerals of intermediate solubility are those with a saturated solution not inhibiting plant growth but their solubility is high enough to contribute significantly to the nutrient composition of the soil solution. The minerals of low solubility contribute to plant nutrient supply only in the long term. Minerals of the silicate, phosphate and oxide groups are almost the end point of the weathering. The fraction of soil mineral matter that contributes to nutrient supply is that which has a moderate degree of water solubility or that with a relatively high specific surface area (surface area / unit weight).

Soil organic matter

Most soils are of mineral origin, but their topsoil contains organic matter that, in spite of its low content, is of great importance to many aspects of soil fertility and plant growth. Soil organic matter (SOM) can range from less than 1 percent in many tropical arid and semi-arid soils of the plains to 5 percent or more in temperate regions or under forest vegetation. The average composition of SOM is 47 percent C, 44 percent O, 7 percent H, 2 percent N and very small amounts of other elements. More than half of SOM consists of carbohydrates, 10–40 percent is the resistant material lignin and the rest consists of compounds of N.

The whole complex of organic matter along with soil organisms and soil flora is of vital importance to soil fertility. SOM contains the well-decomposed fine humus fraction, small plant roots, and members of the plant (flora) and animal (fauna) kingdoms. SOM plays a role far greater than its share of the soil volume. It is a virtual storehouse of nutrients, plays a direct role in cation exchange and water retention, releases nutrients into the soil solution and produces acids that affect the fixation and release of other nutrients.

SOM or "humus" reaches equilibrium during soil formation. Wet and/or cold soil conditions tend to increase the humus content, whereas high temperatures of tropical climates and cropping procedures promote its decomposition. The C:N ratio provides a general index of the quality of SOM, being in the range of 10–15:1 for fertile soils. When organic manures or green manures are added, these become a part of the organic pool of the soil.

Soil pore space

Soil volume that is not occupied by mineral or organic matter is referred to as pore space. This is shared by soil water, soil air and soil life. It has about ten times more CO_2 than the atmosphere. This CO_2 is produced as a result of breathing (respiration) by roots and soil micro-organisms. The ratio of pore space to the volume of solid material in the soil is termed the pore space ratio (PSR). It is an important soil property that determines the dynamics of air, water, temperature and nutrients and also the available root space and ease of working the soil.

SOIL PROPERTIES AND PLANT REQUIREMENTS

Plants need anchorage, water and nutrients from the soil but are sensitive to excesses of growth-impeding substances in the soil. The supply and uptake of nutrients from the soil is not a simple process but requires a suitable combination of various soil properties:

- physical properties (depth, texture, structure, pore space with water and air);
- physico-chemical properties (pH and exchange capacity);
- chemical and biological properties (nutrient status, their transformation, availability and mobility).

A major objective of having the most suitable soil physical, chemical and biological conditions is to provide the most favourable environment for the roots to grow, proliferate and absorb nutrients.

Soil physical properties

Soil physical properties largely determine the texture, structure, physical condition and tilth of the soil. These in turn exert an important influence on potential rooting volume, penetrability of roots, WHC, degree of aeration, living conditions for soil life, and nutrient mobility and uptake. These are as important as soil chemical properties.

Soil depth

Fertile soils generally have a deep rooting zone, which ideally is a minimum of about 1 m for annual crops and 2 m for tree crops. This soil volume should contain no stony or densely compacted layers or unfavourable chemical conditions that impede deep root growth. In addition, the topsoil, which is rich in humus and soil life and the main feeding area for the roots, should be at least 20 cm deep. In practice, many soils have limitations with respect to rooting depth but these can generally be improved by suitable amelioration.

Soil texture

The term texture designates the proportion of different particle size fractions in the soil. Soil texture is primarily represented by its mineral fraction; the organic matter is usually ignored during texture evaluation. Of special importance to soil fertility is the percentage of soil particles of less than 2 mm in diameter, which constitute the fine soil. Fine soil is composed of particles in three size groups: sand, silt and clay. International size units used to classify soil particles in terms of their mean diameter are:

- gravel (> 2.0 mm);
- coarse sand (2.0–0.2 mm);
- fine sand (0.2–0.02 mm);
- silt (0.02–0.002 mm);
- clay (< 0.002 mm).

Based on the relative proportions of these components, soils are classified into different textural classes, such as sandy, loamy or clay soils, and several

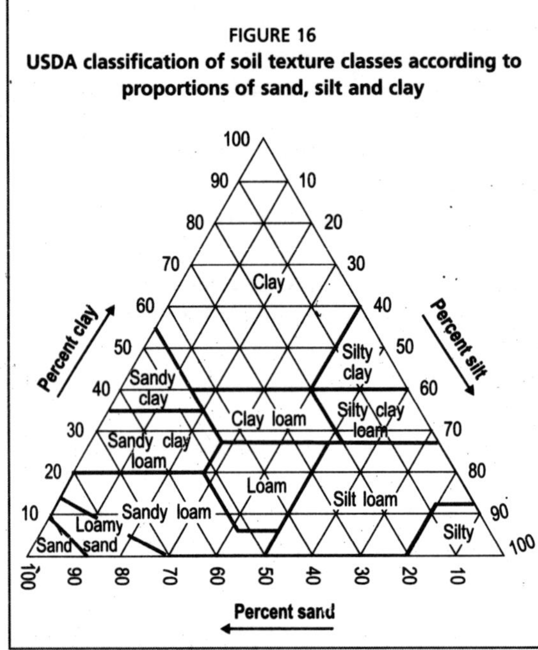

FIGURE 16
USDA classification of soil texture classes according to proportions of sand, silt and clay

intermediate classes (Figure 16). The size of a particle is not related to its chemical composition, e.g. sand may be just silica or nutrient rich feldspar or apatite. Clay particles are of colloidal size and are of special importance because of their mechanical and chemical activity. They consist of plate-like structures that have large external and internal charged surfaces for exchange of nutrient ions, particularly the positively charged cations. Some common types of clays are kaolinite, illite and montmorillonite.

The texture of a soil can be determined approximately by simple "finger rubbing" of moist samples, or precisely by conducting a mechanical (particle size) analysis in a laboratory. Terms used by farmers and sometimes even by researchers such as "light" for sandy soils and "heavy" for clay soils are not based on the actual weight (the volume weight of sandy soils being greater than that of clay soils), but on the practical perception of ease of cultivation as mechanical cultivation requires less energy on a light soil than on a heavy soil.

Soil texture influences to a large extent several components of soil fertility such as the amount of nutrient reserves and their proportion to the available nutrient fraction. It also influences several properties such as aeration, pore space distribution, WHC and drainage characteristics. The broad relation of soil texture with soil fertility can be stated as follows:

- Sandy soils are generally poor in nutrient reserves and have a low WHC, but provide favourable conditions for root growth, soil aeration and drainage of surplus water.
- Clay soils are often rich in nutrient reserves (but not necessarily in plant available forms), have high WHC because of the many medium and small pores, but soil aeration is restricted.
- Loamy soils, like sandy or silty loam, have intermediate properties and are generally most suitable for cropping.

Table 9 lists the effect of various textural classes on different physical properties of a soil. A good soil for plant growth should contain moderate quantities of all the different fractions. As the proportion of any fraction increases, such a soil becomes more suitable for plant growth in some respects and less suitable in other respects. For example, where a soil contains a large proportion of sand, it is well aerated,

TABLE 9
Physical properties of soils related to primary particle size fractions

Physical property	Relative impact of the fraction on the soil property			
	Coarse sand	Fine sand	Silt	Clay
Water holding capacity	Very low	Medium	High	Very high
Capillarity	Very low	Good	Great	Very great
Rate of water movement	Very fast (down)	Fast (down and up)	Slow (down and up)	Very slow (down and up)
Absorption capacity (for gases, water and plant nutrients)	Very slight	Slight	High	Very high
Cohesion and plasticity	None	Slight	High	Very high
Relative temperature	Warm	Fairly warm	Cold	Very cold
Aeration	Very good	Fairly good	Poor	Very poor

excessive moisture drains away easily, and the soil is easy to cultivate. However, such a soil retains too little moisture and plants can quickly suffer from moisture stress. Plant nutrients also leach out of such a soil very easily. Where a soil has a high clay content, it retains moisture and plant nutrients well but such a soil is also poorly aerated, becomes easily waterlogged, and is difficult to cultivate. Thus, it is easy to understand why sandy loams (that contain all the particle size fractions in favourable proportions) are considered the most productive agricultural soils.

Soil structure

The individual particles of the fine soil fraction are usually bound together by organic/inorganic substances into larger aggregates. The process, known as aggregation, results in a vastly increased pore space that is occupied by air and water. The three-dimensional arrangement of the different sizes and shapes of soil aggregates is termed soil structure. In contrast to soil texture, it is a rather variable soil property that, from an agronomic point of view, can improve or deteriorate. There are different types of soil structure, e.g.: single grain and granular structure with good water permeability; blocky and prismatic structure with medium water permeability; and platy and massive structure with slow water permeability. The clay particles, some of which have swelling and shrinking properties (as in black clay soils such as Vertisols) depending on water content are important components of structure formation and, therefore, of pore space distribution.

For agricultural use, the best type is a stable or large granular "crumb" structure with biologically formed sponge-like aggregates of 0.1–1 cm formed by earthworms in combination with string-forming fungal hyphae or gum-producing microbes. Such crumbs are stable against wetting and have a good mixture of different pore sizes, which are desirable characteristics of fertile soils. In contrast, crumbly pieces formed by mechanical tillage are usually much less stable. An important feature of good soil structure is its stability against deteriorating processes such as wetting and pressure. Annual cropping with relatively high disturbance of soils often results in some structural deterioration, which can be reversed to different degrees (as in flooded-paddy soils).

The soil pore system with water and air

The solid soil particles leave large and small holes between them, which make the soil a porous system. The PSR determines the dynamics of air, water, temperature and nutrients and also the available root space and ease of working the soil. Because of the large portion of pores in soils, the volume weight of mineral topsoils is only about 1.5 (1 litre of soil = 1.5 kg).

In a soil, there are a wide range of pore sizes present, and the percentage of the total pore space made up of any particular size varies greatly between soil textural classes. The multiple-shaped pore space is filled by water and air in varying proportions depending on the water content of the soil. Ideally, mineral soils should have a pore volume of almost 50 percent with about one-third of this consisting of large pores. The size of pores determines their function:

- Large pores (10–50 µm diameter or even larger): These contain air or provide drainage.
- Medium pores (0.2–10 µm diameter): These contain the available water.
- Small pores (less than 0.2 µm diameter): These also hold water, but because of the high tension (force) with which it is tightly held, this water is not available to the plants.

Soil water

Soil water added by rain or irrigation is stored up to the WHC of the soil, which is also called the field capacity. The WHC denotes the maximum amount of water that a soil can hold after free drainage has ceased. It is the upper limit of available water. The surplus water is drained by large pores. The WHC is a key soil property because all chemical and biochemical processes require water. The capacity to store plant available water varies greatly among soils depending on their texture, depth, structure and humus content. The loamy/silty soils store the highest amounts of plant available water, whereas coarse sands store very little. Clay soils store considerable amounts of water, but a large portion is not available because it is tightly held in very small pores. Soil water is retained by adsorption and capillary forces, which are measured either in kiloPascals or by its related logarithmic pF value (a pF of 3 indicates an average moist soil). Crops generally use 300–800 litres of water to produce 1 kg of dry matter.

Some practical aspects of soil water are:

- Water is held mainly in medium-sized pores, and medium-textured soils hold the highest amounts.
- Only the free or loosely bound portion of water (15–1 500 kPa) is available to plants. Water drains freely from pores with a diameter exceeding 60 µm. Thus, pores in the range 0.2–60 µm are important in retaining plant available water.
- Nutrient ions travel to the roots as part of the soil water (soil solution).
- Roots can extract available water because of their suction forces.
- Maximum soil water storage against gravitational losses is at field capacity (pF of 2.2 = 15 kPa).

- Dry soils at the wilting point of crops have only non-available water left (pF 4.2 = 1 500 kPa).
- The storage capacity within 150 cm depth varies from 40 to 120 mm of rain for most soils.

Soil air

Soil air is generally similar in composition to atmospheric air except that it has 7–10 times higher concentration of CO_2 than does the atmosphere (0.2 percent compared with 0.03 percent).

As a result of the respiration by roots and micro-organisms, the oxygen in the soil air may be consumed quickly and CO_2 produced, which is unfavourable for both root growth and functions. For most crops, the soil air should contain more than 10 percent oxygen but less than 3–5 percent CO_2. A continuous exchange with atmospheric air, termed soil aeration, is required in order to avoid a deficiency of oxygen. In cropped fields, the breaking of surface compaction can assist in this, but it must be done without destroying soil aggregates. In terms of air–water relations, the two extremes are reprcsented by well-aerated sandy soils (excess air and a shortage of water) and the flooded-rice soils (excess water and a shortage of air).

Soil physical properties and root growth

Crop growth requires that nutrients be present in soil in adequate amounts and in suitable forms for uptake. In addition, the nutrients must be supplied to the root surface at a sufficient rate throughout the growth of the crop so that the crop does not suffer from inadequate nutrient supply. This is particularly important during periods of rapid growth when nutrient demands are high. The physical nature of the soil affects the growth of an established plant through its influence on various factors such as aeration and moisture supply. In addition, such physical properties alter the resistance offered to root elongation and enlargement, proliferation and water uptake, which in turn affect plant nutrition. There are at least three important factors that determine the rate of root elongation. These are: turgor pressure within cells, constraint offered by the cell wall, and constraint offered by the surrounding medium. All of these are affected by the soil physical environment in the vicinity of the elongating root. The requirements of plant roots in soils are:

- deep rooting volume, ease of penetration and no restrictions on root growth;
- adequate available plant nutrients from soil reserves, external inputs or from N fixation;
- sufficient available water to support plants and soil life, for nutrient transformations and for nutrient transport to the roots;
- facility for the drainage of excess water from the rootzone to ensure the right air–water balance (except flooded-paddy fields).
- good soil aeration to meet the oxygen requirements of roots and for the removal of surplus CO_2.

TABLE 10
The effect of moisture and of soil compaction on the growth of maize plants

Treatment		Weight of tops	Weight of roots	Weight of total plant	Top:root ratio
Compaction	Moisture		(g)		
Loose	wet	39.4	14.8	54.2	1:0.38
Loose	dry	27.5	9.3	36.8	1:0.34
Compact	wet	16.0	6.5	22.5	1:0.40
Compact	dry	20.1	11.3	31.4	1:0.56

Root growth and the dynamics of water and air are largely dependent on pore space. Root growth occurs within continuous soil pores, within disturbed zones resulting from macro-organism activity and within the soil matrix itself. Pore size distribution is important for root penetration, water retention and aeration. In general, roots take the path of least resistance as they grow in soil. Root growth is reduced where the pore size is smaller than the root diameter because the plant must spend energy to deform the pore. The existence of sufficient continuous pores of adequate size is an important determinant of root growth. Most of the roots are 60 µm or more in diameter. The first-order laterals of cereals may range in size from 150 to 170 µm. In contrast, the root hairs are much smaller than 60 µm.

Bulk density (in grams per centimetre) is an indirect measure of pore space within a soil. The higher the bulk density, the more compact is the soil and the smaller is the pore space. In addition to absolute pore space, bulk density also affects the pore space distribution (according to size). Soil compaction decreases the number of large pores (> 100 µm) and, as these are the ones through which roots grow most easily, compaction can have an adverse effect on root growth. The effect of bulk density may be altered considerably by changing the moisture content of the soil. As the pore space can be filled with either air or water (containing nutrients) and there is an inverse relationship between these two parameters, an increase in moisture content means a decrease in air-filled pores.

In general, a decrease in soil moisture content reduces root growth even though more space is physically available to roots. Moreover, where the soil moisture content exceeds field capacity, this leads to poor aeration and root growth declines. Table 10 lists the effects of bulk density (compaction) and aeration (moisture) on plant growth. It shows that compaction of the soil under wet conditions can result in a marked decrease in root and top growth through a combination of mechanical impedance and aeration problems. There is a positive response to moisture in loose soil because the large pores drain easily and plant can suffer from a shortage of water. In contrast, adding water to the compact soil reduces root growth because of a lack of air in the soil pores caused by the displacement of air by water.

Organic matter and soil fertility

The effect of SOM on soil fertility far exceeds its percentage share of the soil volume. Organic matter affects soil fertility and productivity in many ways:

- It promotes soil structure improvement by plant residues and humic substances leading to higher WHC, better soil aeration and protection of soil against erosion.

- ➢ It influences nutrient dynamics, particularly:
  - nutrient exchange, thus keeping the nutrients in available forms and protecting them against losses;
  - nutrient mobilization from decomposed organic nutrient sources: N, P, S, Zn, etc.;
  - nutrient mobilization from mineral reserves by complex formation or by changes in pH and redox potential;
  - immobilization of nutrients on a short-term or long-term basis (reverse of mobilization);
  - nutrient gain as a result of N fixation from the air.
- ➢ It influences promotion or retardation of growth through growth hormones.

Organic substances in the soil are important nutrient sources. Moreover, some substances can bring about the mobilization of nutrients from soil mineral reserves by the production of organic acids, which dissolve minerals, or by chelating substances excreted by roots and/or by microbes. Chelates may bind Fe from iron phosphate and, thus, liberate phosphate anions. Organic matter may also have some negative effects, namely the short-term fixation of nutrients such as N, P and S into micro-organisms, which may create a transient deficiency particularly at wide ratios of C with these elements (e.g. C:N, C:P, and C:S ratio). The long-term fixation of these elements into stable humic substances appears to be a loss but it can be beneficial because of its positive effect on soil aggregation and, hence, on soil structure.

Rapid and far-reaching loss of SOM is an important factor in soil degradation. Many of the effects of organic matter are connected with the activity of soil life.

Soil organisms and soil fertility

Soil abounds in the following various types and forms of plant and animal life:

- ➢ animal life (fauna):
  - macrofauna (earthworms, termites, ants, grubs, slugs and snails, centipedes and millipedes),
  - microfauna (protozoa, nematodes and rotifers);
- ➢ plant life (flora):
  - macroflora (plant roots, and macro-algae),
  - microflora (bacteria, actinomycetes, fungi and algae).

Beyond the soil-forming activities of earthworms, termites and other large soil fauna, the multitude of different soil organisms (colloquially also called soil life) contributes significantly to the soil physical and chemical conditions, especially in the transformation of organic matter and plant nutrients. The rate of transformation of most nutrients into available forms is controlled largely by microbial activity. Their huge number represents an enormous capacity for enzyme-based biochemical processes. A special case is N fixation by N-fixing free-living or symbiotic bacteria. Another case relates to the solubilization of insoluble phosphates by several types of soil micro-organisms (Chapter 5).

Soil micro-organisms have similar requirements of soil conditions for optimal activity in terms of air, moisture and pH, as do crops. In general, fungi are more active under acidic conditions, while bacteria prefer neutral–alkaline reaction. Any improvement in soil fertility for crops should also improve conditions for the activity of soil flora and fauna. Microbial activity not only determines soil fertility but it also depends on good soil fertility.

Soil physico-chemical and chemical properties

Three important physico-chemical characteristics of soil fertility are: (i) soil reaction or pH; (ii) nutrient adsorption and exchange; and (iii) oxidation-reduction status or the redox potential.

Soil reaction

The reaction of a soil refers to its acidity or alkalinity. It is an important indicator of soil health. It can be easily measured and is usually expressed by the pH value. The term pH is derived from Latin *potentia Hydrogenii* and is the negative logarithm of the H^+ ion concentration (logarithm of grams of H^+ per litre). Because of the logarithmic scale used, in reality, the actual degree of acidity has enormous dimensions, e.g. the difference in acidity between pH 4 and 5 is tenfold. Thus, a soil of pH 5 is 10 times more acid than a soil of pH 6 and a soil of pH 9 is 10 times more alkaline than a soil of pH 8.

The importance of soil pH is:

- the pH value indicates the degree of acidity or alkalinity of a medium, in this case soil;
- pH 7 is the neutral point, pH of 6.5–7.5 is generally called the neutral range;
- acid soils range from pH 3 to 6.5, alkaline soils from pH 7.5 to 10;
- most soils are in the pH range of 5–8, while the range for plant growth is within pH 3–10;
- the pH of a soil can be altered by amendments and nutrient management practices.

Soil pH is measured in soil/water suspensions. Where dilute calcium chloride solution is used instead of water, the data are lower by 0.5–1.0 of a unit. The pH value obtained is an average of the volume tested. In nature, there is a natural tendency towards soil acidification, the rate of which often increases under leaching, intensive cropping and persistent use of acid-forming fertilizers. Strong acidification leads to soil degradation. However, this can be overcome by the application of calcium carbonate (lime) or similar soil amendments. Unfavourable high pH values, as observed in alkali soils, can be decreased by amendment with materials such as gypsum, elemental S or iron pyrites. Various amendments for acid and alkali soils are discussed in Chapter 5.

Soil reaction is not a growth factor as such but it is a good indicator of several key determinants of growth factors, especially nutrient availability. Soil reaction greatly influences the availability of several plant nutrients. For example, phosphate is rendered less available in the strongly acidic upland soils. The availability of heavy metal nutrients (Cu, Fe, Mn and Zn) increases at lower pH, except for Mo. Although not a nutrient, Al becomes toxic below pH 4.5 (Figure 17). Most plants grow well in the neutral to slightly acid range (pH 6–7) with the dominant cation Ca. Plants are generally more sensitive to strong alkalinity, where the dominant cation is often Na, than to strong acidity where the dominant cation is H. The range of slight and moderate acidity can have special advantages in respect of nutrient mobilization. Soils with very strong acidity (below pH 4.8) contain high levels of soluble Al. Almost no plants can survive below pH 3.

FIGURE 17
Soil pH and nutrient availability

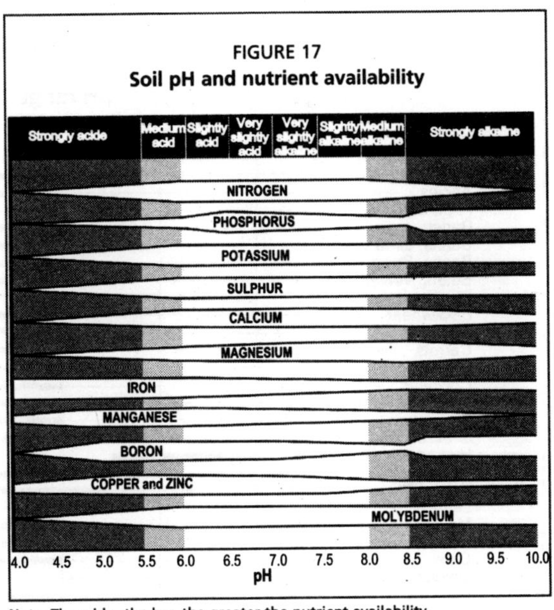

Note: The wider the bar, the greater the nutrient availability.
Source: Brady and Weil, 1996.

The preference of plants for a certain pH range is often determined by aspects of nutrient requirement and efficiency and not because of the pH as such. For example, oats prefer a slightly acid range because of better Mn supply. Tea bushes benefit from an acid environment. This preference could also be caused by the adaptation of a plant species to a certain environment over time (Figure 18). However, this does not mean that the indicated crops cannot be grown outside the depicted pH range.

Exchange capacity and plant nutrition

Only a small percentage of the available nutrients move freely in the soil solution. Most of them are loosely bound on mineral and organic surfaces in exchangeable form. This mechanism acts as a storehouse both for nutrient cations and anions. Clay minerals, especially illitic and montmorillonitic types, have large negatively charged surfaces on which cations like Ca^{2+}, Mg^{2+} and K^{+} are adsorbed and, therefore, protected against leaching. Other particles like oxides and some humic substances also have positive charges and are able to bind anions like phosphate and, to a lesser extent, sulphate. By contrast, nitrate and chloride are hardly bound at all and can be easily lost from the rooting volume by leaching if not taken up by the plant.

The capacity of a soil or any other substance with a negatively charged exchange complex to hold cations in exchangeable form is referred to as its CEC. It is a measure of the net negative charge of a soil and is expressed in me/100 g soil

FIGURE 18
Optimal soil pH for different crops

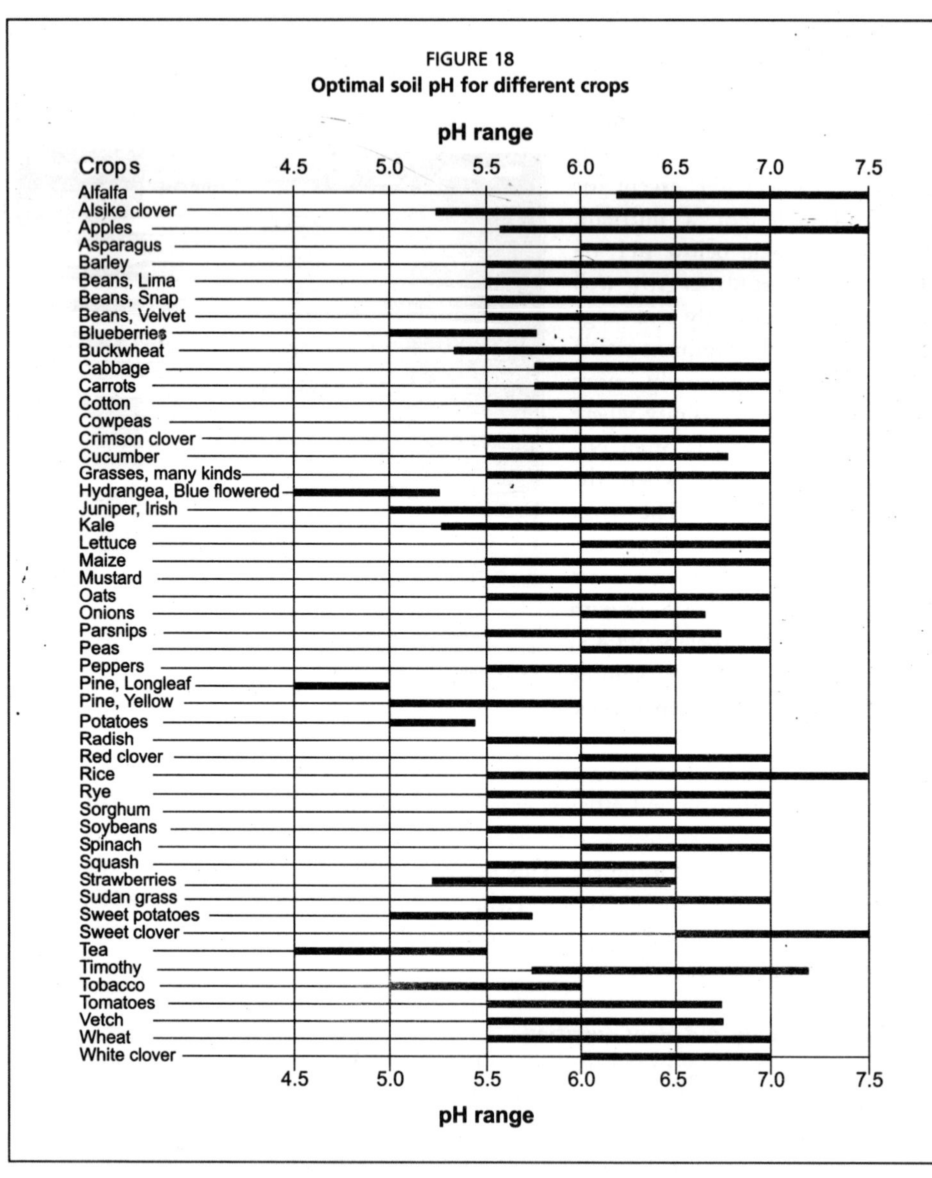

(old term) or cmol (p^+)/kg or cmol/kg (new term). The CEC depends on the type and proportion of clay minerals and organic matter present in the soil. Clay soils have a higher CEC than do sandy soils. Similarly, soils rich in organic matter have a higher CEC than soils that are low in organic matter. Different cations are held on to the exchange sites with differing adsorption affinity or bonding strength (Schroeder, 1984). This determines the ease or the difficulty with which these can be dislodged from the exchange site by cations in the solution. In general, the strength with which different cations are held on the exchange complex is in the following order (which means that Al^{3+} is held most tightly and Na^+ is held most loosely among the cations shown): $Al^{3+} > Ca^{2+} > Mg^{2+} > K^+, H^+, NH^{4+} > Na^+$.

The CEC ranges from less than 10 cmol/kg for sandy soils to more than 30 cmol/kg for clay soils. For average mineral soils, an ideal ratio of cations on the exchange complex would be 75:15:5-3 of Ca:Mg:K. For a soil with an average CEC of 20 cmol/kg, the amounts of exchangeable cations present would be: Ca = 9 000 kg, Mg = 1 000 kg and K = 700 kg/ha. Base saturation refers to the percentage of CEC that is occupied by basic cations. It is an important characteristic of soil fertility. The degree of base saturation is calculated as the ratio:

$$\text{ratio:} \quad \frac{Ca^{2+} + Mg^{2+} + K^+}{Ca^{2+} + Mg^{2+} + K^+ + H^+ + Al^{3-}}$$

If a soil has a CEC of 25 cmol/kg and out of this 3 cmol/kg is occupied by H^+ and the rest by basic cations, then the base saturation will be 88 percent (22/25 × 100). The availability of cations to plants varies according to the strength of bonding with the exchange complex. However, exchangeable cations are generally considered available to plants either directly by contact exchange or more frequently by desorption into the soil solution. In general, soils with a high degree of base saturation are the most fertile ones, provided the exchange complex is not dominated by one particular cation (e.g. Na in sodic soils).

Similarly, the replacement of one anion by another on positively charged soil surfaces or other anion exchange media such as resins is referred to as anion exchange. An example is the exchange of $H_2PO_4^-$ with F^- or of NO_3^- with Cl^-. Anions can enter and exit the soil solution through anion exchange while still remaining in the soil. Anion exchange is of less significance than cation exchange in soil fertility management and plant nutrition. The capacity of a material (soil) to hold anions in exchangeable form is its anion exchange capacity (AEC). It is expressed as centimoles per kilogram of soil or cmol (e^-)/kg. As in the case of cations, different anions are also held on to the exchange complex with varying strength, which is in the order: $PO_4^{3-} > SO_4^{2-} > NO_3^- > Cl^-$.

A decrease in CEC or AEC can result from structural breakdown where clay surfaces become inaccessible or from humus decomposition, but good soil management can reverse this. Table 11 presents a guide to the interpretation of CEC data. The value for each nutrient cannot be considered by itself but in relation to the other ions present. In addition, the rate of its movement to the plant roots is very important. Such rates are determined by soil texture, soil moisture status and nature of the cation, among other factors.

TABLE 11
Ranges of exchangeable cation in soil for the interpretation of cation exchange data

Rating	Exch. Ca	Exch. Mg	Exch. K	Exch. Na	CEC
			(cmol/kg)		
Very high	> 20	> 8	> 1.2	> 2	> 40
High	10–20	3–8	0.6–1.2	0.7–2	25–40
Medium	5–10	1–3	0.3–0.6	0.3–0.7	12–25
Low	2–5	0.3–1	0.2–0.3	0.1–0.3	6–12
Very low	< 2	< 0.3	< 0.2	< 0.1	< 6

Redox potential of the soil

The redox potential is a very important soil property that has a marked influence on the oxidation-reduction state of a soil and, hence, on nutrient dynamics. It is denoted by Eh. The redox potential of a soil varies from -300 mV (strongly reducing condition) to +800 mV in a well-aerated upland soil. The higher the redox potential is, the higher is the oxidizing power of the system. Where the redox potential is high, there is much dissolved oxygen in the soil solution and there is a high proportion of oxidized compounds such as ferric oxide, nitrate and sulphate. Where the redox potential is low, the soil system is deficient in oxygen and there is a greater concentration of reduced forms such as ferrous, ammonium and sulphide. A low redox potential is caused primarily by micro-organisms that break up oxidized compounds and ions in order to obtain oxygen for their metabolism. When the reducing conditions set in, the sequence in which various compounds or ions are set free is: NO_3^-, Mn^{2+}, Fe^{2+}, SO_4^{2-} and, finally, reducible organic matter.

The redox potential is of greatest importance in submerged soils such as under flooded-rice cultivation. As the reducing conditions set in following flooding or ponding, the soil pH moves towards neutrality. In general, the solubility of P increases because of the reduction of iron phosphates; the solubility of Fe and Mn also increases as their less soluble oxidized forms are converted to more soluble reduced forms, e.g. from ferric (Fe^{3+}) to ferrous (Fe^{2+}). Under reduced conditions, the nitrate N is also lost through denitrification.

NUTRIENTS IN SOILS AND UPTAKE BY PLANTS

Soil nutrient sources

Many soils have vast reserves of plant nutrients but only a small portion of these nutrients becomes available to plants during a year or cropping season. Nutrients are present in both organic and mineral forms as stated in Table 6. However, all forms must change themselves to specific mineral ionic forms in order to be usable by plant roots. Thus, in order to become available to plants, nutrients must be solubilized or released from mineral sources and mineralized from organic sources including SOM. Although nutrient mobilization is a rather slow process, it increases sharply with temperature. A temperature increase of 10 °C doubles the rate of chemical reactions. Consequently, the 20–30 °C higher temperature in tropical areas results in chemical transformations (e.g. nutrient mobilization or humus decomposition) at 4–6 times higher the rate in temperate areas.

About 1–3 percent of SOM is decomposed annually and this is a key determinant of N supply. If a fertile soil contains 8 000 kg N/ha in the organic matter (e.g. 2 percent), this corresponds to 160 kg of N transformed from organic N into ammonia, which may then be converted into nitrate. From this amount, crops may utilize about 50 percent, some is taken up by micro-organisms and some lost by leaching, denitrification and volatilization.

As crop yields have increased over the years as a result of technological changes, few soils are able to supply the amounts of nutrients required to obtain higher yields without external inputs. An ideal soil is rich in mineral and organic sources of plant nutrients. In addition, it has the following characteristics:

- It has a strong capacity to mobilize nutrients from organic and inorganic sources.
- It stores both mobilized and added nutrients in forms that are available to plant roots, and protects them against losses.
- It is efficient in supplying all essential nutrients to plants according to their needs.

Practical importance of nutrients

A number of plant nutrients are of large-scale practical importance for successful crop production in many countries. Prominent among these are N, P, K, S, B and Zn. This means that their deficiencies are widespread and external applications are necessary to augment soil supplies for harvesting optimal crop yields while minimizing the depletion of soil nutrient reserves.

N deficiency is widespread on almost all soils, especially where they are low in organic matter content and have a wide C:N ratio. Rare exceptions are soils with very high N-rich organic matter content during the first years of cropping, e.g. after clearing a forest. Widespread N deficiency is reflected in the fact that out of the 142 million tonnes of plant nutrients applied worldwide through mineral fertilizers, 85 million tonnes (60 percent) is N. In addition, substantial external N input is received through organic manures, recycling and BNF.

P deficiency was serious before the advent of mineral fertilization because the native soil phosphate was strongly sorbed in very acid soils or precipitated as the insoluble calcium phosphate in alkaline soils. P deficiencies continue to be a major production constraint in many parts of the world. External input through mineral fertilizers alone was 33.6 million tonnes P_2O_5 in 2002.

K deficiency is most strongly expressed in acid red and lateritic soils or on organic soils that have few K-bearing minerals. Soils rich in 2:1-type clays and those in arid or semi-arid areas are generally better supplied with K than soils in humid regions because of lower or no leaching losses in the former. Ca supply is abundant in most neutral–alkaline soils and, hence, field-scale Ca deficiencies are rare. Where a Ca deficiency occurs, it is mainly in acid soils or because of insufficient uptake and transport of Ca within the plant. Mg deficiency can be widespread in acid soils as a consequence of low supply and leaching losses.

S deficiency was of little practical importance decades ago because of considerable supply from the atmosphere, and widespread use of S-containing fertilizers such as single superphosphate (SSP) and ammonium sulphate. However, S deficiency has developed rapidly in recent years as the atmospheric inputs have declined and high-analysis S-free fertilizers have dominated the product pattern used. The problem has become significant and soil S deficiencies have been reported in more than 70 countries. About 9.5 million tonnes of S are currently applied as fertilizer worldwide (Messick, 2003).

Micronutrient deficiencies are common because of certain soil conditions and have developed at higher yield levels and on sensitive crops. On a global scale, the deficiencies of Zn and B are perhaps of greatest importance. Fe and Mn deficiencies frequently occur on calcareous soils or on coarse-textured soils with neutral or slightly alkaline reaction and rarely on acid soils. In certain areas, Mo deficiencies can even impede the establishment of legume pastures and lessen the potential gains from BNF. On strongly acid soils, there may even be problems of micronutrient toxicities.

Available nutrients in soils

Out of the total amount of nutrients in soils, more than 90 percent is bound in relatively insoluble compounds or is inaccessible within large particles and, therefore, is unavailable for crop use. Only a very small proportion is available to plants at any given point of time. To assess the nutrient supply to crops, it is important to know the amount of available soil nutrients either actually present or likely to be accessible to the plant during a cropping season. All available nutrients must reach the rootzone in ionic forms that plant roots can take up. In order for plants to acquire available nutrients, plant roots must intercept them in the soil or they must move to the root either with the water stream or down a chemical concentration gradient.

Moreover, available nutrients in soils are not a specific chemical entity or a homogeneous pool, but consist of three fractions. In terms of decreasing availability, these are:

- nutrients in the soil solution;
- nutrients adsorbed onto the exchange complex;
- nutrients bound in water-insoluble forms but easily mobilizable nutrient sources.

While the first two fractions are easily available and can be determined by fairly accurate and precise methods, the third fraction comprises a range of substances with varying availability and, therefore, is difficult to assess (described below).

Nutrients in the soil solution

The soil solution is the substrate from which roots take up nutrients. It is comparable with the nutrient solution in hydroponics. Soil solution means soil water containing small amounts of dissolved salts (cations/anions) and some organic substances that is mainly held in medium to small pores. The concentration of these nutrients is very different and varies considerably in time.

The solution of fertile soils may contain 0.02–0.1 percent salts in wet soils but a higher concentration in dry soils. In a neutral soil, the dominant nutrient is generally nitrate (30–50 percent), followed by Ca (20–30 percent), and Mg, K and sulphate-S (about 10 percent each). Ammonium is less than 2 percent and phosphate-P is considerably less than 0.1 percent. In saline and alkali soils, there are large concentrations of Na, chloride and sulphates.

FIGURE 19
Fractions of major nutrients in the soil

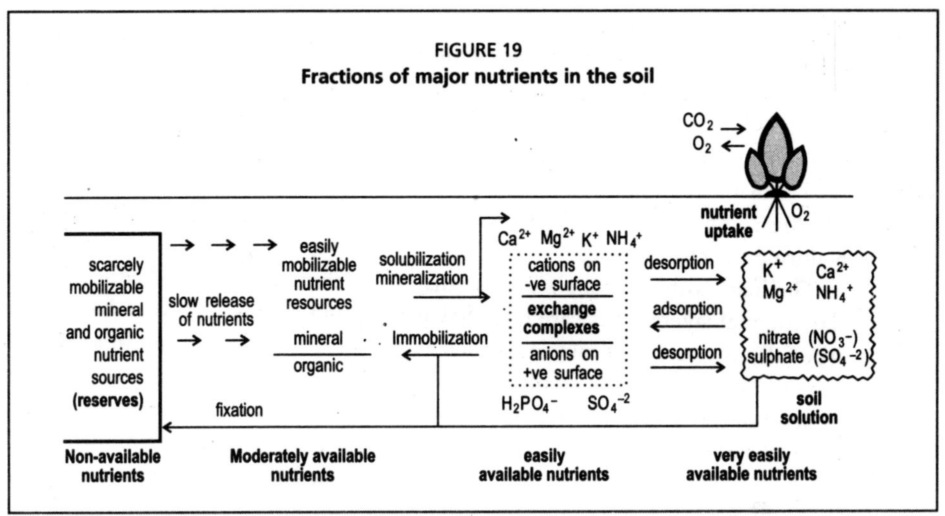

Compared with the daily nutrient requirement of high-yielding crops, the amounts of nutrients in the soil solution are very low and can only meet plant needs for a short period. This is especially true for phosphate and micronutrient cations. At high rates of nutrient uptake, the soil solution soon becomes depleted where it is not replenished from other fractions of available nutrients and unavailable forms, just as well-water is replenished by the groundwater resources as the water is drawn from the well (Figure 19). Adsorbed forms and other potentially available nutrient forms (the capability factor) continuously replenish the soil solution, which represents the intensity factor. Adequate fertilization ensures such replenishment. Where this cannot take place, nutrient deficiencies are very likely to occur.

Nutrients on the exchange complex

The fraction of exchangeable nutrients is much larger than their amount in the soil solution. In fertile soils of neutral soil reaction, about 75 percent of the adsorbed cations are of Ca and Mg and only 3–5 percent are of K. The adsorbed nutrients (cations of Ca, Mg, K, etc.) and anions, especially phosphate, are protected to a considerable extent against leaching loss and yet are easily available. An undesirable reality is that large amounts of nitrate are not adsorbed at all and, hence, can be easily lost through leaching.

Nutrients on the exchange complex must first be desorbed, exchanged or released into the surrounding soil solution before they can be taken up by plants. These replenish the soil solution. There is a steady exchange between nutrients on the exchange complex and those in the soil solution. A cation exchanges places with a cation and an anion can only exchange with another anion. Desorption dominates where the solution is diluted by nutrient uptake or addition of water, whereas adsorption dominates after input of water-soluble nutrient sources or with increasing dryness. Plant roots contribute to the release of adsorbed cations by the production of hydrogen ions (H^+), which may replace other exchangeable cations.

For some nutrients, the exchangeable fraction is a fairly good indicator of the total available pool of a nutrient in many soils. This is especially so in the case of K and Mg except for soils that can release significant amounts of non-exchangeable K during the crop season.

The inaccessible cations are those that are within the close-packed interior of clay mineral layers. These are regarded as fixed and mostly non-available. This is especially the case with nutrients such as K^+ and NH_4^+. K fixation in soils with certain clay minerals can result in severe K deficiencies. However, recent studies of several soils, specially the illite-dominant alluvial soils of India, indicate a very substantial contribution of non-exchangeable or "fixed" fraction of K to K uptake by crops (Subba Rao, Rupa and Srivastava, 2001). This calls for a change in thinking regarding the practical importance of non-exchangeable K for crop nutrition.

Moderately available nutrients

These nutrients are bound within different insoluble mineral and organic sources but are released during the cropping season. They can be easily mobilized by dissolving agents produced by micro-organisms or by plant roots. Some phosphate may be mobilized by organic acids and by mycorrhizae while some micronutrient cations by organic complexing agents known as chelates. The non-exchangeable K referred to above can also be considered in this category of moderately available nutrients.

It is difficult to distinguish this group from the much larger and partly similar pool of non-available nutrient sources as they are in a continuum. For example, typical P-containing compounds are calcium, aluminium and iron phosphates, but whether they belong to the moderately available nutrients depends on several factors. Moderately available P comes from freshly precipitated surface layers of amorphous material of small particle size, which facilitates their dissolution by dilute mineral and organic acids or by complexation. Examples of non-available forms of P are the occluded forms and tricalcium phosphates in alkaline soils. The same phosphates can become moderately available in a strongly acid soil. Insoluble iron phosphates can become available in reduced paddy soils where the ferric form becomes reduced to the soluble ferrous form of Fe.

Thus, the borderline between available and non-available nutrients in chemical compounds is arbitrary. Therefore, the amount of nutrients released into easily available forms during a cropping season is difficult to assess very accurately via practical approaches (discussed below).

Available versus actually used nutrients

Not all available nutrients in the soil are taken up by the roots even where there is a shortage. As root volume occupies only a small proportion of the soil volume and nutrients move relatively short distances to plant roots, they can only be utilized if they are within the reach of roots. Because of the small distances over which nutrients can move, many nutrients must be intercepted by growing plant roots. This is why physical conditions for root growth (soil structure) are very important and why plants with an extensive root system will generally be more efficient in nutrient uptake (discussed above).

The transport of most nutrients to the roots is mainly restricted to the small soil layer surrounding the roots (the rhizosphere). The mechanisms of transport valid for all nutrients, but to a different degrees, are:

- Mass flow: Nutrients flow passively with the water towards the root surface, a movement resulting from the active suction forces of the plant. Most of the nitrate and a part of other nutrients move this way.
- Diffusion: Here, nutrients move along a concentration gradient towards the root surface where the nutrient concentration is reduced because of uptake. Transport by diffusion is caused by random thermal agitation of the ions. Most of K, phosphate and micronutrient cations move by diffusion.

DYNAMICS OF PLANT NUTRIENTS IN SOILS

The content of available nutrients and their degree of availability and accessibility is not a static condition for all situations but ever-changing and very dynamic because of the various inorganic and biochemical processes that take place in soils. These depend on temperature, water content, soil reaction, nutrient uptake, input and losses, etc. Most forms of a nutrient (in solution, adsorbed, fixed, sparingly soluble, etc.) are in a dynamic equilibrium. External applications only cause temporary changes in the relation between different fractions, but the basic nature of the equilibrium remains intact over time.

An increasing water content (e.g. with rains or irrigation) causes a dilution of the soil solution (less nutrient per volume of water but with relatively more monovalent cations such as K^+), a stronger sorption of divalent cations and an increase in the mobilization rate. With increasing dryness, the soil solution becomes more concentrated and contains relatively more divalent cations (such as Ca^{2+}), but, most important, with dryness there is an increased immobilization of nutrients into only moderately available forms.

A decrease in pH from the neutral range results in a smaller proportion of exchangeable Ca and Mg. In the case of phosphate, there is initially a greater mobilization of calcium phosphate, but later a strong immobilization or even fixation into aluminium and iron phosphates. The availability of some micronutrients, especially of Fe, Mn and Zn, is increased strongly, and can even reach toxic levels. An increase in pH by liming can reverse the situation.

Nutrient uptake by plants, biological activity of soil organisms and external nutrient input can result in large or small fluctuations among the nutrient fractions, resulting in an ever-changing soil fertility status. To a certain extent, this can and should be controlled by appropriate management practices (Chapters 6 and 7). As a result of nutrient transformation and dynamics, when a nutrient ion reaches the root surface, the plant cannot distinguish whether this nutrient has come from soil reserves, mineral fertilizers or organic manures. In all probability it does not make any difference to crop nutrition.

DYNAMICS OF MAJOR NUTRIENTS

Nitrogen dynamics

The dynamics of N in soils are quite complex. These are depicted in many ways in the scientific literature. One example of N dynamics is provided in Figure 20.

Many factors affect the level of plant available inorganic N. Soil N is primarily in the organic fraction. The N in the organic matter came initially from the atmosphere via plants and micro-organisms that have since decomposed and left resistant and semi-resistant organic compounds in the soil during development. As the bulk of the organic matter is in the upper horizons, most of the soil N is also in the topsoil.

Inorganic ionic forms of N (NO_3^- and NH_4^+) absorbed by roots usually constitute less than 5 percent of total soil N. In normal cropped soils, where ammonium is added through fertilizers or released from organic matter/organic manures/crop residues by mineralization, it is usually nitrified rapidly to nitrate. N added in the amide form (NH_2) as in urea is first hydrolysed to NH_4^+ with the help of urease enzyme. It can then be absorbed by roots as such or converted to nitrate and then absorbed. Where urea is left on the soil surface, particularly on alkaline soils, some of it can be lost through ammonia volatilization. Such ammonia can return to the soil with rain.

The ratio of NH_4^+ to NO_3^- in soil depends on the presence of satisfactory conditions for nitrification, which is inhibited by low soil pH and anaerobic

FIGURE 20
The nitrogen cycle

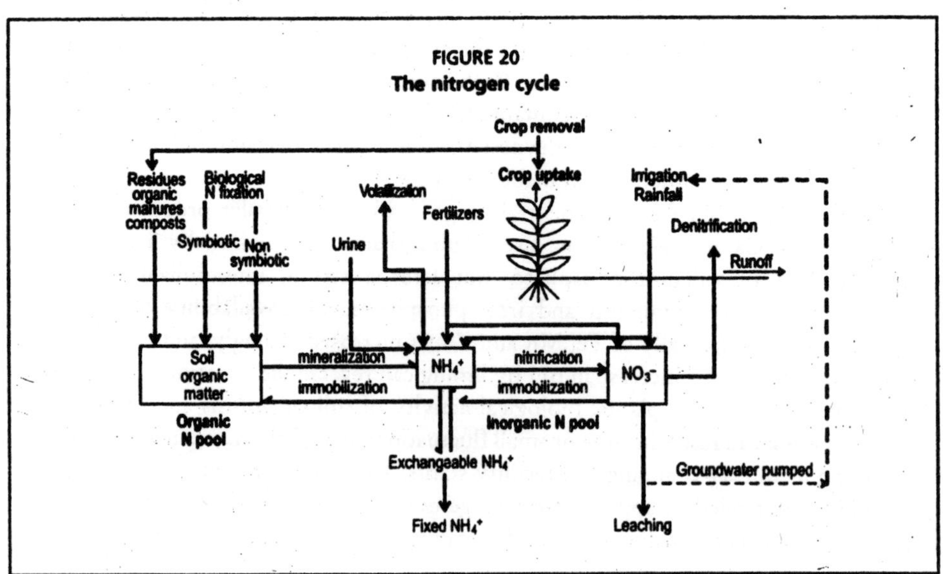

conditions. The nitrate thus formed can be absorbed by roots, immobilized by micro-organisms and become part of organic N, move down with percolating waters and leach out of the rootzone or be denitrified (lost) under anaerobic conditions. In cases where groundwaters are pumped for irrigation, the leached nitrate can re-enter the soil with irrigation water. Certain 2:1-type clays such as vermiculite and illite can fix ammonium ions in a non-exchangeable form.

Gaseous forms of N include dinitrogen (N_2) from the atmosphere or denitrification or nitrous oxides (N_2O, NO) from denitrification. N in the soil atmosphere can only be used by symbiotic N-fixing bacteria such as *Rhizobium* or non-symbiotic N-fixing bacteria such as *Azotobacter, Azospirillum, Cyanobacteria* (blue green algae) and *Clostridium*. Several nitrogen gases that escape from the soil after denitrification or volatilization return to the soil with rain (precipitation). The nitrogen gas itself can return to the soil through biologically or industrially fixed N.

Phosphate dynamics

Chemically, P is one of the most reactive plant nutrient. Thermodynamic principles dictate that P compounds will tend to transform to less soluble and increasingly stable (and unavailable) forms with the passage of time. Hence, P is one of the most unavailable and immobile nutrient elements. One of the indicators of this is that, barely 15–20 percent of the P added through fertilizer is recovered by the crop. It exists in the soil in a variety of forms. The dynamics of phosphate in soil present special problems because of the low solubility of most P compounds (Figure 21).

P added through soluble fertilizers first enters the soil solution, but much of it is converted into adsorbed P within a few hours. Very little of the added P stays

FIGURE 21
Phosphate dynamics in the soil

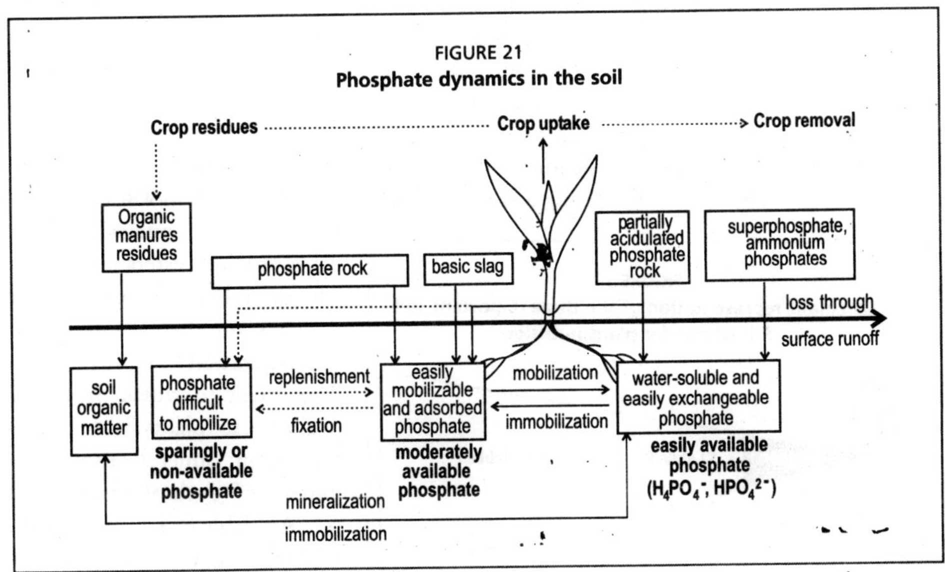

in the soil solution except in very sandy soils or soils lacking in the P-adsorbing agents (clays, oxides of Al or Fe, and carbonates of Ca or Mg). P added through PR is first solubilized by soil acids, after which it participates in various dynamic reactions as in case of soluble P.

Some important aspects of P dynamics are:

- Soil solution P: It is present in very small amounts and ranges from 0.01 to 0.50 mg/litre. In comparison, the concentrations of nutrients such as Ca, Mg and K are of the order of 400, 60 and 40 mg/litre, respectively. The relative distribution of anionic forms of P is dependent on soil pH. In the common pH range for soils, the dominant ionic form is $H_2PO_4^-$, which is also the most common form absorbed by plants. As the pH increases, the relative proportion of $H_2PO_4^-$ decreases while that of HPO_4^{2-} and PO_4^{3-} increases. In addition to ionic P in solution, some solution P may be present as soluble organic compounds, particularly in soils containing appreciable quantities of organic matter.
- Adsorbed P: In acid soils, the reactive phosphate ion is adsorbed onto the surfaces of iron and aluminium hydrous oxides, various clay minerals (e.g. illite and kaolinite) and aluminium–organic matter complexes. In neutral and alkaline soil, inorganic P may be adsorbed onto the microsurface of calcium and magnesium carbonates, iron and aluminium hydrous oxides, various clay minerals and calcium–organic matter complexes. Adsorbed P is a major source of P extracted by reagents used to estimate available P. It is in a dynamic equilibrium with solution P and replenishment as the P from solution is used up.
- Mineral P: These are mainly minerals of P combined with Ca, Al and Fe. In soils above pH 7, calcium and magnesium phosphate are dominant, while iron and aluminium phosphates are the dominant forms in acid soils. The amorphous forms can contribute to plant nutrition, but the crystalline forms are more stable and less reactive.
- Organic P: One-half or more of the total soil P may be present as organic P, the amount depending on the content and composition of organic matter. The major P-bearing organic compounds in soil are inositol phosphate, phospholipids, nucleic acids and others such as esters and proteins. The net release of this P to plants depends on the balance between mineralization and mobilization. The extent to which organic P is available to plants is not certain, but upon mineralization, it can enter the P cycle as adsorbed P.
- Available P: The replenishment of P into the soil solution following P uptake by plant roots is dependent

FIGURE 22
Schematic representation of the three important soil P fractions for plant nutrition

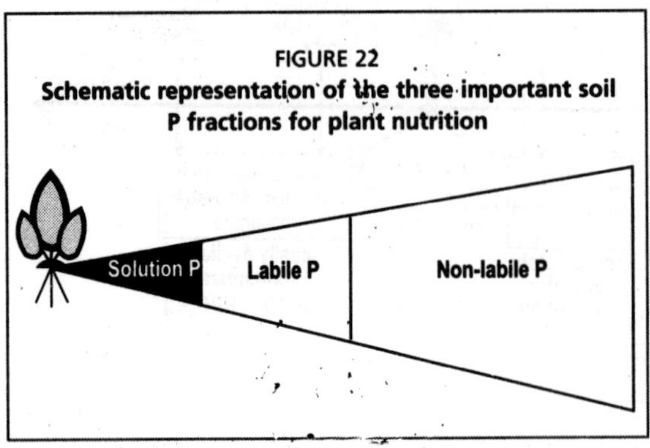

on the quantity of P in the adsorbed and sparingly soluble mineral phase, as well as inorganic P in plant residues (Figure 22). Together, these constitute the reservoir of "readily available P" for soil solution replenishment. This is also known as labile P and is usually defined as the quantity of P that is in equilibrium with the soil solution during the life of the plant. The ability of a soil to maintain its P concentration in solution as plant roots absorb the P, or as P is added by fertilization, is known as the P-buffering capacity. The higher the buffering capacity is, the larger is the proportion of P in the solid phase relative to the solution phase.

Potassium dynamics

Among cations, K^+ is absorbed by plants in the largest amount. The four important forms of K in soil are: (i) mineral K; (ii) non-exchangeable K, or K fixed in between clay plates; (iii) exchangeable K; and (iv) K present in the soil solution. The abundance of different K fractions is usually as follows: 90–98 percent of total K is in mineral form, which is relatively (but not ultimately) inaccessible to a growing crop; 1–10 percent of total K is in the non-exchangeable (fixed) form, which is slowly available; and 1–2 percent of total K is in the exchangeable and water-soluble forms. Of this 1–2 percent, about 90 percent may be exchangeable and 10 percent in the soil solution. All these are in a dynamic equilibrium.

K dynamics are determined by the rate of K exchange from the clay and organic matter surfaces and the rate of release from soil minerals. Except for sandy soils, K^+ is stored on the surface of negatively charged clay minerals. This easily replaceable supply provides the soil solution with additional K when the soil-solution K concentration decreases as a result of crop uptake (Figure 23). Most traditional soil test measurements of available K include the exchangeable and the water-soluble fractions. There is increasing evidence that a part of the non-exchangeable fraction or fixed K should also be included in soil test measurements for making meaningful K fertilizer recommendations.

The K in common fertilizers (KCl, and K_2SO_4) is water soluble. On addition to the soil, the fertilizer dissolves in the soil water and dissociates into the cation (K^+) and the anion (Cl^- or SO_4^{2-}). The cation K^+ is largely held on to the exchange complex as an exchangeable cation and a small amount is present in a freely mobile form in the soil solution. Thus, the K added to soils can be transformed into three fractions apart from any incorporation into the organic matter.

FIGURE 23
Potassium dynamics in the soil

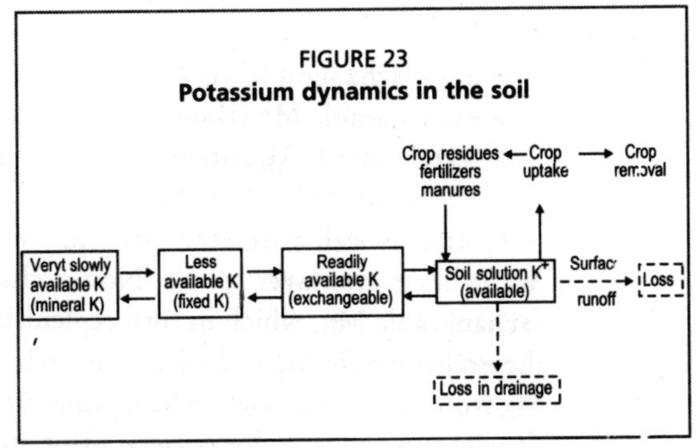

Potassium dynamics in the soil
Source: After A. E. Johnston (www. Kemira-growhow.com).

In very sandy soils, particularly under high rainfall, K can also leach out of the rootzone. Similarly, a significant amount of available K can be added to the soil through crop residues, leaf fall and irrigation water in specific cases. Much of this K is soluble and can again be adsorbed on the exchange complex.

Calcium dynamics

The Ca content of different soils varies greatly, depending on the minerals from which the soils originate and on the degree of weathering. Ca content is lowest in acid peat soils and in highly weathered soils of the humid tropics where, owing to high acidity and heavy leaching losses, as little as 0.1–0.3 percent Ca may be left. On the other hand, calcareous soils sometimes contain more than 25 percent Ca, mostly in the form of calcium carbonate (FAO, 1992). In general terms, Ca in the soil occurs in the following forms:

- Ca-bearing minerals such as calcium aluminium silicates (e.g. plagioclase), calcium carbonates (calcite and dolomite), calcium phosphates (apatite), and calcium sulphate (gypsum);
- exchangeable Ca, adsorbed on negatively charged surfaces of organic and inorganic soil colloids as part of the CEC of the soil;.
- water-soluble Ca.

As is the case with other cations, the different forms of Ca in the soil are in a dynamic equilibrium. Plant roots take up Ca from the soil solution. When the Ca concentration (or activity) in the solution decreases, it is replenished by the exchangeable Ca. An exchange complex dominated by Ca and adequately provided with Mg and K is a favourable precondition for good crop yields. In humid temperate regions, the ratio of the nutrients is considered well-balanced when about 65 percent of CEC is saturated by Ca, 10 percent by Mg, 5 percent by K and the remaining 20 percent by others (H, Na, etc.).

Magnesium dynamics

From the viewpoint of plant nutrition, the Mg fractions in the soil can be considered in a similar manner to those of K. Mg on the exchange complex and in the soil solution is most important for plant nutrition. Mg saturation of the CEC is usually lower than that of Ca and higher than that of K. The major forms of Mg are:

- non-exchangeable Mg (more than 90 percent of total Mg);
- exchangeable Mg (about 5 percent of total Mg);
- water-soluble Mg (about 1–10 percent of exchangeable Mg).

Non-exchangeable Mg is contained in the primary minerals such as hornblende or biotite, as well as in secondary clay minerals such as vermiculite, which hold Mg in their interlayer spaces. Non-exchangeable Mg is in equilibrium with the exchangeable Mg, which in turn replenishes the water-soluble or solution Mg. If the exchangeable Mg is depleted, as under exhaustive cropping, plants will utilize Mg from originally non-exchangeable sites at the clay minerals. However, its rate of release is too slow for optimal plant growth.

Sulphur dynamics

The S content of soils is usually lower than that of Ca or Mg. It is in the range of milligrams per kilogram or parts per million rather than percent. In the soil, S occurs in organic and inorganic forms. A generalized picture of S dynamics is presented in Figure 24. In many soils, the organic S may be 75 percent or more of total S. Tropical soils generally contain less S than do soils in temperate regions because of their lower organic matter content and its rapid rate of decomposition. In a survey of S in the tropics, an average value of 106 mg/kg S for a wide range of tropical upland soils has been reported (Blair, 1979), which is well below the 200–500 mg/kg S reported for non-leached temperate soils in the United States of America.

FIGURE 24
The sulphur cycle

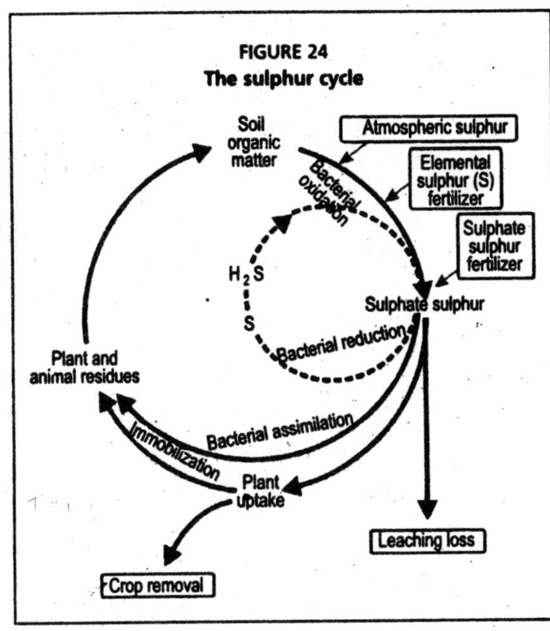

In the soil, inorganic S exists mostly as sulphate. This is either readily soluble or adsorbed on soil colloids. In calcareous soils, sulphate may also be present as cocrystallized impurity with calcium carbonate (FAO, 1992). Some soils have gypsic horizons that are enriched with the sulphates of Ca and Mg. As plant roots take up S from the soil solution in the form of sulphate, only the soluble and the adsorbed SO_4^{2-} are readily available. In many soils, these fractions represent not more than 10 percent of the total S.

S present in SOM or S added through organic manures becomes available for plant use only after conversion to the sulphate form through mineralization. Conditions most favourable for the mineralization of organic S are: high temperatures (the optimum is around 40 °C); a soil moisture status that is about 60 percent of field capacity; conditions favourable for high microbial population; and a minimum S content of 0.15 percent in the organic matter.

When water-soluble sulphate fertilizers such as ammonium sulphate (AS) and single supersulphate are added, the sulphate enters soil water. From there, it can either be moderately adsorbed (stronger than chloride much weaker than phosphate) or immobilized in soil organic water after absorption by micro-organisms. A small portion remains in the form of sulphate ion in the soil solution. Where S is added in elemental or sulphide forms, as in pyrites, these undergo oxidation in the soil to furnish sulphate ions for plant use. This transformation is affected by soil pH, moisture status, aeration and particle size (fineness) of the S carrier.

Dynamics of micronutrients

Micronutrients in the soil are present in a number of forms and fractions (Katyal and Deb, 1982). In a general way, these can be described as follows:

- in the soil solution (soluble): immediately available;
- on the exchange complex (exchangeable): available pool, replenishes soil solution;
- adsorbed on soil surfaces and in inorganic compounds: available pool, replenishes soil solution;
- in organic matter and in living organisms: available on mineralization;
- in primary and secondary minerals: potential medium-long-term sources.

The nutrient ions in the soil solution are available immediately. As in the case of most nutrients, this very minute quantity by itself is unable to meet the total crop needs. It is in a dynamic equilibrium with the exchangeable and adsorbed forms, which sustain the soil solution and replenish it continuously as the nutrient are taken up by plants roots. The dynamics of individual micronutrients are outlined below:

Boron dynamics

The total B content of soils can range from 2 to 100 µg/g. Generally, coarse-textured soils contain less B than do fine-textured soils. Different forms of B in soils are water-soluble, hot-water-soluble, leachable and acid-soluble B. Adsorbed B plays a major part in determining the amount of B available for plant use. B adsorption is affected markedly by soil pH and is maximum at pH 8–9. B is also associated with organic matter, which on mineralization can furnish available B (Shorrocks, 1984).

Water-soluble B is concentrated mostly in the surface layers of well-drained soils. B is susceptible to leaching in coarse-textured soils and its movements in clay soils can be restricted severely. Liming reduces the availability of B, as is the case with several micronutrients except Mo. In saline sodic soils or fields irrigated with B-rich waters, the concentration of soluble B can reach toxic levels, which must be avoided.

Copper dynamics

The total Cu content of soils can vary from 2 to 300 µg/g with an average of 26 µg/g. The five pools of soil Cu are: soluble forms in the soil solution (usually very low concentration); stable organic complexes in the humus; Cu sorbed by hydrous oxides of Mn and Fe; Cu adsorbed on the clay–humus colloidal complex; and the crystal lattice-bound Cu in soil minerals, which is only released on weathering (Shorrocks and Alloway, 1988). Cu is complexed very strongly by organic matter. Cu in the soil solution is immediately available for plant uptake, and that in exchangeable form along with weakly complexed Cu is also quite labile.

The major factors controlling Cu dynamics are: soil pH, carbonate content, organic matter and clay content. The solubility of Cu decreases with an increase in soil pH, as following liming of acid soils. Both adsorption and complexation of Cu are strongest under alkaline conditions, and Cu deficiency is often more pronounced in such soils, although it is also encountered in acid soils.

Iron dynamics

Fe is the fourth most abundant element in the earth's crust, constituting about 5.6 percent. The total Fe content of soils can range from less than 1 percent to more than 25 percent. As is the case with most nutrients, total Fe in the soil is of little value in estimating the available Fe content. The various forms of Fe in soils are: immediately available Fe; the available pool including exchangeable Fe; Fe available on decomposition; and potential medium- to long-term sources (Katyal and Deb, 1982). An assessment of available Fe in soils remains a challenge.

In spite of high total Fe in soils, its adequate availability to crops is a major problem in many upland soils, particularly those of alkaline and calcareous nature. This is because crops take up Fe as Fe^{2+} while in upland soils Fe^{2+} is oxidized to the unavailable Fe^{3+} form. Soil pH is a major factor governing the solubility and availability of Fe to plants. In contrast with upland soils, the transformation in submerged soils is of greater significance for the Fe nutrition of rice. The concentration of Fe^{2+} increases upon reduction in flooded-rice soil. Decomposition of organic matter and green manures is also known to increase the pool of available Fe.

Manganese dynamics

The average content of Mn in soils is about 650 µg/g but it can range from a few to more than 10 000 µg/g (1 percent). Mn in soils is present in various forms: water soluble, exchangeable, easily reducible, complexed or organically bound, occluded within sesquioxides, and Mn present in crystalline minerals. Active or available Mn consists of the water-soluble, exchangeable and easily reducible fractions.

Available Mn constitutes 1–15 percent of the total soil Mn depending on climate conditions, soil type and the extraction method used. Bioavailability of Mn in soils depends upon:

- valence of Mn in the weathering solids;
- nature of the primary minerals;
- redox conditions (Eh and pH);
- organic complex formations;
- microbial activity;
- environmental factors and management practices.

Available soil Mn decreases with increases in soil pH and calcareousness.

Molybdenum dynamics

The average content of Mo in the soil surfaces is 1.5 µg/g, varying from traces to 12 µg/g. The molybdate ion (MoO_4^{2-}) is strongly adsorbed by soil minerals and colloids at pH less than 6. With extensive weathering, the secondary minerals formed may trap Mo. Hydrous aluminium silicate can also fix Mo strongly. Unlike other micronutrients, the availability of Mo increases with increases in soil pH.

Zinc dynamics

The total Zn content in normal soils can range from 10 to 1 000 µg/g. Highly weathered coarse-textured laterite and red soils are poor in Zn. In soil, Zn exists in several forms such as water soluble, exchangeable, complexed, organically bound and acid soluble. Where a fertilizer such as zinc sulphate is added to soils, its Zn dissolves in water and enters one or more of these fractions. Zn present in water-soluble, exchangeable and complexed forms can be considered as available to plants.

Plant available Zn is usually less than 1 percent of total soil Zn. The organic-matter-bound Zn decreases with increase in pH regardless of soil texture. Soil submergence also reduces Zn availability, which is linked with changes in pH and formation of certain relatively insoluble products of Zn. Overliming acid soils can reduce Zn availability drastically and result in its severe deficiency (Gupta, 1995).

ASSESSMENT OF AVAILABLE NUTRIENT STATUS OF SOILS AND PLANTS

The evaluation or assessment of soil fertility is perhaps the most basic decision-making tool for balanced and efficient nutrient management. It consists of estimating the available nutrient status of a soil for crop production. A correct assessment of the available nutrient status before planting a crop helps in taking appropriate measures for ensuring adequate nutrient supply for a good crop over and above the amounts that the soil can furnish. The techniques used include soil testing and plant analysis, the latter including related tools such as total analysis of the selected plant part, tissue testing, crop logging and the diagnosis and recommendation integrated system (DRIS), as described below. The objective of all these techniques is to assess the available nutrient status of soils and plants so that corrective measures can be taken to ensure optimal plant nutrition and minimum depletion of soil fertility.

Soil testing

Soil testing is the most widely used research tool for making balanced and profit-maximizing fertilizer recommendations, particularly for field crops. Soil testing can be defined as an acceptably accurate and rapid soil chemical analysis for assessing available nutrient status for making fertilizer recommendations. Soil testing as a diagnostic tool is useful only when the interpretation of test results is based on correlation with crop response and economic considerations to arrive at practically usable fertilizer recommendations for a given soil–crop situation.

The amount of a nutrient estimated as available through soil testing need not be a quantitative measurement of the total available pool of a nutrient but a proportion of it that is correlated significantly to crop response. Soil testing does not measure soil fertility as a single entity but the available status of each nutrient of interest is to be determined. Based on a high degree of correlation between the soil test value of a nutrient and the crop response to its application, the probability of a response to nutrient input can be predicted. This serves as a basis for making practical fertilizer recommendations, which should be adjusted for nutrient additions expected to be made through BNF and organic manures. Soil testing has to be done for each individual field and for each nutrient of interest. It may be repeated every 3–4 years.

The major steps in practical soil testing for a relatively uniform field of up to 1 ha are:

- representative soil sampling of the fields;
- proper identification and labelling of the sample;
- preparation of soil sample;
- extraction of available nutrients by an appropriate laboratory method;
- chemical determination of extracted nutrients;
- interpretation of soil analysis data – soil test crop response correlation is the key issue.

The usefulness of soil testing depends on a number of factors, such as representative and correct soil sampling, analysis of the sample using a validated procedure, and correct interpretation of the analytical data for making practical recommendations. A sound soil testing programme requires an enormous amount of background research on a continuing basis to cater to changing needs such as the development of new crop varieties, better products and agronomic practices. Such research also helps to determine:

- the chemical forms of available nutrients in soils and their mobility;
- the most suitable extractants for accurately and rapidly measuring such forms;
- the general health and productivity of the soils for various crops;
- norms for field soil sampling and sample processing techniques;
- the response of crops to rates and methods of fertilizer application;
- the effect of season on nutrient availability;
- interactions of a nutrient with moisture and other nutrients.

Soil sampling and sample preparation

The quality of soil testing depends largely on reliable sampling, otherwise the results, even if analytically accurate, are worthless. The test sample (which is only about 0.00001 percent of the topsoil weight) must be representative for the field or the part of the field being tested. Figure 25 provides the suggested sampling procedure for a small field (Finck, 1992) and for a large field (Peck and Melsted, 1967). The best time for sampling is before sowing or planting and certainly

FIGURE 25
Representative soil sampling for small and large fields

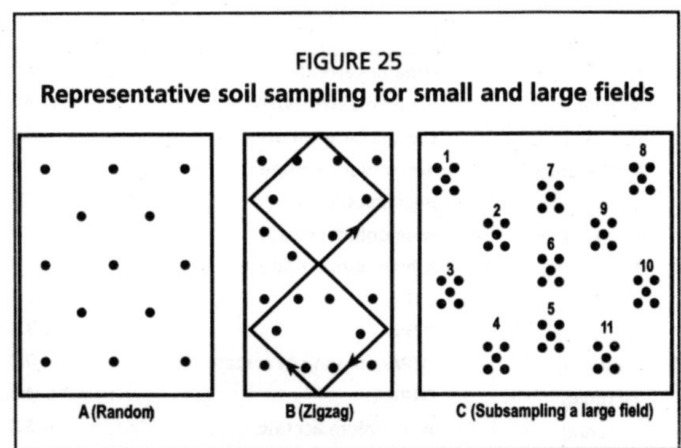

Note: A and B are for fields of up to 1 ha; C is a recommended sampling plan for a field of about 15–16 ha with each spot indicating the site of a subsample.
Sources: Finck, 1992; Peck and Melsted, 1967.

before fertilizer application. Each sample should have an information sheet with field identification, farmer's name and address, previous crops, the crop for which nutrient recommendation is sought, availability of irrigation, and previous fertilization, etc.

A soil sample should cover identifiable soil areas within a field. Abnormal soil patches, areas near a fence or used for storing animal manure or crop residues should be disregarded or sampled separately. For soil sampling, special augers with a core diameter of 1–2 cm are convenient, but small spades can also be used. In any case, a uniform slice of soil should be taken from top to bottom of the desired sampling depth. About 20 cores are taken from a field of 1 ha. Sampling depth can be 20–30 cm on arable land from the plough layer and 0–10 cm on grassland. For better interpretation, a few samples from deeper soil layers should be taken.

The individual cores or slices are then collected in clean containers and mixed well. From this, a final sample of about 0.5 kg is drawn for analysis. The moist samples should be air dried as soon as possible and sieved. Only the fine soil (less than 2 mm) is used for analysis. Most soil sampling is still done by hand, but special sampling machines have been developed in order to facilitate this on large areas. As large fields often have large variations in soil type, they must be divided into subunits of different soils.

Analysis for available nutrients

The concentrations of nutrient in the extracted solutions are determined by standard methods; the results are expressed as nutrient concentrations of air dry soil, using standard units: percent or g/kg or mg/100 g or mg/kg = μg/g = ppm; for organic soils, g/litre of soil are used. Nutrients may be indicated as oxides, e.g. K_2O, or as elements, e.g. K, but it must be stated very clearly whether oxide or elemental

TABLE 12
General soil test limits used for classifying soils into different fertility classes

Nutrient	Method/extractant	General fertility class*		
		Low	Medium	High
N (% organic C)	Organic carbon	< 0.5	0.5–0.75	> 0.75
N (kg/ha)	Alkaline permanganate	< 280	280–560	> 560
P_2O_5 (kg/ha)	Sodium bicarbonate	< 23	23–56	> 56
K_2O (kg/ha)	Ammonium acetate	< 130	130–335	> 335
S (kg/ha)	Heat soluble, $CaCl_2$	< 20	20–40	> 40
Ca (% of CEC)	Ammonium acetate	< 25		
Mg (% of CEC)	Ammonium acetate	< 4		
Zn (μg/g)	DTPA	< 0.6	0.6–1.2	> 1.2
Mn (μg/g)	DTPA	< 3.0		
Cu (μg/g)	DTPA/ammonium acetate	< 0.2		
Fe (μg/g)	DTPA	< 2.5-4.5		
Fe (μg/g)	Ammonium acetate	< 2.0		
B (μg/g)	Hot water	< 0.5		
Mo (μg/g)	Ammonium oxalate	< 0.2		

* Very general limits based on several published Indian sources (Tandon, 2004).

forms are used in order to avoid confusion and errors. For example, 1 unit of P becomes 2.29 units of P_2O_5, and 1 unit of K becomes 1.2 units of K_2O.

The choice of a suitable extractant for available nutrients is very important because of the different amounts of nutrient measured and the degree of their correlation with crop response. Most soil testing methods are based on chemical extractants, as summarized above and listed in Table 12.

For N, the water-soluble fraction (nitrate) is most suitable but the capacity to mobilize organic N can provide additional information. For the nutrient cations (Ca, K and Mg), the exchangeable portion is representative of the whole available fraction in most case and, therefore, determined by suitable extractants such as ammonium acetate and barium chloride. In soils that can release non-exchangeable K, a measurement of exchangeable fractions only does not provide a complete picture.

For phosphate, the choice is more difficult as soil reaction influences the solubility of P. A typical standard method for more acid soils is Bray and Kurtz extractant No. 1 popularly known as the P_1 test (0.03 N NH_4F + 0.025 N^- HCl), whereas for neutral and alkaline soils the Olsen method (0.5 N Na bicarbonate solution of pH 8.5) is more suitable. Among the major nutrients, the soil test methods for P are relatively more reliable.

For most micronutrient cations, diethylenetriamine pentaacetic acid (DTPA) is now widely used as an extractant. In the case of anions, the most commonly used extractant is hot water for B, and Grigg's reagent (ammonium oxalate of pH 3) for Mo.

Although a wide variety of extractants are used, most countries use standard extraction procedures, calibrated for specific soil and climate conditions in order to generate locally applicable recommendations. Soil testing needs to be simple, cheap and relevant to local conditions.

In addition to standard soil testing, simple kits are also available for use by farmers or advisers in the field with immediate but rather doubtful results. There is no good substitute for a valid soil test supplemented by appropriate plant analysis information. As a diagnostic tool, particularly for N in standing crops, the leaf colour chart (LCC) is finding use in rice, maize and some other crops.

Interpretation of analytical data for fertilizer requirement

After soil analysis, the concentrations of available nutrients measured must be interpreted into ranges of nutrient supply and then into the nutrient amounts required to reach a certain yield level. In general, the lower the soil fertility status (soil test value) is, the greater is the need for external nutrient application. For example, a very low soil test indicates a large deficiency of the nutrient in question and, accordingly, a large amount of external application to correct that deficiency and supply adequate amount of nutrient for optimal yield. Similarly, a very high soil test value indicates surplus nutrient supply, hence, no external addition of the nutrient is usually needed. In some agriculturally advanced regions, even on P-rich soils, P application is recommended to compensate for crop removals so that

TABLE 13
Interpretation of soil test data for some nutrients in soils with medium CEC

	Available (extractable) nutrients			Expected relative yield without fertilizer
	P	K	Mg	
Soil fertility class	(mg/kg soil)			(%)
Very low	< 5	< 50	< 20	< 50
Low	5–9	50–100	20–40	50–80
Medium	10–17	100–175	40–80	80–100
High	18–25	175–300	80–180	100
Very high	> 25	> 300	> 180	100

Source: FAO, 1980.

soil fertility is not depleted. Many farmers in developing countries cannot afford to follow such a strategy.

For macronutrients, the data are generally classified into categories of supply, e.g.: very low, low, medium, high, very high. From these categories, the nutrient amounts required for an optimal or stated yield level are estimated. For micronutrients, a critical level is generally used to decide whether an application of that nutrient is needed. Table 13 provides a generalized idea of the relation of available nutrient status to expected yields (without external addition) for a soil of medium CEC (10–20 cmol/kg). The values in the final column of the table indicate the approximate yield level that the existing soil fertility level could support.

In most cases, soil nutrient status is stated as low, medium or high. This needs to be done for each nutrient. For nutrients other than N, P and K, a single critical level is usually designated below which a soil is considered to be deficient in that nutrient, hence requiring its application. As an example, the most commonly used methods and values employed for delineating soils according to their available nutrient status have been provided in Table 12 for India, where more than 500 soil testing laboratories have been established. Most of these laboratories test soil samples for texture, pH and status of available N, P and K. These figures represent general norms but can vary widely with the type of soil, crop and method used. Therefore, only locally developed fertility limits should be used for specific soils and crops, even within a country or region.

On the basis of soil testing, nutrient supply maps can be drawn for farms, larger regions and countries. Such maps provide a useful generalized picture of the soil fertility of an area. However, the extent to which soil fertility maps can be used for planning nutrient management strategies depends on how thorough, recent and representative the soil sampling has been on which such maps are based. Macrolevel maps are more useful as an awareness and educational tool rather than for determining out nutrient application strategies.

Plant diagnosis

The nutrient status of plants can by assessed on a qualitative basis by visual observation and, more accurately, on a quantitative basis by analysing the mineral composition of specific parts of growing plants.

Visual plant diagnosis

A healthy dark-green colour of the leaf is a common indicator of good nutrient supply and plant health. The degree of "greenness" can be specified in exact terms for each crop using Munsell's Plant Colour Chart or other such charts. Any change to light green or a yellowish colour generally suggests a nutrient deficiency where other factors are not responsible such as cold weather, plant diseases and damage caused by sprayings, air pollution, etc.

A deviation from the normal green colour is easily detected. However, it is by no means always caused by N deficiency, as usually assumed. For example, it is a common but questionable practice to always relate light-green late foliar discolouring with N deficiency, but it can often be caused by other deficiencies. Even where the colour is more or less "satisfactory", there may be a latent deficiency ("hidden hunger") that is often difficult to establish from visual observation but can still cause yield reduction. It usually requires chemical plant analysis for conformation.

Fully developed deficiency symptoms can be a useful means for detecting nutrient deficiencies. However, they are only reliable where a single nutrient and no other factors are limiting. Chapter 3 has covered deficiency symptoms by nutrient. However, some general guidelines for the appearance of nutrient deficiencies in cereals are:

- ➢ Deficiencies indicated by symptoms appearing first on older leaves:
  - chlorosis starting from leaf tips, later leaves turn yellowish-brown: N;
  - reddish discoloration on green leaves or stalks: P;
  - leaves with brown necrotic margins, wilted appearance of plant: K;
  - stripe chlorosis, mainly between veins, while veins remain green: Mg;
  - spot necrosis: greyish-brown stripe-form spots in oats (grey-speck disease): Mn;
  - dark-brown spots in oats and barley, whitish spots in rye and wheat: Mn.
- ➢ Deficiencies indicated by symptoms appearing first on younger leaves:
  - completely yellowish-green leaves with yellowish veins: S;
  - yellow or pale yellow to white leaves with green veins: Fe;
  - youngest leaf with white, withered and twisted tip (oats and barley). Cu;
  - yellowish leaves with brownish spots (part of acidity syndrome): Ca.

Chlorosis refers to a condition in which the leaves appear with a light green-yellowish tinge, but the tissue is still largely intact. Necrosis means a brownish dark colour with irreversibly destroyed tissue. The easiest way of visually diagnosing nutrient deficiency symptoms is their identification with good-quality colour photographs of the specific crops. Even with these, farmers are advised to seek professional help and plant analysis as needed before taking corrective measures.

Plant analysis

The nutrient concentration of growing plants provides reliable information on their nutritional status in most cases, expect in the case of Fe. It reflects the current state of nutrient supply and permits conclusions as to whether a supplementary nutrient application is required. Plant analysis generally provides more current plant-based information than soil testing but it is more costly and requires greater efforts in sampling, sample handling and analysis. Ideally, both tools should be used as they complement each other. The key features of plant analysis are:

- Sampling: Representative sampling should be done of specific plant parts at a growth stage that is most closely associated with critical values as provided by research data. Sampling criteria and the procedure for individual samples is similar to that for soil testing in that it should be representative of the field. The composite sample should be about 200–500 g fresh weight.
- Sample preparation: The collected sample should be washed as soon as possible first with clean water and then with distilled water. It should be air dried followed by oven drying at 70 °C. Finally, it has to be carefully ground, avoiding contamination, and the powder mixed well.
- Analysis: After dry or wet ashing and complete dissolution, the determination of nutrients by standard analytical methods is carried out. The results are expressed as a concentration on a dry-matter basis (percent or mg/g for major nutrients; µg/g = mg/kg = ppm for micronutrients).
- Interpretation: Interpretation of plant analysis data is usually based on the total concentrations of nutrients in the dry matter of leaves or other suitable plant parts, which are compared with standard values of "critical nutrient concentrations" ("critical values") and grouped into supply classes. This will determine whether immediate action such as foliar spraying is needed to correct a deficiency. Conclusions can also be drawn on whether the amount of fertilizer applied at sowing time was sufficient or should be increased for the next crop. Where the concentration is in the toxicity range, special countermeasures are required but no application is needed.

For some nutrients, such as Ca and Fe, the "active" (mobile) nutrient content of plants should be considered because immobilization can make the total concentrations misleading. The nutrient concentrations of green (fresh) material or of plant sap can be used as a suitable basis for interpretation in some situations. Because of many interactions between nutrients and other inputs, more sophisticated indicators than just individual concentrations have been suggested, such as simple or complex nutrient ratios, e.g. the DRIS method (discussed below).

Critical values

Between the nutrient concentrations of the deficiency range and those of adequate supply, there is the critical nutrient range as described in Figure 11. The critical level is that level of concentration of a nutrient in the plant that is likely to result in 90 percent of the maximum yields. The plant nutrient concentrations required depend on the cultivar and expected yield level. Standard tables of plant analysis interpretation are based on requirement levels for very high yields (maximum yield) or on those for the more practical and realistic medium to high yields corresponding to critical values of 90 percent.

The main advantage of critical values, once properly established, is their wide applicability for the same crop. Their disadvantage is that they only provide a yes or no type of information and do not cover the entire range over which nutrient supplies need to be managed. Table 14 presents some critical data for a range of crops based on various sources. In most cases, these correspond to 90 percent of maximum yield but in some cases to maximum yield. These are approximations compiled from various sources. Specific situations require further refinement. For example, critical concentrations in the case of oil-palm are different for young palms and for older palms (Fairhurst and Hardter, 2003). A selection of critical plant nutrient concentrations for many crops has been compiled by the International Fertilizer Industry Association (IFA, 1992) among others.

TABLE 14
Critical nutrient concentrations for 90-percent yield for interpretation of plant analysis data

Element	Wheat & rice	Oilseed rape	Sugar cane	Alfalfa (Lucerne)	Grass*	Citrus
				(%)		
N	3.00	3.50	1.50	3.50	3.00	2.50
P	0.25	0.30	0.20	0.25	0.40	0.15
K	2.50	2.50	1.50	2.00	2.50	1.00
Mg	0.15	0.20	0.12	0.25	0.20	0.20
S	0.15	0.50	0.15	0.30	0.20	0.15
						(μg/g)
Mn	30.0	30.0	20.0	30.0	60.0	25.0
Zn	20.0	20.0	15.0	15.0	50.0	20.0
Cu	5.0	5.0	3.0	5.0	8.0	5.0
B	6.0	25.0	1.5	25.0	6.0	25.0
Mo	0.3	0.3	0.1	0.2	0.3	0.2

Note:
Growth stage and plant parts used: wheat/rice: 1–2 nodes, whole shoots; oilseed/rape: pre-flowering, youngest mature leaf; sugar cane: 5–7 months, third leaf blade from top; alfalfa (Lucerne): pre-flowering, whole shoots; grass: pre-flowering, whole shoots; citrus: 5–7 months old leaves from mid of non-fruiting branch.
* Not critical but optimal concentrations for cows producing 15 litres of milk per day.

Field experiment

This is a proven and effective method for assessing nutrient needs. It is also essential for the calibration of soil test and plant analysis results with crop response. Fertilizer experiments provide direct information on the amounts of nutrients required for a crop in a specific environment at a given yield level. However, because of the considerable expertise, effort and cost involved, they are not suitable for individual farmers. These are used mainly by researchers for generating background data needed for making practical recommendations for farmers.

Fertilizer experiments are required to reach conclusions on soil nutrient supply, on optimal rates and combinations of nutrients, best timing, application techniques, etc. The experiments are usually carried out under field conditions both at research stations and on the fields of cooperating farmers. Greenhouse experiments are less expensive but provide only limited information of practical value. Field experiments can be simple test plots or complex replicated trials.

Control test plot

This is the simplest field experiment (e.g. strip technique) where a small plot in the field is treated differently from the main field. For a farmer new to fertilizer use, the strip can be used to demonstrate the need for fertilizer. In this case, the strip receives optimal balanced fertilizer and the remaining field is unfertilized or treated as per the farmer's normal practice. The technique can be used also to determine whether a particular nutrient is likely to be deficient and should be included in the fertilizer programme. For example, to determine the need for S, the strip is fertilized with NPK + S while the remaining area receives NPK. By comparing the colour, growth and yield of the crop in the selected strip and the remaining field, a conclusion can be made about the changes required in existing fertilization practices. This can be quite convincing as it is carried out on the farm. However, adequate supervision is needed in order to ensure that the selected strip is not disturbed. This technique serves the purpose of a simple experiment as well as an on-farm demonstration.

Complex experiments

The design and scope of field experiments varies from simple studies of nutrient needs (e.g. whether crops respond to N, P or K), to increasing nutrient rates for determining the optimal rate and, finally, to complex experiments with several nutrients and other factors. For successful experiments, a number of rules have to be observed. Prerequisites for good field experiments are:

- The experimental field should have a uniform soil, minimum variations and a statistically designed layout.
- The layout of the experiment can range from simple blocks in a single row to more complicated multivariate designs such as Latin square in order to better eliminate soil variation. The total area of an experiment is usually limited by practical management considerations.
- The number of replications required depends on the purpose of the experiment (normally, four replications are required for statistical analysis).
- The size of plots depends largely on the area available and the crop in question. For small cereals, plots of about 20–25 m^2 may be sufficient, whereas larger plots are required for wide row crops such as oilseed rape, sugarcane, tobacco or maize.
- Rates of nutrient application should preferably be graded into equal steps from zero (control plot) to the maximum yield (or above), e.g. 0–50–100–150. A larger number of smaller steps or increments give more information but costs more.
- The total number of plots is determined by the number of different treatments and the number of replications, e.g. five rates of N application (say, 0–40–80–120–160 kg/ha) with four replications requires 20 plots. With three nutrients (say, N, P and K), each with three rates of application, there will be 27 combinations, and with four replications, the experiment will need 108 plots. If this number is too large, it can be reduced to about ten important combinations, such as $N_0P_0K_0$, $N_1P_1K_1$, $N_2P_1K_1$, $N_2P_2K_2$, etc.
- Good management practices including effective measures for weed control and crop protection must be followed in order to guarantee good crop growth.
- Special attention should be given to the control plot (plot without the test fertilizer or nutrient) as this determines the base production level. Occasionally, the test fertilizer or nutrient is added to the control plots

rendering the whole experiment useless, especially if there are no replications. This is more often the case where experiments are conducted on the fields of farmers.

- The results must be analysed by appropriate statistical methods, e.g. analysis of variance, and expressed in a manner indicating the statistical significance of differences in yield or composition. Statistical significance of experimental results does not imply automatically that they are true in a scientific sense. Where significance contradicts common experience, a plausibility check is required because all kinds of mistakes may occur in field experiments.
- Such field experiments usually have to run for at least three years in order to obtain valid conclusions and eliminate season-to-season variations.

The interpretation of field experiments is a standard procedure using statistical methods, and the results are directly applicable. The extrapolation of experimental results requires special care. They are strictly valid only for the crop grown, for the annual growth conditions (unless from long-term experiments), for the soil area of the experimental plot, etc. However, they can often be extrapolated to the surrounding area or similar growth conditions elsewhere through modelling.

IMPACT OF SOIL FERTILITY ON CROP PRODUCTIVITY

The basic aim of sound soil fertility management is to enhance crop productivity, to sustain it, and to keep the soils in good health – physically, chemically and biologically.

Improving soil fertility – the foundation of high-input cropping

The high yield levels obtained are a result of suitable crop growth conditions, optimal and balanced nutrient management (keeping in view the initial soil fertility) and adoption of best management practices. When taken year after year, such steps lead to an improvement in soil fertility, which makes the production of high yields sustainable. After more than a century of mineral fertilization, along with organic supplements available to the farmers, there has been a considerable increase in organic and mineral contents of major nutrients (N, P and K) in the soil in many parts of the world. At the same time, in large areas, soils continue to be mined of their nutrient reserves and are becoming depleted. Such soils are losing their ability to sustain high levels of crop productivity and safeguard food security.

Because of improved soil fertility, cereal yields on many highly productive soils remains high for some time even where left unfertilized. They may remain at 8–10 tonnes/ha in the first year, decrease to 5–7 tonnes/ha after some years, and further to about 3–4 tonnes/ha. Finally, after a longer unknown period, they can probably decline to 1–3 tonnes/ha. Wherever high yields are to be obtained on a sustained basis, the crop requires access to adequate amounts of all essential plant nutrients. Wherever the fertility of a soil is unable to furnish such amounts, soil fertility has to be improved through external additions of required plant nutrients. This is best accomplished through INM.

Degradation and improvement of soil fertility

Soil degradation leads to a deterioration in soil quality, resulting in yield decline. Soil degradation lowers the actual or potential soil productivity in different ways:

- loss of the fertile topsoil components through erosion by water and wind;
- physical degradation (poor structure, compaction, crusting and waterlogging);
- chemical and biological degradation, e.g. decrease in organic matter and soil bioactivity, loss of nutrients through various routes, soil acidification or salinization with their accompanying problems of nutrient deficiencies, toxicities and imbalances.

Soil degradation is widespread in many parts of the world. The basic causes of soil degradation are the result of human activities such as deforestation, overgrazing and poor soil management. Factors that cause soil degradation are interrelated. About 1 200 million ha worldwide are considered to be affected by soil degradation, mostly by erosion. It has been estimated that human-induced soil degradation has affected 46 million ha in Africa and 15 million ha in Asia (FAO, 2000a). Out of these, 25 percent of such soils in Africa and 67 percent in Asia are moderately to severely affected.

Degradation of soil fertility

Soil fertility is not a stable property but a dynamic one. There are widespread problems of soil fertility degradation under many cropping systems even on soils with good initial soil fertility. The result of such a decline is a reduced nutrient supply, which reduces crop yields.

From plant nutrition considerations, chemical degradation of the soil, particularly its fertility status, is of greatest concern. Losses of nutrients from soil can be caused by soil erosion, leaching, crop removal or in the form of gases (as in case of N and to a lesser extent S). Nutrient removal by crop products compared with external nutrient inputs can be similar, higher or lower. Negative nutrient balances result where nutrient removals exceed nutrient additions. These are a cause of soil fertility depletion or nutrient mining. Positive nutrient balances indicate a buildup or improvement in soil fertility.

Nutrient mining or depletion is a widespread problem in low- and medium-input agriculture. This is a major threat to productive sustainable farming. It is accelerated by imbalanced fertilization. Nutrient mining can cause the exhaustion of any nutrient required in moderate to large amounts. It can be particularly severe in the case of N, P, K and S depending on soil nutrient reserves and the amounts replenished. A negative balance can be acceptable for a short period, but, where prolonged, it will lead to soil deterioration. It is expensive to improve depleted soils.

Experience from Africa shows that, on poor soils, 33 percent of the total soil N may be lost within 10 years and 33 percent of P within 20 years, even at grain yields of 2 tonnes/ha. In comparison, K losses are relatively smaller and those of

Zn are very small. In Africa, several countries have a negative nutrient balance of more than 60 kg of total nutrients annually (FAO, 2000a). Negative nutrient balances are quite common also in many Asian countries. For example, in India, the annual gap between nutrients removed by crops and those added through fertilizers has been in the range of 8–10 million tonnes of N + P_2O_5 + K_2O for several years (Tandon, 2004).

Improving soil fertility

In many situations, soil degradation can be reversed by required inputs and improved management. However, once the topsoil has been lost, the damage has been done and there is little or no possibility of restoring it. Loss of topsoils is one of the worst forms of soil degradation. Some generally suggested measures for improving soil fertility/productivity are:

- Physical factors:
  - shallow main rooting zone (deeper cultivation where possible);
  - hard layers in subsoil (mechanical destruction of such layers);
  - very sandy soil (use of organic manure on priority);
  - poor structure (addition of organic matter, mulches, amendments).
- Chemical factors:
  - strong acidity (application of limestone, avoiding acid-forming fertilizers);
  - strong alkalinity (apply amendments such as gypsum and pyrites, green manuring);
  - strong salinity (leaching with non-saline water, growing salt-tolerant crops, green manuring);
  - nutrient toxicities (use of suitable amendments, drainage, tolerant crops);
  - low nutrient status (application of deficient nutrients through mineral, organic and biological sources);
  - nutrient fixation (application of suitable amendments, placement of fertilizers).
- Biological factors:
  - low organic matter (application of organic manures, compost, green manure);
  - poor microbial activity (improvement of aeration, drainage, correction of pH, organic inputs).

Even under conditions of low input cropping and with nutrient depleted soils, fertility degradation can be reversed in acid soils. The first step should be a better P supply with phosphate fertilizers, possibly with some lime application and N input via N fixation by legumes, resulting in a spiralling upwards process.

While soil improvements may result in 50 percent higher yields at a low-input level, more impressive results can be obtained at a high-input level. A good example is that of the formerly degraded and low-yielding, but now highly productive soils of Western Europe with present wheat yields of 8–10 tonnes/ha. The original cereal yield ranged from 0.5 to 1.5 tonnes/ha, a yield that can still be observed in unfertilized control plots of old field experiments and on the fields of millions of farmers in many parts of Asia, Africa and Latin America.

FERTILITY MANAGEMENT OF SOILS IN DIFFERENT CLIMATE REGIONS

The wide range of soils provides vastly different situations for the application of fertilizers and other sources of plant nutrients. The better the fertility of a soil is understood, the more correctly it is possible to develop and adopt nutrient management strategies. Only very few soils are ideal for plant growth by nature and supply nutrients in adequate amounts for high yields. In fact, most soils are in the wide medium-fertility range and many must be considered as poor. There may be many soils with high natural fertility, but in practical agriculture, these must be seen in the context of specific requirements of the crops to produce high yields. From a practical point view, most soils can be considered as requiring some degree of intervention and amelioration.

Soils of the temperate regions

The discussion here is restricted to the Podzolic and transitional soils as these are the main types on sandy substrates. These are predominantly former forest soils and cover large areas. Their organic matter content may be low or high, but the humus forms are generally poor in N and have undesirably high C:N ratios. These soils benefit from organic manures and mulch applied to improve their WHC and nutrient supply. Their fertility ranges from low to medium and their natural yield potential is often low. However, in humid rainy climates, improvement in their nutrient status offers considerable possibilities for yield increases.

On many Podzols, the initial amelioration required consists of removal of root-impeding conditions and better drainage. The often very high soil acidity and damage caused by it must be reduced or overcome by liming. This brings about several improvements, ranging from increase in activity of soil organisms to improvement of nutrient supply. Sandy soils are very sensitive to overliming and the optimal soil reaction (pH 6) is often exceeded. Higher pH values often cause secondary problems, of which Mn deficiency is the most frequent.

The general low status of nutrients, especially in sandy Podzols requires relatively high application of N, P and K. Because of their low nutrient storage capacity, split applications of N are beneficial (to reduce leaching). Because of the low pH, easily mobilized phosphate forms are generally more suitable than water-soluble ones. As Mg deficiency is common on these soils, liming materials containing Mg such as dolomite should be used. Deficiencies of micronutrients are also common. The deficiencies of Mn and B are frequently caused by high pH. Cu deficiency is characteristic of partly boggy sandy soils that have Cu-fixing humus forms. For the health of grazing animals, Cu and Co deficiencies need to be corrected.

Climate-stress factors that need to be taken into account are cold spells in spring, which primarily damage poorly nourished young plants, and dry periods in spring, which increase nutrient immobilization. Thus, full utilization of the production potential of these soils requires a comprehensive plant nutrient management programme with relatively large inputs, which offers large possibilities for yield increases.

Soils of the humid tropical regions

Red loam and red earth soils

Red loams and red earth soils (Latosols, Ferralsols, etc.) cover large areas on both sides of the equator under tropical rain forests and moist savannahs. Their average fertility ranges from low to medium, except for some red earths that have especially high fertility. Even where these soils support abundant natural vegetation, crop yields are frequently low. However, there are considerable possibilities for yield improvement. The introduction of intensive high-yielding production systems is possible because of favourable climate conditions. Sustainability of such systems in these soils can be achieved by using lime and a nutrient management programme that addresses the problems caused by production-limiting soil properties and the humid tropical climate.

Common properties of humid tropical soils relevant to nutrient supply and crop nutrition are:

- limited usable soil depth owing to hard layers (iron oxide concretions, stone layers);
- low natural soil reaction, often between pH 4 and 5, resulting in deficiencies of P, K and Mg and in Al toxicity;
- low nutrient reserves in old mineral soils (partly compensated by a higher mobilization rate than in temperate regions), in organic soils, the contents of P, K and Mg are often very low;
- low humus and N content and a rapid mineralization of organic matter under warm and more or less permanent moist conditions;
- low CEC of kaolinitic clay minerals in these soils is responsible for high losses of K, Ca and Mg by leaching under high rainfall in annual cropping;
- strong adsorption (fixation) of P on some soils with high contents of active iron and aluminium oxides;
- stable soil structure tends to deteriorate under prolonged cropping, which leads to poor accessibility by roots to otherwise available nutrients;
- danger of soil erosion by water.

For improving the status and availability of nutrients, a basic measure is proper liming. As with other soils, the optimal reaction of these soils also depends on soil texture. However, as the structure of these soils is usually relatively stable owing to iron oxides, a moderately acid pH range between 5 and 6 is frequently optimal. Problems may arise where there are no natural lime supplies within a reasonable distance, as is the case in some large tropical forest regions. Plant-ash from burning trees may serve as a substitute to some extent. However, even if the pH is increased above the toxicity level of Al (pH 4.8), disappointing results may be obtained because of unaccounted-for yield-limiting factors that have to be overcome in order to obtain satisfactory yields.

Acid sulphate soils
The acid sulphate soils, also known as cat clays, have a soil pH as low as pH 3 when drained. This is caused by the presence of dilute sulphuric acid resulting from sulphide oxidation. Some soils are also salt-affected. They often contain toxic amounts of Fe, Al and sulphate but are generally poor in most major nutrients. Such soils often produce very low yields, sometimes less than 1 tonne/ha of grain. However, even small amounts of lime, in combination with PR may increase the yield to 2–3 tonnes/ha. After complete amelioration, which is not easy, these soils can produce yields as high as normal soils.

The best way to manage such soils is amelioration with lime, PR and a general increase in nutrient levels, which requires substantial capital input. However, these soils are generally used in their natural state for growing paddy rice as flooding increases soil reaction. Where the main problem is Fe toxicity, using cultivars with high tolerance to Fe can solve the problem to a certain extent.

Soils of the subtropical arid regions

Arid soils
These soils cover large parts of the arid regions. Soil types at the zonal level often have a light-brownish colour. These soils belong to a variety of soil types depending on the parent material and climatic conditions. However, many of them have similar fertility status. Most of these soils are characterized by a neutral to slightly alkaline soil reaction. They usually have a good soil structure, are not subject to leaching and are rather well supplied with K, S, Ca, Mg, B and Mo. Their major limiting factor is inadequate water. Because of the lack of water, they generally support only sparse vegetation, or none at all. However, more than water is needed in order to make them productive. Their production potential is often considerable except in very sandy and stony soils.

The special problems of nutrient management in arid soils are:

- almost always a severe shortage of water, so that irrigation is usually required;
- sometimes little usable profile depth because of hardened lime crusts in the subsoil;
- low to very low organic matter content;
- low storage capacity for mineral nutrients in sandy soils;
- low content of available and mobilizable nitrogen (owing to low organic matter);
- phosphate deficiency in sandy soils;
- frequent Fe and Zn deficiencies because of fixation under alkaline reaction;
- low biological activity;
- occasional excess of soluble salts, adsorbed Na and B;
- high susceptibility to wind erosion.

As for major nutrients, fertilization with N is almost always necessary. All common N fertilizers act relatively quickly because of the high transformation capacity of these warm soils. On coarse-textured soils, fertilization with P and K is often necessary in order to obtain high yields, but it is less so on medium- and fine-textured soils. Losses of N as gaseous ammonia can occur in the neutral to alkaline range where N fertilizers (particularly urea) are applied to the soil surface and left unincorporated.

Owing to the high soil pH, micronutrient availability poses several problems. In particular, acute Fe deficiency occurs frequently in the form of lime chlorosis. This can be caused by the immobilization of Fe in the soil and also its immobilization in the roots and in leaves. The calcareousness of these soils plays an additional detrimental role. Zn deficiency is frequent, but Cu deficiency is rare. A certain lowering of the high soil pH by acid-producing fertilizers, e.g. ammonium sulphate, is desirable for mobilizing Fe, Mn and Zn.

Vertisols

An important soil type in subtropical zones is the dark clay soils, called Vertisols. With irrigation, their high productivity potential places them among the most productive soils. Vertisols can be shallow to very deep and usually have a neutral to alkaline soil reaction. Their large montmorillonitic clay content (more than 50 percent) makes cultivation of the soil difficult, reduces the permeability to water and reduces aeration.

Vertisols are well supplied with most mineral nutrients and have a high CEC. Their main soil fertility problem is the low level of total and available N. Among nutrients, the application of N, P and Zn is most frequently needed. In the case of N, while leaching losses are small, fertilizers such as urea can be subjected to ammonia volatilization where these are broadcast applied on the dry soil surface. Because of their neutral alkaline reaction, water-soluble sources of P are preferred to untreated PR. Band or spot placement of P is more effective than surface broadcast application. These soils are generally well supplied with K, Ca and Mg.

Fe and Zn deficiency can be a problem in crops grown on such soils, particularly in sorghum, groundnut, sugar cane, upland rice nurseries and fruit trees. The nutrient dynamics in flooded Vertisols change markedly where such soils are submerged for growing rice. Upon submergence, the solubility and availability of P increases, but nitrate in the topsoil can move down into the anaerobic zone and become denitrified if not absorbed by the plant.

Organic soils

These soils contain more than 20 percent organic matter throughout the solum (the upper part of the soil profile consisting of A and B horizons). Such soil can occur independently of climate zones. Organic or bog soils are created where decaying moss and other vegetable matter sinks to the bottom of a lake or pond. True bog soils contain more than 30 percent organic matter by weight. Because of their low specific weight, all data concerning nutrients are referred to in volumes, e.g. nutrient contents per 100 ml of soil.

The WHC of bogs is high. While much of it remains unavailable to the plant, water supply to plants is generally adequate. Poor aeration may limit crop growth where drainage is inadequate. Except for some specific cropping, e.g. by rice, organic soils must be drained for agricultural use. Organic amendments are not required on bog soils, except as nutrient sources. In order to define management strategies for bog soils, they are best divided into raised bog and fen soils.

Raised bog soils

Raised bog soils are highly acidic and poor in minerals. They often require liming before fertilization, but the soil pH should only be raised to a limited extent (unlike mineral soils). Soil supplies of N, P and K are low because of the absence of mineral reserves. Thus, adequate fertilization is necessary. K deficiency is often severe. Relatively small amounts of micronutrients are present, but their availability is generally good (except for Mo). Fertilizer-use efficiency is generally high because immobilization is insignificant as there are no Fe or Al compounds for sorbing phosphate. The deficiency of mineral matter can often be corrected by amelioration with sand. However, this is extremely expensive. Formerly, nutrients were also mobilized by burning bog layers.

Fen soils

Fen soils often occur in large or small depressions in the landscape with mineral soils. They are often rich in lime and mineral nutrient reserves. Therefore, they do not require liming, but acidifying fertilizers should be applied. An abundance of N is characteristic of many fens, providing adequate N for the plants. Their high lime content causes immobilization of phosphate and some micronutrients (especially Mn). An important climate stress is caused by cold conditions in the early stages of plant growth. This increases the importance of proper supplies of nutrients such as K, Mn and Cu in order to improve the tolerance of plants to cold.

Tropical bog soils

Tropical bog soils originating from forest bogs, papyrus marshes, etc. are usually more fertile after drainage than are corresponding bog soils in temperate zones. This is because of the high rate of decomposition of organic matter resulting in an accumulation of minerals in the drained layer. The principal problem in such soils after drainage is their extreme acidity (pH 2–3), often caused by the formation of sulphuric acid as a result of the oxidation of iron sulphide (FeS_2). In this respect, these soils are similar to acid sulphate soils. Such extreme acidity prevents any plant growth either directly or by mobilizing toxic quantities of metal ions. Liming is essential for any practical use, even in less extreme cases. Cropping with drainage causes a substantial loss of organic matter.

Chapter 5

Sources of plant nutrients and soil amendments

A large number of diverse materials can serve as sources of plant nutrients. These can be natural, synthetic, recycled wastes or a range of biological products including microbial inoculants. Except for microbial inoculants (biofertilizers), all of these contain one, two or several plant nutrients in readily or potentially available forms. A certain supply of mineral and organic nutrient sources is present in soils, but these often have to be supplemented with external applications for better plant growth. In practical farming, a vast variety of sources can find use in spite of large differences in their nature, nutrient contents, forms, physico-chemical properties and rate of nutrient release. These are not mutually exclusive but can be used together as components of INM.

Nutrient sources are generally classified as organic, mineral or biological. Organic nutrient sources are often described as manures, bulky organic manures or organic fertilizers. Most organic nutrient sources, including waste materials, have widely varying composition and often only a low concentration of nutrients, which differ in their availability. Some of these, such as cereal straw, release nutrients only slowly (owing to a wide C:N ratio) while others such as the N-rich leguminous green manures or oilcakes decompose rapidly and release nutrients quickly.

Residues from processed products of plant or animal origin are increasingly important as nutrient sources and lead to nutrient saving by recycling. In addition, a very wide range of products obtained from the recycling of crop, animal, human and industrial wastes can and do serve as sources of plant nutrient. A significant amount of N is made available through BNF by a number of micro-organisms in soils either independently or in symbiosis with certain plants. The inocula of such micro-organisms are commonly referred to as biofertilizers, which are used to enhance the N supply for crops.

The majority of nutrient input to agriculture comes from commercial mineral fertilizers. Organic manures are considered to play a significant but lesser role in nutrient contribution, leaving aside their beneficial effects on soil physico-chemical and biological properties. Such a conclusion could be due in part to inadequate data on the production and consumption of organic sources as compared with mineral fertilizers. Appreciable amounts of nutrients can also be brought in with rain (e.g. atmospheric deposition of nitrate and sulphate) and with irrigation water. This chapter describes common sources of plant nutrients. The last section deals with various soil amendments. Chapters 7 and 8 provide guidelines for the application of various nutrients through different sources.

MINERAL SOURCES OF NUTRIENTS (FERTILIZERS)

Definition, classification and general aspects

Definition

The term fertilizer is derived from the Latin word *fertilis*, which means fruit bearing. Fertilizer can be defined as a mined, refined or manufactured product containing one or more essential plant nutrients in available or potentially available forms and in commercially valuable amounts without carrying any harmful substance above permissible limits. Many prefixes such as synthetic, mineral, inorganic, artificial or chemical are often used to describe fertilizers and these are used interchangeably. Although organic fertilizers are also being prepared and used, they are not yet covered by the term fertilizers, largely owing to tradition and their generally much lower nutrient content. Strictly speaking, the most common mineral fertilizer, urea, is an organic compound that releases plant available N after transformation in the soil. In this section, the term fertilizer is used in a more narrow sense and widest acceptability.

Fertilizer grade is an expression used in extension and the fertilizer trade referring to the legal guarantee of the available plant nutrients expressed as a percentage by weight in a fertilizer, e.g. a 12–32–16 grade of NPK complex fertilizer indicates the presence of 12 percent nitrogen (N), 32 percent phosphorous pentoxide (P_2O_5) and 16 percent potash (K_2O) in it. On a fertilizer bag, the NPK content is always written in the sequence N, P_2O_5 and K_2O.

Synthetic fertilizers are sometimes referred to as being artificial or chemical fertilizers, implying that these are inferior to those termed natural (mainly organic) products. However, fertilizers are neither unnatural nor inferior products. Many fertilizers are finished products derived from natural deposits, either made more useful for plants (e.g. phosphate fertilizer) or separated from useless or even harmful components (e.g. K fertilizer). Although most N fertilizers are indeed produced artificially, i.e. synthesized in chemical factories, their N is derived from atmospheric air and their components such as nitrate, ammonia or urea are identical with the substances normally occurring in soils and plants. The primary source of all P in fertilizers is PR, a natural mineral that has to be mined, refined and solubilized in order to be useful.

Classification

Fertilizers have been traditionally classified as follows:

- Straight fertilizers: These contain one of the three major nutrients N, P or K. This is a traditional term referring to fertilizers that contain and are used for one major nutrient as opposed to multinutrient fertilizers. For secondary nutrients, these include products containing elemental S, magnesium sulphate, calcium oxide, etc. In the case of micronutrients, borax, Zn and Fe chelates and sulphate salts of micronutrients are straight fertilizers. However, the term is not often used for micronutrient carriers. This is not a very accurate term because many straight fertilizers also contain other essential plant nutrients, such as S in ammonium sulphate. These can also be termed single-nutrient fertilizers. The term focuses on the most important nutrient for which a product was traditionally used disregarding other valuable constituents. In a strict sense, the term is justified only for products such as urea, ammonium nitrate (AN), and elemental S.

- Complex/compound fertilizers: These contain at least two out of the three major nutrients. They are produced by a chemical reaction between the raw materials containing the desired nutrients and they are generally solid granulated products. These include both two-nutrient (NP) and three-nutrient (NPK) fertilizers. These are also referred to as multinutrient fertilizers, but do not include fertilizer mixture or bulk blends as no chemical reaction is involved. The term is rarely used for multimicronutrient fertilizers or fortified fertilizers containing both macronutrients and micronutrients or for liquid fertilizers. The term multinutrient fertilizers is more appropriate as it includes both major nutrients and micronutrients. Moreover, it does not restrict itself to a particular production process. Multinutrient fertilizers can be further classified into: (i) complex/compound fertilizers; (ii) mixtures and bulk blends; (iii) multimicronutrient carriers; and (iv) fortified fertilizers.

A brief historical overview

The use of fertilizers started in the early nineteenth century when saltpetre and guano where shipped from Chile and Peru to the United Kingdom and Western Europe, respectively. The first "artificial fertilizer", namely SSP, was produced in 1843 in the United Kingdom, to be followed by many SSP factories throughout Europe. Production of potash fertilizers started in 1860 in Germany and of that N fertilizers from ammonia (derived from coal) in about 1890. A significant advance in the production technology of N fertilizers came with the production of synthetic ammonia by the Haber-Bosch process in Germany in 1913. Production and use of urea as a fertilizer started from 1921. Since then, a large variety of solid and liquid fertilizers containing one, two or several plant nutrients have been produced and used. The fertilizer scene is dominated by products containing N, P and K in many chemical and physical forms and their combinations in order to meet the need for their application under different conditions throughout the world.

General aspects

In most countries, the effectiveness and safe use of substances to be registered as fertilizers is ensured by law. Recently, in developed countries, there has been a trend towards regulating some aspects of fertilizer application in respect of pollution.

The nutrient concentration of fertilizers is traditionally expressed in terms of N, P_2O_5, K_2O, etc. For example, an NPK fertilizer 15–15–15 contains 15 percent each of N, P_2O_5 and K_2O, or 45 percent total nutrients. The percentage composition of a fertilizer refers mostly to the total concentration of a nutrient, but sometimes only to its available portion. For solid fertilizers, the percentage generally refers to the weight basis, e.g. 20 percent N means 20 kg of N in 100 kg of product. For

TABLE 15
Five leading countries in terms of the consumption of mineral fertilizers, 2002–03

Country	Consumption				kg/ha of arable area
	N	P_2O_5	K_2O	Total	
	(million tonnes)				
China	25.200	9.854	4.162	39.216	275.0
United States of America	10.878	3.875	4.545	19.298	109.6
India	10.474	4.019	1.602	16.095	99.7
Brazil	1.816	2.807	3.059	7.682	130.2
France	2.279	0.729	0.960	3.968	215.1
World	84.746	33.552	23.273	141.571	100.8

Source: FAOSTAT, 2005.

liquid fertilizers, both weight and volume percentages are used, e.g. 20 percent by weight of N of a solution with the specific weight of 1.3 corresponds to 26 percent by volume (260 g N/litre).

In scientific literature, the nutrients are expressed mostly in elemental form whereas the industry, trade and extension services continue to express P and K in their oxide forms. The fact is that neither N nor P exists in soils, plants or fertilizers in elemental form. In any case, owing to the mismatch between the forms in which plant nutrients are expressed in research, extension and trade literature, care is needed when converting research data into practical values. Where the optimal application rate is reported as 26 kg P/ha in a research document, this translates into 60 kg P_2O_5/ha.

From small beginnings in the nineteenth century, the use of fertilizers has grown dramatically. The total consumption of NPK through fertilizers is now almost 142 million tonnes at an average rate of 100 kg of nutrients (N + P_2O_5 + K_2O) per hectare of arable area (Table 15). Five countries (China, the United States of America, India, Brazil and France) account for 61 percent of the total fertilizer consumption, while more than half of total consumption takes place in China, the United States of America and India.

The nutrient consumption rate in different countries varies from very high to extremely low (Figure 2). Even more than 150 years after the beginning of fertilizer use, there are still large areas of the world where no or very little fertilizer is used.

Fertilizers containing nitrogen

Origin

All N in fertilizers originates from the nitrogen gas (N_2) in the atmosphere, which contains 79 percent N by volume. Above every hectare of land at sea level, there are 78 000 tonnes of N_2. This is the N that is converted into ammonia in the fertilizer factories, and this is also the N that is fixed biologically into ammonium by various micro-organisms. Thus, there are abundant supplies of N for the production of nitrogenous fertilizers. Only a small amount of fertilizer N is still obtained from natural deposits such as Chile saltpetre and guano. As the nutrient

N is captured from the air, N fertilizer production is primarily a matter of available energy, which is mainly derived from oil or natural gas reserves.

Production of N fertilizers

The main features of the production of N fertilizers are:

- Ammonia: It is the starting point and basic intermediate for the production of N fertilizers. It is synthesized by the Haber-Bosch reaction which combines the very stable molecule of atmospheric N_2 with hydrogen, e.g. from natural gas, under a pressure of 200 atmospheres at 550 °C:

  air + natural gas + water → ammonia + carbon dioxide

  $O_2 + N_2 + CH_4 + H_2O \rightarrow NH_3 + CO_2$
- Nitrate fertilizers: In this case, nitric acid (HNO_3) is produced by the oxidation of ammonia and then neutralized with materials such as calcium carbonate ($CaCO_3$) to produce calcium nitrate $Ca(NO_3)_2$. Nitrate fertilizers may also be derived from other sources such as Chile saltpetre.
- Ammonium nitrate (AN) fertilizers: These are produced by neutralizing nitric acid (derived from the oxidation of ammonia) with ammonia. The solid granulated fertilizer is obtained by spraying the highly concentrated solution in cooling towers.

  $HNO_3 + NH_3 \rightarrow NH_4NO_3$

  nitric acid + ammonia → ammonium nitrate (solution)
- AN with lime: It is produced: (i) by mixing AN with calcium carbonate to obtain calcium ammonium nitrate (CAN); and (ii) by the reaction of calcium nitrate with ammonia and CO_2.
- Urea: It is produced by the reaction of NH_3 and CO_2 at 170 atmospheric pressure and a temperature of 150 °C. Care is needed during drying to ensure that the biuret formed is minimum and within the permissible limits set out in fertilizer-quality standards.

Consumption of N fertilizers

The annual consumption of N through fertilizers is almost 85 million tonnes of N (2002–03 data). Out of this total, more than 50 million tonnes of N is consumed in five countries (China, United States of America, India, France and Brazil). China, India and the United States of America each consume more than 10 million tonnes of N through fertilizers annually. The number of N-containing fertilizers is large. Straight N fertilizers are listed in Table 16 and the major ones are described below. Multinutrient fertilizers containing N are discussed in a later section.

Anhydrous ammonia

Gaseous ammonia can be used directly as a fertilizer. It has a pungent odour and is toxic to plants and humans when concentrated but harmless in dilute form. When liquefied under pressure for transportation, it is referred to as liquid or anhydrous ammonia (containing 82 percent N). It is injected as a gas by special equipment into the soil, where it reacts rapidly with water to form ammonium hydroxide.

TABLE 16
Common straight N fertilizers

Fertilizer	Percent N
Ammonium fertilizers	
Anhydrous ammonia NH_3	82
Aqua ammonia NH_4OH	26
Ammonium sulphate $(NH_4)_2\ SO_4$	21(also 24% S)
Ammonium bicarbonate NH_4HCO_3	17
Ammonium chloride NH_4Cl	25
Nitrate fertilizers	
Calcium nitrate $Ca(NO_3)_2$	16 (also 20% Ca)
Sodium nitrate (Chile saltpetre) $NaNO_3$	16
Ammonium + nitrate fertilizers	
Ammonium nitrate NH_4NO_3	35
Calcium ammonium nitrate $NH_4NO_3 + CaCO_3$	27
Ammonium nitrate sulphate $NH_4NO_3 + (NH_4)_2\ SO_4$	26 (also 15% S)
Amide fertilizers	
Urea $CO(NH_2)_2$	46
Calcium cyanamide $CaCN_2$	22
Urea ammonium nitrate fertilizers	
Urea ammonium nitrate solution	28
Slow-release N fertilizers	Variable
Several products, e.g. CDU, S-coated urea, polymer-coated products, oxamide, IBDU	

Because of its low price, and in spite of its high application cost, it accounts for a large part of N consumption in some countries, e.g. the United States of America. Special safety precautions are needed during its transportation, handling and application. It is also the major intermediate for the production of other N fertilizers, both straight and complex.

Aqua ammonia

Aqueous ammonia is a solution containing water and ammonia in any proportion, usually qualified by a reference to ammonia vapour pressure. For example, aqua ammonia has a pressure of less than 0.7 kg/cm^2. Commercial grades commonly contain 20–25 percent N. It is used either for direct application to the soil or in the preparation of ammoniated superphosphate. It is easier to handle than anhydrous ammonia, but because of its low N concentration, it involves higher freight costs per unit of nutrient.

Ammonium sulphate (AS)

AS is the oldest synthetic N fertilizer. It contains about 21 percent N (all as ammonium) and 23–24 percent S (all as sulphate). It is an acid-forming fertilizer and is highly soluble in water. It can be produced through various processes and used directly or as an ingredient of fertilizer mixtures. It is used as part of the basal dressing or as top-dressing to provide both N and S. In S-deficient soils, it works as an N + S fertilizer. AS should not be mixed with PR or urea.

Ammonium nitrate (AN)

AN is produced by neutralizing nitric acid with ammonia. Fertilizer-grade AN has 33–34.5 percent N, of which 50 percent is present as ammonium and 50 percent as nitrate. It is usually in a granular or prilled form and coated with a suitable material to prevent absorption of moisture and caking in storage. It is a valuable N fertilizer, but also a dangerous explosive, hence, its trade and use as fertilizer is forbidden in many countries. It can be rendered harmless by mixing it with calcium carbonate to produce CAN. It is also used to produce liquid fertilizers. AN leaves behind an acidic effect in the soil.

Calcium ammonium nitrate (CAN)

CAN is a mixture of AN and finely pulverized limestone or dolomite, granulated together. It contains 21–26 percent N, half in the form of ammonium and the rest in the form of nitrate. Its use does not make the soil acid by virtue of the carbonate in it.

Sodium nitrate

Also known as Chilean nitrate of soda or Chile saltpetre, it was the first mineral N fertilizer to be used. It is obtained by refining the crude nitrate deposits called Caliche found in Chile. It contains about 16 percent N, all as nitrate. Natural saltpetre from Chile is still used as a fertilizer. The product also contains 0.05 percent B, which makes it particularly suitable for fertilizing sugar beets.

Urea

Urea is the most important and widely used N fertilizer in the world today. It is a white, crystalline, non-protein, organic N compound made synthetically from ammonia and CO_2. Urea contains 46 percent N, all in amide (NH_2) form and it is readily water soluble. It is the most concentrated solid N fertilizer that is produced as prills or granules of varying sizes. It is hydrolysed in the soil by the enzyme urease to furnish ammonium and then nitrate ions. During the manufacture of urea, a small amount of biuret (NH_2-CO-NH-CO-NH_2) is also produced. Urea should not contain more than 1.2 percent of the toxic biuret for soil application and not more than 0.3 percent where sprayed on leaves. It is used as a solid N fertilizer for soils, for foliar application, as an ingredient of liquid fertilizers and in NP/NPK complexes. Urea leaves behind an acidic effect in soils. However, this is much smaller than the acidic effect of AS.

Others

N is also provided through a number of liquid fertilizers or fertilizer solutions. One example is the aqueous ammonia discussed above. Another is urea ammonium nitrate solution, which contains 28–33 percent N. Liquid N fertilizers can be high-pressure solutions or low-pressure solutions.

Slow-release fertilizers are of particular importance for special applications and they increase the efficiency of N. These have been developed to better adapt the rate of N release to the N demands of plants, reduce the number of splits required, improve nitrogen-use efficiency and reduce N losses.

There are a large number of slow-release fertilizers and their mixtures, with N-release rates extending from short to long periods. Some examples of slow-release fertilizers are crotonylidene urea (CDU), isobutylidene diurea (IBDU), combinations of formaldehyde and urea, and oxamide (diamide of oxalic acid). Polymer-coated urea has been shown to be an effective N source. However, like the other slow-release products, the cost is high. Different degrees of release can be distinguished by analytical methods with fractions soluble in hot water acting more slowly than those soluble in cold water, and fractions insoluble in hot water acting extremely slowly. Soil microbes gradually liberate the N in these slow-release fertilizers with the decomposition rate depending largely on temperature. They are expensive in terms of per unit of N and are, therefore, restricted mainly to commercial and special applications.

Fertilizers containing phosphorus

Phosphatic fertilizers contain P, mostly in the form of calcium, ammonium or potassium phosphates. The phosphate in fertilizers is either fully water soluble or partly water soluble and partly citrate soluble, both being considered as plant available. Citrate-soluble P dissolves slowly and is relatively more effective in acid soils. The concentration of P (usually indicated as percent P_2O_5) refers either to the available or the total portion of phosphate.

Origin and reserves

The primary source of phosphate in fertilizers is the mineral apatite, which is primarily tricalcium phosphate [$Ca_3(PO_4)_2$]. It is the major constituent of PR, the basic raw material for the production of phosphatic fertilizers. These phosphate-containing rocks are found in special geological deposits and some phosphate-containing iron ores or other P compounds. PRs consist of various types of apatites. Depending upon the dominance of F, Cl or OH in the apatite crystal structure, it is known as fluorapatite, chlorapatite or hydroxyapatite. Weathering processes over long periods of time resulted in the accumulation of primary apatites or apatite-containing bones, teeth, etc. of animals of earlier geological periods. Many such deposits occur near the earth's surface, from where they are obtained by opencast mining and utilized either directly or after beneficiation for fertilizer production.

Large deposits of PR exist in several parts of the world, for example:

- North Africa (Morocco, Algeria, Tunisia, etc.) in the form of organogenic phosphorite, either as more or less hard rocks or as soft earth phosphate;
- the United States of America, e.g. Florida apatite, which is in the form of moderately hard pebbles and the teeth and bones of sea animals.
- Russian Federation, in the form of hard earth, coarsely crystalline apatite, e.g. magmatic Kola apatite.

It is not always realized that phosphate is a scarce raw material, probably the most critical one. Global reserves (actual and probable) with more than 20 percent P_2O_5 content seem to be in the range of 30–40 000 million tonnes, amounting to about 10 000 million tonnes P_2O_5. With a future annual consumption of 40–50 million tonnes P_2O_5, these reserves would last less than 200 years, or may be 100 years assuming an increased rate of consumption. In the past 100 years, phosphate has been discovered at a rate that exceeds the rate of P consumption (Sheldon, 1987). One source of future phosphate production is offshore deposits, which occur on many continents. None of these deposits is currently being mined because ample reserves exist onshore.

Production of P fertilizers

Superphosphate, or rather SSP, was the first mineral fertilizer to be produced in factories in the 1840s in the United Kingdom. There are two principal ways of producing P fertilizers from PRs:

- ➢ Chemical solubilization of PR into fully or partially water-soluble form by:
  - Sulphuric acid resulting in SSP:
    $Ca_3(PO_4)_2 + H_2SO_4 \rightarrow Ca(H_2PO_4)_2 + CaSO_4$
    tricalcium phosphate + sulphuric acid → [monocalcium phosphate + gypsum] = SSP
  - Phosphoric acid resulting in triple superphosphate (TSP) as follows:
    $Ca_3(PO_4)_2 + H_3PO_4 \rightarrow Ca(H_2PO_4)_2$
    tricalcium phosphate + phosphoric acid → [monocalcium phosphate] = (TSP)
  - Partial solubilization of PR with lesser amounts of sulphuric acid to produce what are known as partially acidulated phosphate rocks (PAPRs).
- ➢ Mechanical fine grinding of reactive PR for direct application as fertilizer.
- ➢ For the commercial evaluation of PRs, their total P content is determined using strong mineral acids. Most P fertilizers are evaluated by the "reactive" or "available" portion of their total phosphate content. This is based on chemical solubility, which is supposed to correspond to plant availability. Several solvents are employed for the extraction of the "available" portion of P fertilizers:
- ➢ Water: for SSP, TSP, etc.; extraction of water-soluble phosphate.
- ➢ Neutral ammonium citrate for SSP, PR, etc. is used in some countries to determine quick-acting phosphate. In some cases, the first extract is discarded and the second extract taken for evaluation of PR. High solubility in citrate (> 17 percent) indicates high reactivity.
- ➢ Citric acid (2 percent) for nitrophosphates and Thomas phosphate.
- ➢ Formic acid (2 percent) for PR in some countries. High solubility (> 55 percent) indicates high effectiveness.

Consumption of P fertilizers

The world consumption of phosphate fertilizers is 33.6 million tonnes P_2O_5, accounting for 24 percent of total nutrient usage (Table 15). Almost 63 percent of the global P_2O_5 consumption in 2002–03 occurred in China, India, the United States of America, Brazil and France. China alone accounts for almost 10 million tonnes P_2O_5 consumption through fertilizers. The consumption in terms of arable area ranges from negligible in several countries to 109 kg P_2O_5/ha in Japan, with a world average of 24 kg P_2O_5/ha.

The nutrient composition of major phosphate fertilizers is summarized in Table 17. This is followed by a brief description of common P fertilizers. Ammonium phosphates are discussed under complex fertilizers.

TABLE 17
Some common phosphate fertilizers

Fertilizer	P_2O_5	P
	(%)	
Single superphosphate (SSP): $Ca(H_2PO_4)_2$ + $CaSO_4{\cdot}2H_2O$	16–18	7–8
Enriched superphosphate (ESP) is a special form of SSP	27	12
Triple superphosphate (TSP): $Ca(H_2PO_4)_2$ + $CaHPO_4$	46–50	20–22
Partly acidulated phosphate rock (PAPR). About 40% water soluble + 30% citric acid soluble P, giving 70 percent "available" portion, contains 20% gypsum	23	10
Basic slag (Thomas phosphate): citric-acid-soluble concentration contains Ca phosphate silicate (75 percent), CaO (5 percent), some Fe, Mn, etc.	10–15	4–7
Phosphate rocks, finely ground (< 0.16 mm): evaluated according to solubility in citrate or formic acid	23–40	10–17

Superphosphates

Single superphosphate (SSP) is the oldest commercially produced synthetic fertilizer and the most common among the group of superphosphates. The prefix "super" probably refers to its superiority over crushed animal bones when it was first produced in the 1840s. SSP is a mixture of monocalcium phosphate [$Ca(H_2PO_4)_2$] and calcium sulphate or gypsum ($CaSO_4{\cdot}2H_2O$). It contains 16 percent water-soluble P_2O_5, 12 percent S in sulphate form and 21 percent Ca. As is clear from its composition, it is known as a straight or single-nutrient (P) fertilizer only for historical and traditional reasons. Its bulk density is 96.1 kg/m^3, critical relative humidity is 93.7 percent at 30 °C and angle of repose is 26°. It is commonly used as part of basal dressing either as such or as part of fertilizer mixtures. Its S component comes from the sulphuric acid used during its manufacture. The Ca component of SSP is particularly valuable for crops such as groundnut during pod formation. SSP should not be mixed with CAN or urea unless the mixture is applied immediately and not stored.

TSP is obtained by treating PR with phosphoric acid. It contains about 46 percent P_2O_5, mainly in water-soluble form. Unlike SSP, it contains very little S.

Basic slag

Basic slag is a by-product of the steel industry. It is considered to be a double silicate and phosphate of lime [$(CaO)_5P_2O_5SiO_2$]. It contains 10–18 percent P_2O_5 (part of which is citrate soluble), 35 percent CaO, 2–10 percent MgO and 10 percent Fe. Basic slag can be used as a fertilizer-cum-soil conditioner because it contains lime and citric-acid-soluble P. The steel slags are very hard – their use in agriculture is possible only where they are ground to a fine powder.

Thomas phosphate, a type of basic slag, is a by-product of the open-hearth process of making steel from pig iron. It may contain 3–18 percent P_2O_5 depending on the P content of the iron ore. Thomas phosphate (14–18 percent P_2O_5) was a popular phosphate fertilizer in Europe. It is a dark powder and its slow action is well-suited to maintaining soil P levels. The standard specification of Thomas slag is that 70–80 percent of the material should pass through 100 mesh. It has some liming effect. The availability of this fertilizer is decreasing and it is unimportant in much of the world.

Phosphate rock (PR)

PR can also be used directly as a fertilizer. It contains 15–35 percent P_2O_5. The quality of PR as a fertilizer depends on its age, particle size, degree of substitution in the crystal structure and solubility in acids. PR also contains several micronutrients. Their average contents are 42 mg/kg Cu, 90 mg/kg Mn, 7 mg/kg Mo, 32 mg/kg Ni and 300 mg/kg Zn. Their Cd content varies from 1 to 87 mg/kg of PR. In PRs for direct application, the Cd content should preferably not exceed 90 mg Cd/kg P_2O_5 (27 mg/kg of PR).

Reactive PRs can also be used directly as P fertilizer in acid soils with or without any pre-treatment. Such PRs can be used in acid soils and for long-duration crops. Their suitability depends on the reactivity of the rock, its particle size, soil pH and type of crop. Their suitability for direct application can be estimated by dissolving the PR in certain extracting solutions. The most common solutions are neutral ammonium citrate, 2-percent citric acid and the preferred 2-percent formic acid. The effectiveness of PRs is not only related to the reactive "available" portion but it also depends on the P-mobilization capacity of the soil, which is related to pH, moisture status and biological activity. This means that the final evaluation of PR must be based on field experiments. Several aspects of PR for direct application have been dealt with in detail in publication produced by FAO (2004b).

Partially acidulated phosphate rock (PAPR)

PAPR is obtained by the partial acidulation of PR to convert only a part of its P into water-soluble form, as compared with complete acidulation, where fertilizers such as SSP or TSP are produced. The degree of acidulation is usually referred to in terms of the percentage of acid required for complete acidulation, e.g. to produce SSP. Where only 30 percent of the acid needed to make SSP is used for preparing PAPR, it is referred to as PAPR 30 percent H_2SO_4. It is an intermediate kind of product between SSP and PR. It can serve as an effective phosphate fertilizer in neutral to alkaline soils that are not highly deficient in P and where long-duration crops are grown. These are widely used in Europe and South America (FAO, 2004a, 2004b).

Others

Dicalcium phosphate ($CaHPO_4$) is a slow-acting product used as a component of multinutrient fertilizers but it is rarely used as a fertilizer by itself in present times. Other P fertilizers are polyphosphates and diluted phosphoric acid (H_3PO_4), which can be used in hydroponics or for preparing liquid fertilizers. The problem of low P-utilization efficiency and the desire to obtain products suitable for fertilizer solutions and fertigation has led to a range of new P fertilizers, such as condensed phosphates (polyphosphates, metaphosphates and ultraphosphates), all with high P concentrations. They are partly water soluble and rapidly hydrolyse in the soil, i.e. convert into the plant available orthophosphate form. Phosphates coupled with sugars (glycido-phosphates) have been found to be useful for fertigation.

There are also liquid fertilizers based on phosphoric acid that may have several other nutrients such as N and micronutrients along with P.

Phosphate fertilizers can also be derived from the processing of municipal wastewaters, namely iron and aluminium phosphates. Where practically free of toxic impurities, these are valuable although slow acting and are likely to gain greater importance in the future.

Fertilizers containing potassium

Potash fertilizers are predominantly water-soluble salts. For historical reasons, their K concentration is generally still expressed as percent K_2O, particularly by the industry, trade and extension. As such, the nutrient K does not exist as K_2O in soils, plants or in fertilizers. It is present as the potassium ion K^+ in soils or plants and as a chemical compound (KCl, K_2SO_4) in fertilizers.

Origin and reserves

Large deposits of crude K salts were first found in Germany in the mid-1850s. In recent times, deposits in several countries, especially in Canada, have been mined and utilized for the production of potash fertilizers. Canada and the countries of the former Soviet Union have 90 percent of the known potash reserves (IFA, 1986). These deposits were formed millions of years ago during the process of drying up of seawater in former ocean basins. Layers of common salt (NaCl) were overlain by smaller layers of K minerals, which hardened to rock under pressure. Crude K salts are thus natural seawater minerals, which are now mined from great depths. World K reserves are large and more are expected to be discovered.

Production and consumption

The first potash fertilizers were ground crude K salts containing 13 percent K_2O. These are still used to some extent for fertilization of grassland in order to supply K and Na. They are also accepted in biofarming as a natural fertilizer. The main K fertilizers used at present are purified salts.

The production of potassium chloride (KCl) or MOP involves grinding of the salt rocks, which consist of minerals such as kainite (19 percent K_2O) and carnallite (17 percent K_2O). The unwanted components such as Na, Mg and Cl are then separated, which involves heating (dissolution of salts) followed by crystallization of KCl upon cooling. In the newer flotation process, KCl crystals are coupled with organic agents, floated to the surface and removed. Electrostatic methods separate solid crystals of KCl from other compounds.

Potassium sulphate is produced by the chemical reactions of different crude salts as also by the reaction of KCl with sulphuric acid. Besides the salt deposits, there are K-containing industrial waste products, e.g. dust from cement production, that can serve as a K fertilizer.

World consumption of K through fertilizers was 23.3 million tonnes K_2O in 2002–03. This amounted to about 16 percent of the total nutrient consumption through fertilizers. Almost 62 percent of total potash consumption takes place in five countries (the United States of America, China, Brazil, India and France) with the United States of America, China and Brazil accounting for 50 percent of the total potash consumption. Unlike most countries, potash consumption exceeds phosphate consumption in large-consuming countries such as the United States of America, Brazil and France while it is well below phosphate consumption in India and China. At the global level, potash consumption ranges from negligible in many areas to 107 kg K_2O/ha of arable area in the Republic of Korea, with a world average of 16.6 kg K_2O/ha.

Potassium chloride (MOP)

Potassium chloride (KCl), also called muriate of potash (MOP), is the most common K fertilizer. It is readily soluble in water and is an effective and cheap source of K for most agricultural crops. Grades of MOP vary from 40 to 60 percent K_2O. Fertilizer containing 60 percent K_2O is almost pure KCl containing about 48 percent Cl. MOP comes as powders or crystals of varying colours and hues from white to pink but these differences have no agronomic significance. Its critical relative humidity is 84 percent at 30 °C and it has a higher salt index than potassium sulphate. It is used either directly as a fertilizer or as an ingredient of common NPK complexes.

Potassium sulphate (SOP)

SOP is actually a two-nutrient fertilizer containing 50 percent K_2O and 18 percent S, both in readily plant available form. It is costlier than MOP but is particularly suitable for crops that are sensitive to chloride in place of KCl. It has a very low salt index (46.1) as compared with 116.3 in case of MOP on material basis. It also stores well under damp conditions. SOP should not be mixed with CAN or urea.

Others

Other important sources of potash such as potassium magnesium sulphate and potassium nitrate are discussed under multinutrient fertilizers in a later section. As there may be some salinity damage with high K applications, particularly as MOP (especially in gardening), slow-acting K fertilizers such as less soluble double salts, fritted K containing glass and soluble-coated K salts have been developed. Special rock powder, e.g. from potassium feldspar, is an extremely slow-acting K fertilizer, even after fine grinding.

Fertilizers containing sulphur

Most S-containing fertilizers are in fact sulphate salts of compounds that also contain other major nutrients or micronutrients. S-containing fertilizers such as AS, SSP and SOP have been discussed above under the respective sections on fertilizers containing N, P or K. Multinutrient fertilizers including NP/NPK complexes containing S as also liquid fertilizers (e.g. ammonium thiosulphates) are discussed in a later section. The only truly single-nutrient S fertilizers are the elemental S products.

Some sources of S and their approximate S content are:

- ammonium sulphate $(NH_4)_2 SO_4$: contains 24 percent S;
- ammonium sulphate nitrate $(NH_4)_2SO_4.NH_4NO_3$: contains 12 percent S;
- SSP: contains 12 percent S;
- ammonium phosphate sulphate: contains 15 percent S;
- potassium sulphate (K_2SO_4): contains 18 percent S;
- potassium magnesium sulphate ($K_2SO_4.2MgSO_4$): contains 22 percent S;
- magnesium sulphate monohydrate ($MgSO_4.H_2O$): contains 22 percent S;
- magnesium sulphate heptahydrate ($MgSO_4.7H_2O$): contains 13 percent S;
- gypsum/phosphogypsum ($CaSO_4.2H_2O$): contains 13–17 percent S;
- elemental S products: contain 85–100 percent S;
- sulphur bentonite: contains 90 percent S;
- pyrites (FeS_2): contains 18–22 percent S;
- sulphate salt of micronutrients: contain variable amounts of S.

Formulations containing S in elemental form are increasingly finding use as S fertilizers (Messick, de Brey and Fan, 2002). Elemental S products are the most concentrated source of S. The elemental S in them has first to be oxidized to sulphate in the soil by bacteria (*Thiobacillus thiooxidans*) before it can be absorbed by plant roots. The rate of S oxidation depends on the particle size of the fertilizer, temperature, moisture, degree of contact with the soil, and level of aeration. To facilitate oxidation from S to SO_4^{2-}, elemental S sources are usually surface applied a few weeks ahead of planting.

Fertilizers containing calcium

Raw materials for Ca fertilizers are abundant as whole mountains consist of calcium carbonate ($CaCO_3$) and there is no shortage of gypsum ($CaSO_4.2H_2O$) either as a mineral or as a by-product (phosphogypsum) of the wet-process phosphoric acid production. Common Ca fertilizers are:

- calcium oxide (CaO): contains 50–68 percent Ca (Ca × 1.4 = CaO);
- slaked lime [$Ca(OH)_2$]: contains 43–50 percent Ca;
- agricultural limestone ($CaCO_3$): contains 30–38 percent Ca;
- dolomite ($CaCO_3.MgCO_3$): contains 24–32 percent Ca,
- CAN: contains 7–14 percent Ca;
- calcium nitrate [$Ca(NO_3)_2$]: contains 20 percent Ca;
- calcium chloride ($CaCl_2.6H_2O$): 15–18 percent Ca;
- SSP: contains 18–21 percent Ca;
- gypsum ($CaSO_4.2H_2O$): contains 23 percent Ca;
- calcium chelates: variable.

Calcium nitrate contains about 15 percent N and 28 percent CaO. It is a good source of nitrate N and water-soluble Ca and is particularly used for fertilizing horticultural crops and for fertigation. Calcium nitrate is suitable only where N application may also be required. Water-soluble Ca fertilizers such as calcium chloride or calcium nitrate may be applied as foliar sprays. A component of several commercial leaf sprays, calcium chloride solutions with 10 percent Ca are used for spraying fruits such as apples.

Gypsum, with its moderate water solubility, is a very useful Ca fertilizer for soil application, but few soils need it to increase Ca supply. The main role of mineral gypsum is on alkali (sodic) soils for the removal of toxic amounts of Na and to supply S in deficient situations. The same is true of phosphogypsum, where it is not contaminated with heavy metals such as Cd.

Fertilizers containing magnesium

Natural reserves of Mg are very large, both in salt deposits ($MgCl_2$, $MgCO_3$, etc.) and in mountains consisting of dolomite limestone ($CaCO_3.MgCO_3$). There are several commercially available materials of acceptable quality that can be used to provide Mg to soils and plants. There are two major groups of Mg fertilizers, namely, water soluble and water insoluble. Among the soluble fertilizers are magnesium sulphates, with varying degree of hydration, and the magnesium chelates. The sulphates can be used both for soil and foliar application whereas the chelates, such as magnesium ethylenediamine tetraacetic acid (Mg-EDTA), are used mainly for foliar spray. Some sources of Mg are:

- magnesium oxide (MgO): contains 42 percent Mg (Mg × 1.66 = MgO);
- magnesite ($MgCO_3$): contains 24–27 percent Mg;
- dolomitic limestone ($MgSO_4.CaSO_4$): contains 3–12 percent Mg;
- magnesium sulphate anhydrous ($MgSO_4$): contains 20 percent Mg;
- magnesium sulphate monohydrate ($MgSO_4.H_2O$): contains 16 percent Mg;
- magnesium sulphate heptahydrate ($MgSO_4.7H_2O$): contains 10 percent Mg;
- magnesium chloride ($MgCl_2.6H_2O$): contains 12 percent Mg;
- potassium magnesium sulphate ($K_2SO_4.2MgSO_4$): contains 11 percent Mg.

Magnesium sulphate is the most common Mg fertilizer. In anhydrous form, it contains 20 percent Mg. As a hydrated form, $MgSO_4.7H_2O$ (Epsom salt), it contains 10 percent Mg. It is readily soluble in water, has a bulk density of 1 g/cm^3 and an angle of repose of 33°. It can be used for soil application and for foliar application. Kieserite is the monohydrate form of magnesium sulphate ($MgSO_4.H_2O$). It contains 16 percent Mg and is sparingly soluble in cold water but readily soluble in hot water. Its bulk density is 1.4 g/cm^3 and its angle of repose is 34°. It is used as a fertilizer for soil or foliar application to provide Mg as well as S.

Among the insoluble or partially water-soluble sources are magnesium oxide, magnesium carbonate and magnesium silicates. The insoluble or partially soluble materials are used more often as liming materials. However, in acid soils, they can also be used as Mg fertilizers. Magnesium carbonate, the major component of the mineral magnesite, is also used as a raw material for the production of magnesium sulphate.

Fertilizers containing nitrogen and phosphorus (NP)

These are not only the starting materials for the production of NPK fertilizers but they are also used for the simultaneous supply of two major nutrients (N and P) required in many cropping systems. They are produced by different processes and their nutrient concentration is indicated in percent N + P_2O_5.

The main solid types of NP fertilizers are mono-ammonium phosphate (MAP), di-ammonium phosphate (DAP), nitrophosphates, urea ammonium phosphates and ammonium phosphate sulphates. NP solutions consist of ammonium phosphate and polyphosphates with a specific gravity of about 1.4 and nutrient concentrations about 10 percent N + 34 percent P_2O_5. Special-purpose NP types are ultrahigh concentration fertilizers that are not phosphates but phosphonitriles or metaphosphate with a composition of 43 percent N + 74 percent P_2O_5 as an example (sum of nutrients > 100 percent if based on P_2O_5), but actually 43 percent N + 33 percent P.

Mono-ammonium phosphate (MAP)

MAP ($NH_4H_2PO_4$) is produced by reacting phosphoric acid with ammonia. It contains 11 percent N and 55 percent P_2O_5. It can be used directly as an NP fertilizer for soil application or as a constituent of bulk blends. It can also be fortified with S to make it more effective on S-deficient soils.

Di-ammonium phosphate (DAP)

DAP [$(NH_4)_2HPO_4$] is an important finished fertilizer as well as an intermediate in the production of complex fertilizers and bulk blends. It is produced by treating ammonia with phosphoric acid. It typically contains 18 percent N + 46 percent P_2O. About 90 percent of the total P is water soluble and the rest is citrate soluble. In some countries, efforts are underway to fortify DAP with the needed micronutrients.

Ammonium nitrate phosphate (ANP)

ANP is produced by reacting PR with nitric acid. Several grades are produced and a typical grade contains 20 percent N and 20 percent P_2O_5. Also known as nitric phosphates or nitrophosphates, all of them contain 50 percent of the total N in nitrate form and 50 percent as ammonium. Part of the total phosphate (30–85 percent) is water soluble, the rest being citrate soluble. Products with less water-soluble P are more efficient in acid soils or soils that are at least of medium P fertility, particularly for long-duration crops. In neutral to alkaline soils, particularly for short-duration crops, 60 percent or higher levels of water-soluble P_2O_5 content are generally preferred.

Ammonium phosphate sulphate (APS)

These are in reality three-nutrient fertilizers containing N, P and S, all in water-soluble, plant available forms. APS can be seen as a complex of AS and ammonium phosphate. Both the common grades (16–20–0) and 20–20–0) also contain 15 percent S, which comes from the AS portion.

Urea ammonium phosphates (UAPs)

UAPs are produced by reacting ammonia with phosphoric acid to which urea is also added in order to increase the N content in the product. The most common example of this type of NP complex is 28–28–0 (the first UAP to be commercially produced in the world). As the name suggests, it contains part (68 percent) of its N in the amide (urea) form and the rest (32 percent) in ammonium form. All its nutrients are readily soluble in water and in available form, amide N being available after conversion into ammonium.

Fertilizers containing nitrogen and potassium (NK)

Of the fertilizers containing N and K, potassium nitrate is perhaps the most important. It typically contains 13 percent N and 44 percent K_2O (37 percent K). It is a good source of K and N for crops that are sensitive to chloride. It finds greatest use for intensively grown crops, such as tomatoes, potatoes, tobacco, leafy vegetables and fruits, and in greenhouses. It has a moderate salt index (between that of MOP and SOP) and is also less hygroscopic. It is useful for normal application and also for fertigation.

Fertilizers containing nitrogen and sulphur (NS)

Fertilizers containing N and S have already been mentioned under nitrogenous fertilizers. Common types are AS, ammonium sulphate nitrate and combinations of urea with ammonium sulphate. S-coated urea is a slow-release fertilizer. Fertilizers such as AS are ideal for top-dressing a growing crop where S deficiency has been detected and an N application is also required. They combine two important nutrients for crops with high S demand.

Ammonium thiosulphate is a liquid NS fertilizer containing 12 percent N and 26 percent S (thio refers to sulphur). Fifty percent of its S is in the sulphate form and the rest is in elemental form. It can be used directly or mixed with neutral to slightly acid P-containing solutions or aqueous ammonia or N solutions to prepare a variety of NPK + S and NPKS + micronutrient formulations. It can also be applied through irrigation, particularly through drip and sprinkler irrigation systems.

Fertilizers containing nitrogen, phosphorus and potassium (NPK)

Theoretically, with 6 major nutrients, there are 20 possible combinations of three nutrient fertilizers. The most prominent ones of these are NPK fertilizers. These can be complex/compound fertilizers, mixtures or bulk blends. In fact, even some so-called single-nutrient or straight fertilizers such as superphosphate can belong to this group as they contain P, Ca and S.

There are a large number of standard-type NPK fertilizers with different nutrient ratios. Their nutrient concentrations are indicated as percentage of N + P_2O_5 + K_2O, the individual nutrient concentrations ranging from about 5 percent to more than 20 percent. While a different fertilizer for every crop and field may appeal to sophisticated farmers, the majority of growers use a limited number of standard types. Most NPK types are produced by the acid decomposition of PR with incorporation of ammonia, thus producing an NP fertilizer to which a K salt, usually MOP or SOP, is added. These can be solid or liquid fertilizers.

Solid NPK fertilizers

More than 50 types are available on the market, with the N and P components being present in one or several forms. Thus, even in NPK fertilizers with the same grade or nutrient ratio, a given nutrient can be present in several chemical forms (Table 18). In most NPK complexes, the K component is often derived from MOP, but some types contain K through SOP, which makes them suitable for many chloride sensitive plants and horticultural crops. Some NPK fertilizers contain Mg as an additional component. This is often through magnesium sulphate, which makes them suitable for crops with high Mg requirements. This actually results

TABLE 18

Forms of nitrogen and phosphate in various NP/NPK fertilizers

	Percent N as			Percent P_2O_5 as	
Fertilizer (grade)	NH_4	NO_3	NH_2	WS[1]	CS[2]
Di-ammonium phosphate (18–46–0)	18.0	0	0	41.0	46.0
Ammonium phosphate sulphate (16–20–0)	16.0	0	0	19.5	20.0
Ammonium phosphate sulphate (20–20–0)	20.0	0	0	17.0	20.0
Ammonium nitrate phosphate (20–20–0)	10.0	10.0	0	12.0	20.0
Ammonium nitrate phosphate (23–23–0)	11.5	11.5	0	18.5	23.0
Ammonium nitrate phosphate (23–23–0)	13.0	10.0	0	20.5	23.0
Urea ammonium phosphate (28–28–0)	9.0	0	19.0	25.2	28.0
Urea ammonium phosphate (24–24–0)	7.5	0	16.5	20.4	24.0
Mono-ammonium phosphate (11–52–0)	11.0	0	0	44.2	52.0
Ammonium polyphosphate (10–34–0) (liquid)	10.0	0	0	22.1	34.0
Nitrophosphate with K (15–15–15)	7.5	7.5	0	4.0	15.0
NPK complex (15–15–15)	12.0	0	3.0	12.0	15.0
NPK complex (17–17–17)	5.0	0	12.0	14.5	17.0
NPK complex (17–17–17)	8.5	8.5	0	13.6	17.0
NPK complex 18–18–18 (100 % ws[1])	8.2	9.8	0	18.0	18.0
NPK complex (19–19–19)	5.6	0	13.4	16.2	19.0
NPK complex 19–19–19 (100 % ws[1])	4.5	4.0	10.5	19.0	19.0
NPK complex 20–20–20 (100 % ws[1])	3.0	4.9	12.1	20.0	20.0
NPK complex (10–26–26)	7.0	0	3.0	22.1	26.0
NPK complex (12–32–16)	9.0	0	3.0	27.2	32.0
NPK complex (22–22–11)	7.0	0	15.0	18.7	22.0
NPK complex (14–35–14)	14.0	0	0	29.0	35.0
NPK complex (14–28–14)	8.0	0	6,0	23.8	28.0
NPK complex (20–10–10)	3.9	0	17.1	8.5	10.0
NPK complex 13–5–26 (100 % ws[1])	6.0	7.0	0	5.0	5.0
NPK complex 6–12–36 (100 % ws[1])	1.5	4.5	0	12.0	12.0
Calcium nitrate (15.5 % N, 18.8 % Ca)	1.1	14.4	0	0	0
Mono-ammonium phosphate (12–61–0) (100 % ws[1])	12.0	0	0	61.0	61.0
Monopotassium phosphate (0–52–34) (100 % ws[1])	0	0	0	52.0	52.0
Potassium nitrate (13–0–45)	0	13.0	0	0	0

[1]Water soluble; [2]Citrate soluble.
Source: Tandon, 2004.

into a fertilizer containing four major nutrients. NPK fertilizers are granulated for uniform distribution. Their colour is often greyish but, in order to be better recognized by farmers, some fertilizers are specially coloured in some countries, e.g. red may indicate a composition of 13–13–21, yellow of 15–15–15, and blue of 12–12–20 with K as sulphate.

Liquid NPK fertilizers

For more accurate and convenient application of fertilizers on large farms, liquid fertilizers offer certain advantages. Farmers do not need to carry fertilizer bags, they simply rely on pumping. Spraying machines used for crop protection can be used but suspensions require special nozzles. There are two different types of liquid fertilizers:

- Fertilizer solutions: These are clear liquid fertilizers of low to medium nutrient content. In most of these, the sum of nutrients adds up to 30 percent and they have a specific gravity range of 1.2–1.3. Their common components are urea, ammonium, nitrate, ammonium phosphate and a K salt.
- Suspensions: These are saturated solutions with fine crystals in a stabilized condition in which the sum of nutrients can be up to 50 percent. Their specific gravity is about 1.5. Their components are urea, ammonium, nitrate, polyphosphates and other phosphates, and a K salt.

For both types, the nutrient ratios vary in a wide range from 5:8:15 up to 25:6:20 ($N:P_2O_5:K_2O$).

The optimal nutrient ratio in NPK fertilizers

On the question of optimal nutrient ratios in NPK fertilizers, theoretical considerations and the actual trend are not in agreement. Strictly speaking, nutrient ratios should be fine tuned to every cropped field. However, in practice, this is neither possible nor necessary. Farmers want to handle as few fertilizers as possible.

A practical approach to the optimal nutrient ratio is derived from nutrient removal data. Decades ago in Western Europe, average rotations removed nutrient from the fields in an $N:P_2O_5:K_2O$ ratio of 1:0.5:1.2. This figure was corrected for the different utilization ratios, which resulted in a final ratio of 1:1:1.6. This was the basis for the common NPK fertilizer of 13:13:21. In recent decades, the ratio has become increasingly dominated by N with a tendency towards 1:0.5:0.5. This is partly explained by greater the buildup of P and K in the soils over the years and the consumers' emphasis on N supply.

In India, which is the world's third-largest user of fertilizers, on a macrolevel, balanced nutrient application is represented by the ratio 1:0.5:0.25. This historical ratio has represented the trend of importance given to fertilizer nutrients and the extent to which these are qualitatively deficient in Indian soils. This ratio bears no relationship to the ratio in which plant nutrients are absorbed by crops or the ratio in which these are removed with the harvest. The overall ratio in which nutrients are removed by crops in India is 1:0.45:1.75. Although a large number of NPK complexes with a wide range of nutrient ratios are produced and used in India, there is no such thing as an ideal ratio that can be applied over large areas. Even within a given region, the optimal nutrient ratio can never be the same for diverse crops (grains, fodders, fruits, sugar cane, tea, etc.).

At present, the nutrient ratio of global fertilizer consumption is about 1:0.4:0.3. Differences in ratios among countries are as large as between regions within the same country. The search for a single optimal ratio or a few ratios is thus futile for large countries with diverse soils and cropping systems. With increasing emphasis on precision farming and site-specific nutrient management (SSNM), it is best that the optimal ratio be determined by the soil, the crop and the growth conditions.

Fertilizers containing other combinations of major nutrients

Fertilizers containing N and Mg are suitable for supplying these two nutrients in the growing season. They contain AS or AN combined with magnesium sulphate or magnesium carbonate (as dolomite). Micronutrients may be added, such as 0.2 percent Cu for grassland. Potassium magnesium sulphate is a unique three-nutrient fertilizer without N. It typically contains 11 percent Mg, 22 percent K_2O and 22 percent S. Potassium magnesium sulphate is used where the application of S and K is also required. It contains less than 1.5 percent Cl. It has a neutral effect on soil reaction but should not be mixed with urea or CAN.

Micronutrient fertilizers

The importance of fertilizers containing micronutrients has been increasing over the years for several reasons. Decades ago, at medium yield levels, fertilization with micronutrients was restricted to the recovery of acute visible deficiencies that occurred in some areas of sandy, metal-fixing, overlimed or just poor soils. However, on most soils, the natural soil supply of micronutrients was adequate, so that micronutrients were not a large component of fertilization programmes.

With intensive cropping and high yields, the situation has changed considerably (Chapters 4, 6 and 7). For several micronutrients, there are now increasing reports of insufficient soil supplies to meet increased crop requirements. This is affecting both crop yields and produce quality. Increasingly, micronutrients have become yield-limiting factors and are partly responsible for a decreasing efficiency of NPK fertilizers. Therefore, standard NPK-based fertilization must often be supplemented by the deficient micronutrients.

Of the six practically relevant micronutrients, deficiencies of Fe, Mn and Zn tend to occur more on neutral to alkaline soils and under arid and semi-arid conditions. A deficiency of B and Cu is more likely to occur on acid soils in humid climates although large-scale B deficiencies have been reported from many neutral to alkaline soils in east India. Common micronutrient fertilizers are briefly described here. Chapters 7 and 8 provide their application guidelines.

Boron fertilizers

Historically, Chile saltpetre was the first B fertilizer used. Its excellent effect on crops such as sugar beets was not only due to the N but also to the B contributed by the small amount of borax present in it. This B contribution was not recognized during the first 70 years of its use.

Common B fertilizers are sodium tetraborate or borax ($Na_2B_4O_7.10H_2O$) (10.5 percent B), boric acid (H_3BO_3) (17 percent B), Solubor $Na_2B_4O_7.5H_2O$ + $Na_2B_{10}O_{16}.10H_2O$ (19 percent B), and boron frits. Borax, or sodium tetraborate, is the standard B fertilizer. It is a white gritty salt suitable both for soil and foliar application. Boric acid is more soluble but relatively toxic to plants where applied as a foliar spray. The best fertilizers for spraying on leaves are polyborates. For soil application, borax involves the risk of B toxicity to sensitive plants. However, there are slow-acting B fertilizers, such as colemanite or fritted boron silicates (fine glass powder containing B), that are safe. However, they lack a rapid initial supply.

On B-deficient soils, about 1–2 kg B/ha may be needed for high yields. As the actual fertilizer amounts applied are small and difficult to distribute evenly, B is usually supplied together with special combined fertilizers (N or P or NPK with B).

Chlorine fertilizers

The nutrient Cl is often present in the soil in adequate amounts or is incidentally added through chloride-containing fertilizers and in some cases through irrigation water or seaspray in coastal areas. Chloride deficiency is not common. It has been encountered in palms cultivated away from coastal areas. Common fertilizers containing Cl are KCl (47 percent Cl), NP/NPK complexes in which KCl is an input, sodium chloride (60 percent Cl) and ammonium chloride (66 percent Cl).

Copper fertilizers

Cu fertilizers were first used for the treatment of Cu deficiency in boggy soils to correct the "heath-bog disease" of oats or for the "lick disease" of cattle raised on Cu-deficient grassland because humic substances tend to fix Cu in unavailable forms. Some common Cu fertilizers are: copper sulphate $CuSO_4.5H_2O$ (24 percent Cu), $CuSO_4.H_2O$ (35 percent Cu); and copper chelate Na_2Cu-EDTA (12–13 percent Cu).

Copper sulphate ($CuSO_4.5H_2O$) is the oldest and best-known fertilizer. It is a blue salt containing 24 percent Cu or 35–36 percent Cu with less water in its structure. It comes in particle sizes varying from fine powder to granular and is used either in solid form for soil application or as a dilute solution for foliar spraying, which is more effective than soil application. For foliar spraying, copper oxychloride and copper chelate are preferable to the sulphate salts. Cu fertilizers based on metallic oxide and silicate forms can also be used to treat Cu-deficient soils. These substances must first be solubilized in the soils, i.e. converted into Cu^{2+} ions. These are more suitable for long-term Cu supply, in contrast to copper sulphate, which is more suitable for immediate effect. Some fertilizers for grasslands contain both Cu and Zn and even Co.

Iron fertilizers

The majority of Fe fertilizers are water-soluble substances, being either salts or organic complexes (chelates). Common Fe fertilizers are ferrous sulphate $FeSO_4.7H_2O$ (19 percent Fe) and ferrous ammonium sulphate $(NH_4)_2SO_4.FeSO_4.6H_2O$ (16 percent Fe), which is in fact a three-nutrient fertilizer containing N, S and Fe. Other important Fe fertilizers are iron chelates, iron polyflavonoides (10 percent Fe) and iron frits, which have variable Fe content.

Ferrous sulphate ($FeSO_4.7H_2O$) is a common fertilizer but in many countries there is greater acceptability of iron chelates for foliar spraying. Iron chelates are the principal Fe-containing fertilizers for soil and foliar application in many developed countries and becoming popular in other countries as well. Common Fe chelates in use are:

- Fe-EDTA = ethylenediamine tetraacetic acid with 5–12 percent Fe (Fe^{2+});
- Fe-EDDHA = ethylenediamine di(o-hydroxyphenyl) acetic acid with 6 percent Fe (Fe^{3+}).

Fe uptake by the leaves is greater from chelates than from salts. In the soil, the chelates protect the Fe against rapid fixation. Moreover, chelates have a less damaging effect on leaves. For application on Fe-fixing soils, which are generally neutral to alkaline, the stability of the chelate in the soil is important. In this respect, Fe-EDDHA is more stable and effective than Fe-EDTA.

Manganese fertilizer

Important Mn fertilizers are manganese sulphate $MnSO_4.H_2O$ (30.5 percent Mn), manganese oxide MnO (41–68 percent Mn), manganese frits (10–35 percent Mn), and Mn chelates (5–12 percent Mn). Manganese sulphate is a pink salt that is water soluble and can be used both for soil treatment and for foliar application. It is also a constituent of Mn-containing multinutrient fertilizers. As in the case of Fe, Mn chelates are more effective than salts. Other Mn fertilizers for soil application are various manganese oxides, manganese carbonate and manganese phosphate. These can be used mainly for soil application. Manganese oxides are mobilized through bacterial reduction under acid conditions, thus converting unavailable MnO_2 into available Mn^{2+} ions.

Mn fertilization is problematical as Mn deficiency is usually not caused by soil impoverishment but by Mn fixation, which decreases the available Mn. Mn fertilizers are not very effective in Mn-deficient soils and whatever effect they have may be small and not long lasting, because soluble Mn is fixed rapidly. Soil acidifying N-fertilizers can even be more effective than Mn fertilizers.

Molybdenum fertilizers

The standard Mo fertilizer is sodium molybdate ($Na_2MoO_4.2H_2O$) with 40 percent Mo, but ammonium molybdate [$(NH_4)_6$ $Mo_7O_{24}.4H_2O$] (54 percent Mo) is also suitable. Both products are water soluble and quick acting. These are used for soil and for foliar application. Other potential sources of Mo are molybdenum oxide MoO_3 (66 percent Mo) and molybdenum frits.

Zinc fertilizers

Common Zn fertilizers are zinc sulphates, Zn-EDTA chelate (12 percent Zn), zinc oxide ZnO (55 percent Zn), zinc frits (variable Zn content) and natural Zn chelates. Zinc sulphate is the most common fertilizer and it is available either as $ZnSO_4.7H_2O$ (21 percent Zn) or $ZnSO_4.H_2O$ (33 percent Zn). It can be used for soil or foliar application and like all sulphate salts also provides S. It is less suitable for foliar application because of its acidic action, for which zinc sulphate with some lime is preferable, or Zn chelates like Zn-EDTA can be used. Zinc oxide (ZnO) can be used for soil application, for pre-plant dipping of roots of rice seedlings in its slurry and also soaking of potato-cut seed tubers before planting.

Zn mobilization in soil is aided by acid-forming N fertilizers such as AS or other substances, e.g. pyrite (FeS_2), which produce localized areas of sulphuric acid in soil thus solubilizing Zn.

Combinations of micronutrients

On soils deficient in several micronutrients, multiple micronutrient fertilizers are required. However, this principle is more appropriate for soils under horticultural crops than for soils where only one or two nutrients may be limiting as in case of field crops. In horticulture, particularly for fruit trees, slow-release micronutrient fertilizers are required that can provide a continuous supply of all micronutrients without damage caused by excess supply at a given time. Such fertilizers, with several or all micronutrients, are generally partly water soluble but have mainly slow-acting components. Where applied at planting time, they are effective during the whole growth period.

A large number of multimicronutrient formulations have been developed in several countries. These are meant for soil application or for foliar spray. As is the case with all such formulations, there is always a chance that some nutrients are underapplied and some are overapplied. These umbrella-type formulations are sometimes also seen as prophylactic applications. There is a persistent disagreement between the research data on micronutrient deficiencies and the composition of commercial formulations of multimicronutrient fertilizers marketed in a given area.

Fertilizers containing major nutrients and micronutrients

Some fertilizers are more or less "complete" fertilizers in which many if not all nutrients are present. However, their use has remained limited as, under most cropping systems, not all nutrients need to be supplemented. Nevertheless, some complete fertilizers have a special place in agriculture, particularly in gardening. For example, fertilizers containing N, P, K, Mg and S are enriched with micronutrients Mn, Cu and B, resulting in an eight-nutrient fertilizer that has widespread applicability. Similarly, others are based on slow-acting N and permit a complete nutrient supply to potted plants when applied at planting time, or serve as a lawn fertilizer for the whole vegetative period with no problems of toxicity caused by excess supply early in the season.

Aqueous solutions with all or most nutrients have been developed for foliar application and also for crops where the cause of poor growth is unknown. The problem with such products is that rarely do all nutrients need to be applied, and the really deficient nutrients might be added in insufficient amounts while the not so deficient nutrients may be delivered in excess.

Another view, important for intensive high-value cropping, is based on the consideration that, during vegetative growth, a number of nutrients must be added in order to prevent the minimum factors from limiting growth and yields. As, without precise diagnosis, farmers do not know what is limiting, they tend to use combinations of nutrients that are or might be in short supply. There are numerous products containing various combinations of major nutrients and micronutrients on the market. Whether and to what extent each of their components makes a positive contribution to plant nutrition and economic yield gain is extremely difficult to confirm.

Multinutrient (macro plus micro) applications may take care of existing nutrient deficiencies where applied in time at required intervals. Therefore, they have their place in nutrient management in the absence of accurate information about the nutrient status of a given soil and crop. Money spent on nutrients that are not really needed is the price for a lack of precise information and may be as an insurance against unforeseen limiting factors. However, these are no substitutes for a good nutrient supply from the soil, which must be planned before planting the crop with the help of a good soil test.

Fortified and speciality fertilizers

Apart from the conventional fertilizers described above, there are a number of fortified fertilizers and speciality fertilizers that are targeted at specific situations. Many countries have a fertilizer legislation in which the definition and list of approved fertilizers is provided. Strictly speaking, only such fertilizers can be produced, labelled and marketed as fertilizers. In reality, the number of products in a given market is much larger than the number of officially approved fertilizers. Many products containing plant nutrients and non-essential beneficial elements and also other constituents are often sold as soil improvers, plant growth promoters, or yield enhancers in order to bypass the conditions laid down in the fertilizer legislation. However, several of these have a role to play in meeting the nutrient needs of modern high-technology farming.

Fortified fertilizers

Fortified fertilizers are generally common fertilizers to which one or more specific nutrients have been added in order to increase their nutrient content and make them more versatile. These are also useful for applying the very small quantities of some micronutrients. Some examples of fortified fertilizers are:

- zincated urea, containing 2 percent Zn;
- boronated SSP, containing 0.18 percent B;
- DAP and NPK complexes fortified with 0.5 percent Zn or 0.3 percent B;

- SSP fortified with elemental S, containing 20–50 percent S or with 0.05 percent Mo;
- TSP coated with elemental S to contain 10–20 percent S;
- MAP fortified to contain 10–12 percent elemental S.

Speciality fertilizers

Speciality fertilizers are mainly produced to cater to special crop-production or nutrient-delivery systems. These systems include: intensive indoor farming, greenhouse farming, intensive cultivation of speciality crops, and fertigation. Most of the speciality fertilizers are either fully water-soluble formulations, slow-release materials or material containing organic compounds (humates and amino acids). They may contain one, two or several nutrients (macro and micro). Fertilizers for drip irrigation systems have to be fully water soluble so that they do not leave any residue that will clog the nozzles. In several cases, these are purified versions of common fertilizers that give 100-percent water solubility. Some examples of such fertilizers are:

- monopotassium phosphate containing 52 percent P_2O_5 and 34 percent K_2O;
- NPK complexes of various grades that are 100 percent water soluble (Table 18);
- seaweed extracts or granules fortified with mineral nutrients;
- potassium sulphate that is 100 percent water soluble;
- materials containing major nutrients and micronutrients for specific applications;
- special products containing amino acids, vitamins, humic acids, etc.

Fertilizers containing non-essential beneficial elements

Some cropping areas may need supplementation with beneficial mineral nutrients such as Na, Si, Co and Al. Some pastures may need additional nutrients such as the Co and Se required by grazing animals. All these and other materials cannot be sold as fertilizers in many countries because they may not feature in the definition and list of approved fertilizers in fertilizer legislation.

Sodium fertilizers

Na improves the growth of the so-called "Na-liking plants", i.e. sugar beets, spinach, cabbage and barley. The Na concentrations in the leaves of such plants should be 1–3 percent, which is much higher than the Na concentration in cereals. The salt (NaCl) requirements of cattle make Na concentrations of about 0.2 percent in grass desirable. Fertilizers for improving the Na supply are sodium nitrate and multinutrient fertilizers with Na, such as special pasture fertilizer supplemented with 3-percent Na. Sodium chloride (NaCl) is only rarely used.

Silicon fertilizers

Silicate or silicic acid is beneficial to cereals because it improves the stalk stability and, thus, resistance to lodging. Although most soils contain enormous amounts of silicates, its uptake is not always sufficient and may have to be improved by application of soluble silicate, a practice used in flooded-rice cropping in some areas. The quantities applied as Si fertilizers vary within wide limits. Silica fertilizers used are soluble silicic acid or soluble silicates and Si-containing phosphate fertilizers.

Cobalt fertilizers

Cobalt (Co) is beneficial for plants because it is essential for the N-fixing bacteria and blue green algae (BGA). Therefore, legumes and other N-fixing plants require a sufficient supply of Co, which is generally derived from the soil reserves. Co is mainly applied as cobalt sulphate ($CoSO_4$ with 21 percent Co). As the amount required on pastures is very small (50–80 g Co/ha), it is generally applied as an additive to phosphate fertilizers, e.g. 0.5 kg Co/ha can last for a long period. Because of the small amounts required, an alternative to Co fertilization is the direct supply of Co to animals together with ordinary salt.

Aluminium fertilizers

Al appears to be beneficial to only to a few plants, e.g. tea. Tea leaves contain 0.2–0.3 percent Al, which appears to promote growth. Where Al is considered to be deficient, aluminium sulphate [$Al_2(SO_4)_3$] can be added. However, aluminium sulphate acts mainly as a soil-acidifying agent and its favourable effect on some "acid-loving" plants such as blueberries may not be due to an improved Al supply but to the mobilization of some micronutrients as a result of acidification. For most crops, even small amounts of soluble Al ions are toxic.

Fertilizers with mineral nutrients for animals

For animal nutrition, additional elements may be required and these may have to be applied through fertilizer in some areas. Co has already been mentioned above. As Se deficiency has been discovered in animals grazing on pastures on soils that are poor in available Se, fertilizers containing Se have been developed. Generally, addition of Se to fertilizer is not recommended because the optimal supply range of Se is narrow and there may be a danger of toxicity on soils already well supplied. Polymer-coated Se fertilizers are available that reduce this risk. Little is known to date about the required "animal" nutrients Cr or vanadium (V) in soils.

Transportation, storage and mixing of solid fertilizers

The chemical composition and physical condition of a fertilizer as well as climate conditions directly affect its handling, storage, transportation and mixing with other fertilizers.

Effect of humidity

Many fertilizers absorb moisture from the atmosphere. This can adversely affect their physical condition and sometimes their quality. Moisture uptake by fertilizers is indicated by their hygroscopicity coefficient. This coefficient is obtained by deducting the relative humidity of the air above a saturated solution from 100. The coefficient increases with increase in temperature, so that the risk of deterioration in fertilizer quality is greater in tropical than in temperate climate.

Another indicator is the critical relative humidity (CRH), which is the relative humidity at which a material starts absorbing moisture. The CRH is usually stated at 30 °C. The hygroscopicity coefficient and CRH values of some fertilizers as affected by temperature are provided in Table 19. The lower the CRH of a fertilizer is, the more hygroscopic it is. Such materials need special care during storage. CRH in the case of micronutrient fertilizers has not received much attention.

TABLE 19
Moisture absorption by fertilizers from the atmosphere

Fertilizer	Hygroscopic coefficient at		Critical relative air humidity at	
	20 °C	30 °C	20 °C	30 °C
Calcium nitrate	45	53	-	47
Ammonium nitrate	33	41	63	61
Sodium nitrate	23	28	-	72
Urea	20	28	79	74
Ammonium sulphate	19	21	81	81
Potassium chloride	14	16	-	84
Potassium sulphate	-	-	-	96
Di-ammonium phosphate	-	-	-	83

Sources: Finck, 1982, 1992; Tandon, 2004.

Some fertilizers, such as calcium nitrate and CAN, are extremely sensitive to moisture, harden and become liquefied. Only a few nitrogenous fertilizers, e.g. AS, retain their good flow properties at increased air humidity and, therefore, are very suitable for use in the tropics. The undesirable hardening of fertilizers is caused by crystal bridges being formed between the particles after wetting and drying.

Transportation and storage

Fertilizer particles should be spherical because spheres have maximum stability against pressure and make minimum contact with one another. Most fertilizer granules have a diameter of 2–4 mm, and uniformity in granule size is a precondition for good spreading and mixing of fertilizers.

The stability of the fertilizer granules is made vulnerable by the absorption of moisture from the air. Fertilizer granules may be conditioned during the production process to protect them from atmospheric moisture absorption. Coating fertilizer granules with non-hygroscopic conditioning substances such as lime, and diatomaceous earth, prevents granules from sticking together where humidity is high, prevents the collapse of granules under pressure, prevents the liquefaction of the fertilizer as a whole, and keeps the granules free flowing and dispersible during transportation, storage and application.

Fertilizer weight is important in transportation, storage and application. The bulk density (weight of the loosely filled fertilizer per unit volume) of most solid fertilizers is about 1 kg/litre. However, urea is considerably lighter with a bulk density of 0.7 kg/litre. Some fertilizers such as basic slag are exceptionally heavy with a bulk density of 2.0 kg/litre.

Care must be taken during transportation and storage not only to avoid detrimental effects to the fertilizers, but also to avoid any harm or injury to people handling them. Some fertilizers become heated and create a fire hazard when they absorb moisture. Others are potentially explosive (e.g. AN), many are corrosive, and some may release harmful gases. Fertilizers are generally conditioned against such undesirable effects, but such conditioning is only possible to a certain extent. Regulations are generally issued at country level for the proper handling, storage and transportation of various fertilizers, especially in large quantities.

Bags made of plastic and paper (and laminated jute in some areas) are the usual containers for fertilizers. The 50-kg bags prevalent in developing countries often have to be carried manually. However, large farms may use large bags that contain 500–1 000 kg of material and require mechanical handling. Bulk transportation and storage of loose (bulk) fertilizers saves packing and handling labour, but requires suitable equipment for transport and protection against moisture during storage. Large farming enterprises are increasingly moving towards bulk fertilizers.

Mixing of solid fertilizers

As plants need several nutrients, fertilizers can be bought individually and distributed separately or blended together prior to spreading. There are several alternate ways to apply multiple nutrients. Mixing is generally not required when appropriate complex/compound fertilizers are selected. Several fertilizers can be mixed without problems (compatible fertilizers), but there are three chemical reasons for not mixing fertilizers indiscriminately:

- possibilities of losses of N by chemical reactions;
- possibilities of immobilization of water-soluble phosphate;
- possibilities of deterioration of distribution properties due to hygroscopicity.

The compatibility of fertilizers, allowing for these factors, is indicated in Figure 26.

Reactions of ammonium fertilizers after moisture absorption, with alkaline substances such as lime, etc., result in loss of N with ammonia escaping in gaseous form, CAN being an exception. Water-soluble phosphates should not be mixed with lime-containing or alkaline-acting fertilizers because insoluble and less available compounds are formed. Highly hygroscopic fertilizers are conditionally miscible, which

FIGURE 26
Guide for fertilizer compatibility and mixing

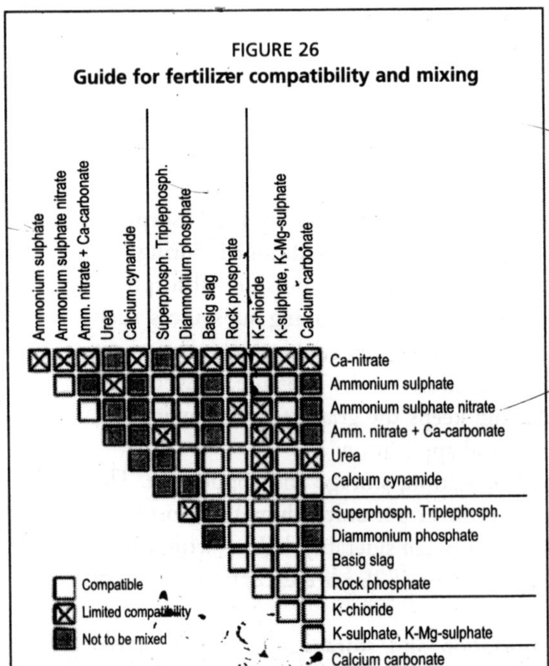

Source: Finck, 2001.

means that they should not be used in a mixture and stored but mixed only in dry weather shortly before application.

Bulk blending is a special type of fertilizer mixing in which only granulated products of fairly uniform size and density are used. Blended fertilizers are prepared by the mechanical mixing of two or more granular materials in defined proportions. Bulk blending originated in the United States of America and now dominates the fertilizer market in many areas. Often, a farmer has a bulk blend prepared according to the soil test report of the particular farm – a tailor-made, ready-to-use mixture. The main advantages to the farmers are:

- nutrients are supplied in ratios to suit the needs of particular soils and crops;
- the cost per unit of plant nutrient is generally low;
- the cost of transportation and spreading is low because of the high analysis of bulk blends.

However, the fertilizers used for mixing must be compatible both chemically and physically. The granules must be dry and strong so that they do not "cake" (stick together) and the granules must be similar in size in order to avoid segregation during mixing, transport and spreading. Common fertilizers used for bulk blending are DAP, MAP, TSP, AN, urea, MOP and special fertilizers to supply S, Mg and needed micronutrients.

The most important issues relate to the size and the density of granules. Granule size ranges from 1–4 mm in the United States of America and from 2–4 mm in Europe. The lower range is mainly caused by cheaply produced prilled urea with an average diameter of 1.5 mm, whereas phosphates and other common constituents exceed 2 mm in diameter. In addition to different granule size, large differences in bulk density may also cause segregation, the main problem being with urea, which has 30 percent lower density than most other fertilizers. Segregation of granules results in uneven distribution and erratic nutrient supply in the field. Another difficulty with bulk blending is mixing small amounts of micronutrients or herbicides with the much larger quantities of major nutrients.

ORGANIC SOURCES OF NUTRIENTS

Definition

Organic sources of nutrients are derived principally from substances of plant and animal origin. Partially humified and mineralized under the action of soil microflora, the organic sources act primarily on the physical and biophysical components of soil fertility. These sources cover manures made from cattle dung, excreta of other animals, other animal wastes, rural and urban wastes, composts, crop residues and even green manures. The term "bulky organic manure" is used collectively for cattle dung, FYM, composts, etc. because of their large bulk in relation to the nutrients contained in them. Concentrated organic manures, such as oilcakes, slaughterhouse wastes, fishmeal, guano and poultry manures, are comparatively richer in NPK.

General aspects

Organic sources of plant nutrients are used to varying extents in all countries. They may be used in the form in which they are obtained from the source or after having undergone varying degrees of processing. In most cases, the kinds of organic manures in use in a region are determined by the organic materials that are locally available or can be generated in the area, except for commercial organic fertilizers. According to surveys conducted by FAO through its various field projects (Roy, 1992), the main nutrient sources (in order of priority) in a number of countries are:

- Bangladesh: animal wastes, BNF (*Rhizobium*), green manuring;
- Burkina Faso: animal wastes, crop residues, BNF (*Rhizobium*);
- Democratic Republic of the Congo: crop residues, leaves of forest trees, BNF (*Rhizobium*);
- Guinea Bissau: crop residues, BNF (*Rhizobium* and *Azolla*);
- Indonesia: BNF (*Rhizobium*), recycling of legume crop residues, rice straw, animal wastes;
- Madagascar: animal wastes, crop residues (particularly rice straw), BNF (*Rhizobium* and *Azolla*).
- Nepal: in hill areas, animal wastes and BNF (*Rhizobium*); in terai areas, BNF (*Rhizobium*) and green manuring;
- Pakistan: animal wastes, BNF (*Rhizobium*), green manuring;
- Rwanda: animal wastes (in Butare and Gitarama regions), BNF (*Rhizobium*), crop residues;
- Sri Lanka: rice straw and legume crop residues, BNF (*Rhizobium*);
- Sudan: animal wastes, crops residues, BNF (*Rhizobium*);
- Thailand: BNF (*Rhizobium*), crop residues, agro-industrial wastes;
- United Republic of Tanzania: BNF (*Rhizobium*), crop residues;
- Zambia: animal wastes (certain areas in southern, western and central provinces), crop residues, BNF (*Rhizobium*).

Crop residues and green manures

Secondary products of crops, or auxiliary plants, are low-grade nutrient and soil-fertility improving resources. Composting can sometimes increase their value as a nutrient resource. Crop residues of legumes are richer in nutrients and have a low C:N ratio, which facilitates their mineralization compared with the residues of cereals. Similarly, processed residues such as oilcakes have a much higher nutrient content than conventional crop residues such as straw and stover.

Crop residues

Crop residues represent the bulk of the crop biomass left after removal of the main produce (grain, fruit, etc.) from the field. Most crops produce a voluminous amount of residues, e.g. straw, stalk, stubble, trash, and husks, which can have varying uses including as sources of plant nutrients either directly or after composting. Straw is produced in about the same and often higher amounts than grain (2–10 tonnes/ha) and can serve several purposes on the farm where not used for fuel, roofing, cattle bedding or sold. Crop residues contain a substantial proportion of plant nutrients (Table 20).

TABLE 20
Average nutrient content of some crop residues

Crop residues	Grain:straw ratio	Nutrient content (oven-dry basis)		
		N	P_2O_5	K_2O
			(%)	
Rice straw	1:1.5	0.58	0.23	1.66
Wheat straw	1:1.5	0.49	0.25	1.28
Sorghum stalks	1:2.0	0.40	0.23	2.17
Pearl millet stalks	1:2.0	0.65	0.75	2.50
Maize stalks	1:1.5	0.59	0.31	1.31
Average pulses	1:1.0	1.60	0.15	2.00
Pigeon pea	1:2.5	1.10	0.58	1.28
Chickpea	1:1.0	1.19	n.a.	1.25
Sugar-cane trash	1:0.2	0.35	0.04	0.50

However, the low N concentration of straw presents a special problem for its decomposition where the soil contains insufficient available N. Cereal straw has a C:N ratio of about 100:1 whereas ratios of below 25.1 are required for microbial decomposition in order to avoid N deficiency in the next crop. Such a growth-retarding effect can be avoided by adding 1 percent of mineral N to cereal straw. In spite of the low concentrations, as much as 125–250 kg K_2O can be added to the soil by 10 tonnes of cereal straw or 5 tonnes of oilseed rape straw. Being easily accessible to the farmer for use on the land, these have traditionally played an important role in maintaining soil productivity.

With some crops, such as sugar beets and sugar cane, large amounts of leaves are left on the field. They represent a large and valuable nutrient source, but their animal feed value is generally too high to be used as manures. Heavy leaf shedding before harvest is characteristic of jute plant and, in the process, large amounts of absorbed nutrients are returned to the soil.

Oilcakes

Oilcakes represent a special type of crop residue. These are the residues left behind after oil has been extracted from an oilseed. Table 21 provides a list of the average nutrient content of common oilcakes. Non-edible oilcakes can be used as manure, while edible oilcakes are used primarily as cattle feed. Oilcakes have a much higher nutrient content, particularly of N and P, than do normal crop residues, such as cereal straw or bulky organic manures. Owing to their low C:N ratio, these decompose at a faster rate in the soil to furnish available nutrients.

Green manures

Green manures represent fresh green plant matter (usually of legumes and often specifically grown for this purpose in the main field) that is ploughed in or turned into the soil to serve as manure. Several legume plants can be used as green manure crops. These are an important source of organic matter and plant nutrients, especially N where the green manure crop is a legume. Where feasible, green manuring is a key component of INM.

Green manure can either be grown *in situ* and incorporated in the field or grown elsewhere and brought in for incorporation in the field to be manured, in which

TABLE 21
Average nutrient content of some oilcakes

Oilcake sources	% N	% P_2O_5	% K_2O	kg N + P_2O_5 + K_2O/tonne of cake
Edible oilseeds				
Groundnut	7.29	1.65	1.33	103
Mustard	4.52	1.78	1.40	77
Rapeseed	5.21	1.84	1.19	82
Linseed[1]	5.56	1.44	1.28	83
Sesame	6.22	2.09	1.26	96
Cotton seed (decorticated)	6.41	2.89	1.72	110
Cotton seed (undecorticated)	3.99	1.89	1.62	75
Safflower (decorticated)	7.88	2.20	1.92	120
Safflower (undecorticated)	4.92	1.44	1.23	76
Non-edible oilseeds				
Castor	4.37	1.85	1.39	76
Neem (*Azadirachta indica*)	5.22	1.08	1.48	59
Mahua (*Madhuca indica*)	3.11	0.89	1.85	59
Karanj (*Pongamia glabra*)	3.97	0.94	1.27	62
Kusum (*Schleichera oleosa*)	5.23	2.56	1.37	92
Khakan (*Salvadora oleoides*)	4.32	2.45	1.24	80

[1] Edible and non-edible.

case it is referred to as green-leaf manuring. Not all plants can be used as a green manure in practical farming. Green manures may be: plants of grain legumes such as pigeon pea, green gram, cowpea, etc.; perennial woody multipurpose legumes, such as *Leucaena leucocephala* (*subabul*), *Gliricidia sepium*, and *Cassia siamea*; and non-grain legumes, such as *Crotalaria*, *Sesbania*, *Centrosema*, *Stylosanthes* and *Desmodium*. Because green manures add whatever they have absorbed from the soil, they in fact recycle soil nutrients from lower depths to the topsoil besides contributing to soil N through N fixation by the legume green manure crop. For major crops, some common green manures are:

- rice: sunnhemp, *Sesbania* and wild indigo (*Indigofera tinctoria*), *Azolla*;
- sugar cane: sunnhemp;
- finger millet: sunnhemp;
- wheat: sunnhemp;
- sorghum: sunnhemp, Leucaena (*Leucaena leucocephala*);
- banana: leaves of *Gliricidia sepium*;
- potato: sunnhemp, cowpea, cluster bean, lupin (*Lupinus albus*).

Green manures can add substantial amounts of organic matter and N as well as other nutrients. The bulk of the N input through leguminous green manures comes from BNF. Using rice culture as an example, this can range from 50 to 200 kg N/ha (Table 22). The nutrient contribution of a green manure crop is greatest where the entire green plant is ploughed in and incorporated in the soil. It is minimum but still appreciable where the grain of the legume is harvested and the straw or stover is ploughed in.

Green manure crops are often sown and incorporated in the field prior to planting a main crop such as rice, potato or sugar cane. Short-duration legumes

TABLE 22
Some green manure crops and their N contribution under optimal conditions

Crop	Scientific name	Suitable soil	Optimal temperature (°C)	Duration in days (days)	N added (kg/ha)
Black gram	*Vigna mungo* L.	Well drained	15–35	70	60
Mung bean	*Vigna radiata* L.	Well drained	20–35	60–65	55
Cowpea	*Vigna unguiculata* L.	Well drained	10–38	45–60	60
Sesbania	*Sesbania rostrata* L.	Poorly drained	15–40	45–50	100
Sunnhemp	*Crotalaria juncea* L.	Poorly drained	12–35	45–50	120
Siratro	*Indigofera hirsute*	Well drained	15–35	100–120	80–90
Sesbania	*Sesbania bispinosa*	Wet to waterlogged	15–38	45–50	80
Cluster bean	*Cyamopsis tetragonoloba*	Marginal	12–35	-	80–90
Ipil-iplil	*Leucaena leucocephala*	Fertile	15–35	-	125[1]
Gliricidia	*Gliricidia sepium*	Acid, low fertility	8–35	-	80–100[1]

[1] N added through 4–5 tonnes of biomass.
Source: Pandey, 1991

can also be used as intercrops along with long-duration crops and used as green manures before or after picking the pods. After a few months of growth, generally at the beginning of flowering, the plants are cut and mixed into the soil. The gains in N with these short-duration legumes are generally of the order of 30–50 kg/ha N. There are limits to the use of green manuring under arid conditions because of the additional water requirement. Green manures and cover crops have an important place in plantations. Where grown on marginal lands and brought to fields, their nutrients can be considered as an external input, which is also the case where "weeds" such as water hyacinths are applied.

Farmyard manure and animal slurry

Farmyard manure (FYM)

FYM refers to the bulky organic manure resulting from the naturally decomposed mixture of dung and urine of farm animals along with the litter (bedding material). Average, well-rotted FYM contains 0.5–1.0 percent N, 0.15–0.20 percent P_2O_5 and 0.5–0.6 percent K_2O. The desired C:N ratio in FYM is 15–20:1. In addition to NPK, it may contain about 1 500 mg/kg Fe, 7 mg/kg Mn, 5 mg/kg B, 20 mg/kg Mo, 10 mg/kg Co, 2 800 mg/kg Al, 12 mg/kg Cr and up to 120 mg/kg lead (Pb). Often, fully or partially air-dried dung is used as FYM. FYMs can be used simply after air drying or after composting. Grazing animals return them directly to the soil as a natural nutrient supply, or the dry dung may be collected, stored and used as fuel or again as a manure in the desired area. A list of the average nutrient content of some organic manures including FYM and other organic manures is given in Table 23. The list includes manures derived from plants, animals and human wastes.

During storage, organic manure is partly decomposed by fermentation, which also produces valuable humic substances. Some losses of N as ammonia occur, but these can be reduced by the addition of about 2-percent water-soluble phosphate. Nutrient concentrations of fermented moist FYM (25 percent dry matter) depend

TABLE 23
Average nutrient content of bulky organic manures and composts

Type of manure	N	P_2O_5	K_2O
		(%)	
Cattle dung	0.3	0.10	0.15
Sheep/goat dung	0.65	0.5	0.03
Human excreta	1.2–1.5	0.8	0.5
Hair and wool waste	12.3	0.1	0.3
Farmyard manure	0.5	0.15	0.5
Poultry manure	2.87	2.90	2.35
Town/urban compost	1.5	1.0	1.5
Rural compost	0.5	0.2	0.5

on feeding intensity, and vary over a wide range. In several tropical and subtropical areas such as South Asia, the FYM is applied preferentially before the rainy-season crops such as rice, maize and pearl millet rather than to wheat in the dry post-monsoon season. FYM is also frequently applied to potato, groundnut, sugar cane and vegetable crops in preference to crops such as wheat.

Animal slurry

In many developed countries, because of the shift towards intensive labour-saving animal production systems, many of which do not require bedding straw, there has also been a large output of animal slurry. In large areas, slurry is now the dominant animal manure although this can hardly be regarded as a desirable feature from an environmental and animal welfare point of view. Slurry from domestic animals consists of dung and urine, partly mixed with a small portion of straw and with small or large portions of water in order to improve its fluidity. It is a semi-liquid nutrient source that can be mechanically collected (pumped up to 12 percent dry matter), stored and distributed. The amounts of slurry produced per year are about 15–20 m^3/cow (7–10 percent dry matter) and about 15 m^3/pig unit (7 pigs) with 5–8 percent dry matter.

In regions with frozen or cold soils, slurry cannot be spread throughout the year. Therefore, it must be stored in large containers for up to several months. During this period, fermentation and conversion of urea to ammonia takes place and ammonia losses occur. Unpleasant odours may also be produced. Nutrient concentrations of fermented slurry with 5–10 percent dry matter are of the following order:

- cow slurry: 0.25–0.5 percent N, 0.3–0.5 percent K, 0.05–0.1 percent P;
- pig slurry: 0.4–0.8 percent N, 0.3–0.4 percent K, 0.1–0.2 percent P.

The main effect of slurry on crops is through its N supply. A large portion of N, about half with pig slurry, is ammonia N derived from decomposed urea. About half of the organic N is slow acting, the K fraction is mineral and the phosphate is mostly organic, but partly in mineral form ($MgNH_4PO_4$). The pH of slurry is about neutral.

Biogas plant slurry

The use of organic wastes for biogas production can be an important source of energy on the farm and also of manure. In India, many small-scale biogas production units have been established (Plate 1). Cattle dung is most commonly used as an input, mainly because of its availability. In addition to the animal and human wastes, plant materials can also be used. Materials with a high C:N ratio could be mixed with those of a low C:N ratio to bring the average ratio of the composite input to a desirable level. In China, as a way of balancing the C: N ratio, it is customary to load rice straw at the bottom of the digester upon which latrine waste is discharged. Similarly, at Machan Wildlife Resort located in Chitawan District, Nepal, feeding the digester with elephant dung in conjunction with human waste enabled a balanced C:N ratio for the smooth production of biogas (Karki, Gautam and Karki, 1994). In the biogas production units, waste materials, are fermented under anaerobic conditions in a closed metal container (about 3 m^3) for a few days. The resulting methane and hydrogen is used as fuel for cooking and lighting, and the residual material in slurry form can be used as manure either directly or as compost. The typical composition of biogas slurry is 1.4–1.8 percent N, 1.1–1.7 percent P_2O_5 and 0.8–1.3 percent K_2O. It is a useful organic manure. Effective small-scale biogas production is restricted to warm climates. It requires capital investment, maintenance and a considerable amount of manual work, but the energy gain can be considerable.

Plate 1
Biogas plant, example from India.

Compost

Although many organic waste products can be added directly into the soil, most of them have a better soil-improving effect after their decomposition through the composting process. The resulting mixed and improved products following decomposition are termed compost (Latin *componere* = mixing). Compost can be defined as an organic manure or fertilizer produced as a result of aerobic, anaerobic or partially aerobic decomposition of a wide variety of crop, animal, human and industrial wastes. Composting has a long tradition almost everywhere in the world. It was a central concept of early Chinese agriculture, but it has also been practised in India and Europe for centuries. Composts are generally classified as:

- Rural compost: This is produced from materials available on the farm and in other rural areas. The raw materials used can be straw, leaves, cattle-shed bedding, fruit and vegetable wastes, and biogas plant slurry. On average, it contains 0.5 percent N, 0.2 percent P_2O_5 and 0.5 percent K_2O. Rural compost primarily finds use on farms as a bulky organic manure.
- Urban or town compost: This refers to compost prepared from urban and industrial wastes, city garbage, sewage sludge, factory waste, etc. Its typical composition is 1.5–2.0 percent N, 1.0 percent P_2O_5 and 1.5 percent K_2O. Commercially prepared urban compost has been reported to contain 1 percent Fe, about 375 mg/kg Cu, 705 mg/kg Zn, 740 mg/kg Mn and small amounts of other micronutrients.

- Vermicompost: This is an important type of compost that contains earthworm cocoons, excreta, beneficial micro-organisms, actinomycetes, plant nutrients, organic matter, enzymes, hormones, etc. It is an organic fertilizer produced by earthworms and contains on average 0.6 percent N, 1.5 percent P_2O_5 and 0.4 percent K_2O. In addition to NPK, it is also a source of micronutrients, containing an average of 22 mg/kg Fe, 13 mg/kg Zn, 19 mg/kg Mn and 6 mg/kg Cu. It helps in cost-effective and efficient recycling of animal wastes (poultry, horse, piggery excreta and cattle dung), agricultural residues and industrial wastes using low energy.

Compost preparation

Composts are prepared through the action of micro-organisms on organic wastes such as leaves, roots and stubbles, crop residues, straw, hedge clippings, weeds, water hyacinth, bagasse, sawdust, kitchen wastes, and human habitation wastes. Virtually any biodegradable organic material can be composted. For making town or urban garbage compost, the organic wastes from households and other establishment should be carefully collected, separated from unsuitable materials and not contaminated with toxic substances. The main problem with compost prepared from urban wastes and garbage is the potential contamination with toxic substances that must be avoided.

A number of composting processes are in vogue in different parts of the world, comprising practices adopted as a convention, and the recently introduced methodologies for expediting the process that entail individual or combined application of treatments, such as: shredding and frequent turning, mineral N compounds, effective micro-organisms, use of worms, cellulolytic organisms, forced aeration and mechanical turnings. Conventional methods generally adopt an approach based on limited aerobic/anaerobic decomposition or one based on aerobic decomposition using passive aeration through measures such as little and infrequent turnings or static aeration provisions such as perforated poles/pipes. These processes take several months. On the other hand, using the recently developed techniques, rapid methods expedite the aerobic decomposition process and reduce the composting period to about four to five weeks. Most of these methods include a high temperature period, and this adds further value to the product by eliminating pathogens and weed seeds (FAO, 2003a).

During compost preparation, special supplements can be used such as some mineral N (1–2 kg N/m^3 in order to obtain a C:N ratio of about 10–15:1, 2–3 kg $CaCO_3/m^3$ for neutralization of surplus acids and possibly some PR for better P supply). By doing so, compost can be enriched and fortified. Phosphocompost is one such type of material where less reactive PR can be utilized effectively and the nutrient content of compost upgraded.

Nutrient content and quality standards

The nutrient content of a compost depends largely on the nutrient content of the wastes composted. The quality of composts varies widely. On average, compost may contain 30–50 percent dry matter, 10–15 percent organic matter and the indicated amounts of plant nutrients. Ideally, compost should be rich in available plant nutrients, contain readily decomposable material and relatively stable humic substances, and have a crumbly structure, similar to a humus-rich topsoil. Composts are not only nutrient sources, but also effective soil amendments.

Quality standards define the composition and characteristics of compost and prescribe the maximum acceptable limits of undesirable elements. Such standards have been emerging gradually in the western world. Several European countries have adopted specific standards (Brinton, 2000). However, such standards are still in the process of development for most developing countries. Sometimes, a total minimum N, P_2O_5 and K_2O content of 5 percent is suggested as a requirement. One example relating to Bangkok is:

- minimum nutrient content: 1–3 percent N, 1.5–3 percent P_2O_5, 1–15 percent K_2O;
- moisture content: should not exceed 15–25 percent;
- organic matter: should be at least 20 percent C;
- C:N ratio: should be between 10:1 and 15:1
- pH: should be around neutral (6.5–7.5).

In garbage compost, harmful substances and pollutants such as toxic metals (e.g. Cd, Cr and Hg) or toxic organic compounds should be below the critical level (CL). Therefore, the compost materials need to be controlled for safe use in order not to endanger soil quality, plant growth, food quality or human health. Assuming that urban compost is used primarily for urban agriculture, the users are well advised to insist on proper compost quality in respect of toxic metals, even if the gain of cheap nutrients appears rather attractive. The principle should be that if the urban areas want to free themselves of waste materials, it is their responsibility to offer useful and safe products.

Recyclable waste products

The utilization of common waste products of plant and animal origin as sources of plant nutrients has been discussed above. In addition, several wastes or by-products of animal, human and industrial origin can also be used as sources of plant nutrients.

Waste products of animal origin other than excreta

A number of wastes derived from the bodies of domestic animals can be used as sources of plant nutrients. Important among these are various types of animal meals including bonemeal, which is a long-established source of phosphate for crop production. A list of the nutrient content of several such manures derived from the animal bodies is given in Table 24.

Animal meal is the common term used for the group of organic manures derived from animal wastes other than dung and urine (Table 24). Bonemeal is rich in P, others are rich in N. Bonemeal is an organic fertilizer derived from bones. Raw bonemeal consists of ground bones without any of the gelatin or glue removed.

TABLE 24
Average nutrient composition of some organic manures derived from the animal wastes

Manure	Nutrient content		
	N	P_2O_5	K_2O
		(%)	
Meatmeal	10.5	2.5	0.5
Bloodmeal	10–12	1–2	1.0
Horn and hoof meal	10–15	1.0	-
Bonemeal (raw)	3–4	20–25	-
Bonemeal (steamed)	2–5	26–28	-
Fishmeal	4–10	3–9	1.8
Leather waste	7.0	0.1	0.2
Hair and wool waste	12.3	0.1	0.3

It contains at least 3 percent N and about 22 percent P_2O_5, of which about 8 percent is citrate soluble (available). It also contains variable amounts of micronutrients. Steamed bonemeal is obtained by treating crushed bones with steam under pressure in order to dissolve part of gelatine and then grinding the residue into a power, which is then passed it through a sieve of 1-mm mesh size. It contains about 28 percent P_2O_5, of which about 16 percent is citrate soluble.

Waste products of human origin

Human excreta composed of faeces and urine along with domestic wastewater carried through sewers to the disposal points/treatment tanks is termed sewage. Sometimes, this may be further contaminated through industrial effluents (high in heavy metals). Sewage sludge is the end product of the fermentation (aerobic or anaerobic) of sewage. It is semi-solid and a useful organic manure. Activated sewage sludge refers to biologically active sewage sludge obtained by repeated exposure of the sewage to atmospheric oxygen, thus facilitating the growth of aerobic bacteria and other unicellular micro-organisms. In the process, it is improved for use on land.

The general composition of sewage sludge is 1.1–2.3 percent N, 0.8–2.1 percent P_2O_5 and 0.5–1.7 percent K_2O. It also contains Na, Ca, S, several micronutrients and toxic heavy metals (e.g. Al) in some cases. The typical nutrient content of activated sewage sludge is 5.8 percent N, 3.2 percent P_2O_5 and 0.6 percent K_2O. It also contains lesser and variable amounts of secondary and micronutrients and toxic heavy metals. Therefore, care has to be taken in deciding the optimal application rates depending on its composition.

Properly treated sewage effluent and processed products such as sewage sludge can serve as irrigation water and manure. The relative number of enteric pathogens in sewage effluent and sewage sludge depends on the type of sewage treatment. Primary treatment (consisting mostly of settling) removes 35–45 percent of pathogens while more than 95-percent pathogen removal is achieved by secondary treatment. Thus, the use of treated sewage for crop production minimizes the health risk. Chapter 7 discusses suggested cropping patterns for irrigation with untreated and treated sewage waters.

Waste products of industrial origin

Several industrial wastes and by-products can be used as sources of plant nutrients or as soil amendments after suitable processing. One such source is press mud or filter cake obtained from sugar factories.

Press mud is a by-product of sugar factories. It is the residue obtained by filtration of the precipitated impurities that settle out in the process of clarification of the mixed juice from sugar cane. The material has 55–75 percent moisture, is soft and spongy, light in weight and amorphous dark brown. It can readily absorb moisture when dry. Depending on the process used in the sugar factory, it can be either sulphitation press mud (SPM) or carbonation press mud (CPM). It contains 1–3 percent N, 0.6–3.6 percent P_2O_5, 0.3–1.8 percent K_2O and 2.3 percent S.

SPM contains about 9 percent gypsum while CPM has 60 percent calcium carbonate. SPM is richer in plant nutrients compared with CPM. Thus, material from factories using a sulphitation process is a good source of S. Press mud from sugar factories using the carbonation process can find use as a liming material. Press mud can also be utilized after it is composted. It can be composted alone or with sugar-cane trash and animal dung. While preparing such compost, a 22.5-cm thick layer of SPM is arranged alternatively with a 22.5-cm thick layer of the yard sweepings consisting of cane trash, cattle dung and urine in pits for composting. It takes 6–8 months for the compost to be ready. The compost thus prepared has good manurial value, containing 1 percent N, 3 percent P_2O_5, 1 percent K_2O and 8 percent CaO on a fresh-weight basis. Preparation of compost from distillery spent wash is also possible.

Commercial organic fertilizers

In their original state, waste products have a wide range of nutrient concentrations and are often difficult to handle. It is only reasonable and for the user's benefit that they should be processed into standardized nutrient sources. Such products are commercial organic fertilizers produced on a large scale, and they are much preferred by commercial growers to the original unprocessed waste materials.

Organic fertilizers can be defined as materials that have been prepared from one or more materials of a biological nature (plant/animal) and/or unprocessed mineral materials (lime, PR, etc.) that have been altered through controlled microbial decomposition into a homogenous product with a sufficient amount of plant nutrients to be of value as a fertilizer. Usually, they must contain a minimum of 5 percent nutrients ($N + P_2O_5 + K_2O$).

The raw materials used are processed through a process of drying, shredding, mixing, granulating, odour removal, pH modification, partial fermentation and composting, and always with proper hygienic control. This process provides standard products with certified concentrations of organic matter, a definite C:N ratio, guaranteed nutrient concentrations, and products without growth-impeding substances or sanitary problems. Finally, they are also easy to store and handle.

The types of commercial organic fertilizers, based on plant and/or animal residues, are often classified as follows:

- organic N fertilizers (at least 5 percent N, often higher);
- organic P fertilizers, mainly from bones (e.g. 25 percent P_2O_5);
- organic NP fertilizers (at least 3 percent N and 12 percent P_2O_5);
- organic NPK fertilizers (at least 15 percent of N, P_2O_5 and K_2O together);

- organo-mineral NP or NPK fertilizers, supplemented by mineral fertilizer or guano (e.g. NP with at least 5 percent each of N and P_2O_5, or NPK with at least 4 percent each of N, P_2O_5, and K_2O);
- organo-mineral fertilizers based on peat, but with nutrient supplements.

All these types of organic fertilizers are widely used, especially in gardening, where low nutrient concentrations and slow-acting N sources are preferred. In agriculture, they are applied mainly to vegetables. Some of these can be important inputs in organic farming.

Other types of organic inputs gaining popularity are those derived from seaweeds. These are red, brown or green algae living in or by the sea. Seaweeds like *Ascophyllum nodosum*, *Laminaria digitata*, and *Facus serratus*, contain gibberellin, auxins, cytokinin, etc. and are being used as liquid organic fertilizer with or without fortification with minerals in many countries. Their role is more of a plant-growth stimulant rather than of a nutrient supplier.

The term guano covers a special group of organic fertilizers derived from the excreta of, usually, small animals and includes materials such as bat guano, Peruvian guano, and fish guano. The general N content of guano can be 0.4–9.0 percent and total P_2O_5 can be 12–26 percent. Guano is found and used in certain areas only.

Application techniques for organic manures are discussed in Chapter 7.

BIOFERTILIZERS (MICROBIAL INOCULANTS)

Definition, classification and general aspects

Definition

Biofertilizer is a broad term used for products containing living or dormant micro-organisms such as bacteria, fungi, actinomycetes and algae alone or in combination, which on application help in fixing atmospheric N or solubilize/mobilize soil nutrients in addition to secreting growth-promoting substances. They are also known as bioinoculants or microbial cultures. Strictly speaking, although widely used, the term biofertilizer is a misnomer. Unlike fertilizers, these are not used to provide nutrients present in them, except in the case of *Azolla* used as green manure.

Classification

Biofertilizers can be grouped into four categories:

- N-fixing biofertilizers: These include the bacteria *Rhizobium*, *Azotobacter*, *Azospirillum*, *Clostridium* and *Acetobacter* among others; BGA or cyanobacteria and the fern *Azolla* (which works in symbiosis with BGA).
- P-solubilizing/mobilizing biofertilizers: These include phosphate-solubilizing bacteria (PSB) and phosphate-solubilizing micro-organisms (PSMs), e.g. *Bacillus*, *Pseudomonas* and *Aspergillus*. Mycorrhizae are nutrient-mobilizing fungi, also known as vesicular-arbuscular mycorrhizae or VA-mycorrhizae or VAM.
- Composting accelerators: (i) cellulolytic (*Trichoderma*); and (ii) lignolytic (*Humicola*).

➢ Plant-growth-promoting rhizobacteria (PGPR): Species of *Pseudomonas*. These do not provide plant nutrients but they enhance plant growth and performance.

General aspects

The most important biofertilizers used in agriculture are those that contain cultures of N-fixing organisms; next in importance are the cultures of P-solubilizing organisms.

BNF involves the conversion of nitrogen gas (N_2) into ammonia through a biological process (in contrast to industrial N fixation). Many micro-organisms (e.g. *Rhizobium*, *Azotobacter* and BGA) utilize molecular N_2 through the help of nitrogenase enzyme and reduce atmospheric N_2 to ammonia (NH_3):

$$N_2 + 6H^+ + 6e^- \longrightarrow 2NH_3$$

BNF is a major source of fixed N for plant life. Estimates of global terrestrial BNF range from 100 to 290 million tonnes of N/year. Of this total, 40–48 million tonnes is estimated to be biologically fixed in agricultural crops and fields. The first commercial *Rhizobium* biofertilizer was produced as Nitragin in the United States of America in 1895. PSMs secrete organic acids that dissolve insoluble phosphate compounds. The first commercial P-solubilizing biofertilizer, Phospho-bacterin, was produced in the then Union of Soviet Socialist Republics.

Only N-fixing micro-organisms bring in net additional supplies of a nutrient (N) into the soil plant system. All other biofertilizers simply solubilize or mobilize the nutrients that are already present in soils. *Azolla* is unique in the sense that it acts as host to the N-fixing cyanobacteria, after which it is used virtually as a green manure. In the process, it adds not only the biologically fixed N but also the other nutrients absorbed from the soil and present in its biomass. While *Rhizobium* is legume specific, BGA and *Azolla* are specific to wetlands and, hence, useful in augmenting the N supply in flooded-rice cultivation.

Nitrogen-fixing biofertilizers

Rhizobium

Bacteria of the genus *Rhizobium* are able to establish symbiotic relationships with many leguminous plants, as a result of which the nitrogen gas (N_2) of the air is "fixed" or converted to ammonium ions that can be utilized by plants. These bacteria survive in the soil as spores. Where a root of a compatible species grows close to the spore, recognition occurs and symbiosis begins. The root hair curls and an infection thread appears from the spore and enters the root cells. The root responds by multiplying cells and these form the nodules on the roots that contain the bacteria. The root nodules act as the site of N fixation. The optimal temperature for their growth is 25–30 °C and the optimal pH is 6–7. Inoculation with *Rhizobium* is recommended for legumes (pulses, oilseeds and forages). On average, yield response to *Rhizobium* inoculation varies from 10 to 60 percent depending on the soil–climate situation and efficiency of the strain.

Not all species of *Rhizobium* can form a symbiotic relationship with all legumes and form nodules. There is generally high specificity between the bacteria and the host plant, called cross-inoculation groups. However, some plants can be infected by a range of *Rhizobium* species and form effective symbiotic association. In contrast to the root-nodule-forming *Rhizobium,* there is also the *Azorhizobium* bacteria, which is capable of forming root nodules as well as stem nodules on the tropical legume *Sesbania rostrata.* It is grouped under *Azorhizobium* in *Rhizobium* classification. The *Rhizobium* species that can form nodules and fix N with specific leguminous plants are:

- *Rhizobium ciceri*: It nodulates chickpea.
- *Rhizobium etli*: It nodulates beans.
- *Rhizobium japonicum* (now known as *Bradyrhizobium japonicum*): It nodulates soybean.
- *Rhizobium leguminosarum*: It nodulates peas, broad beans, lentils, etc.
- *Rhizobium lupini*: It nodulates *Lupinous* sp. and *Ornithopus* sp.
- *Rhizobium meliloti*: It nodulates *Melilotis* (sweet clover), *Medicago* (alfalfa) and *Trigonella* (fenugreek).
- *Rhizobium phaseoli*: It nodulates temperate species of *Phaseolus.*
- *Rhizobium trifolii*: It nodulates *Trifolium* spp.

Most soils contain these bacteria but their population may not be adequate or effective for forming productive associations with the crops sown. In such cases, the organisms must be artificially introduced into the system. This is generally done by mixing a culture/inoculum of the organism with the seed before sowing. Artificially prepared *Rhizobium* culture that is used for seed dressing of legumes before sowing to enhance the supply of N is referred to as the *Rhizobium* inoculant or biofertilizer. It is the most widely used biofertilizer in the world. Inoculation of grain legumes such as pulses is associated with an N gain of 20–40 kg N/ha. Application techniques of biofertilizers are discussed in Chapter 7.

Azotobacter

Azotobacter is a non-symbiotic, aerobic, free-living, N-fixing soil bacterium. It is generally found in arable soils but its population rarely exceeds 10^2–10^3/g soil. Its six species are: *Azotobacter armeniacus, A. beijerinckii, A. chroococcum, A. nigricans, A. paspali* and *A. vinelandi.* Unlike *Rhizobium,* inoculation with *Azotobacter* can be done for a wide variety of crops. Grain yields obtained from plots untreated with fertilizer N but inoculated with N-fixing bacteria are similar to yields obtained from the application of 20–35 kg N/ha.

Azotobacter also synthesizes growth-promoting substances, produces group B vitamins such as nicotinic acid and pantothenic acid, biotin and heteroauxins, gibberellins and cytokinin-like substances, and improves seed germination of several crops. Both carrier-based and liquid-based *Azotobacter* biofertilizers are available. It is recommended as a biofertilizers for cereals and horticultural crops including flowers and vegetables. Its application is usually done through seed treatment, seedling treatment or soil application (described in Chapter 7).

Azospirillum

Azospirillum, a spiral-shaped N-fixing bacteria, is widely distributed in soils and grass roots. Major species of *Azospirillum* are *Azospirillum brasilense* and *Azospirillum lipoferum*. It can fix 20–50 kg N/ha in association with roots. It also produces hormones such as indole acetic acid (IAA), gibberellic acid (GA), cytokinins and vitamins.

Acetobacter

Acetobacter is a rod-shaped, aerobic, N-fixing bacteria. *Acetobacter diazotrophicus* is an N-fixing bacteria found in the roots, stems and leaves of sugar cane with the potential to fix up to 200 kg N/ha. It is capable of growth at pH 3. It can also solubilize insoluble forms of P. Inoculation with *Acetobacter* is recommended for sugar cane.

Blue green algae

BGA are photosynthetic, unicellular, aerobic, N-fixing algae. They are also known as cyanobacteria and are used primarily as a biofertilizer in flooded-rice culture. More than 100 species of BGA are known to fix N. Commonly occurring BGA are *Nostoc*, *Anabaena*, *Aulosira*, *Tolypothrix* and *Calothrix*. These are used as biofertilizer for wetland rice (paddy) and can provide 25–30 kg N/ha in one crop season, or up to 50 kg N/ha/year. The BGA also secrete hormones, such as IAA and GA, and improve soil structure by producing polysaccharides, which help in the binding of soil particles (resulting in better soil aggregation). BGA are also used as a soil conditioner and, through mat formation, they protect the soil against erosion.

Soil pH is the most important factor in determining BGA growth and N fixation. The optimal temperature for BGA is about 30–35 °C. The optimal pH for BGA growth in culture media ranges from 7.5 to 10, and its lower limit is about 6.5–7. Under natural conditions, BGA growth is better in neutral to alkaline soils. BGA need all the plant nutrients for their growth and N fixation. N fertilizers generally inhibit BGA growth and N fixation. Adequate available P should be present in the floodwater as P enhances BGA growth and N fixation. Consequently, P deficiency causes drastic reduction in BGA growth and, hence, in N fixation. Mo is another essential nutrient for the growth and performance of BGA.

The inoculum of BGA can be prepared in the laboratory or in the open fields. The open-air soil culture method is simple, less expensive and easily adaptable by farmers. BGA are multiplied in shallow trays or tanks with 5–15 cm standing water in 4 kg soil/m^2. A thick BGA mat is formed on the soil surface in about 15 days and the tray is allowed to dry in the sun. BGA flakes are collected and stored for use (described in Chapter 7).

Azolla

Azolla is another N-fixing biofertilizer of specific interest in rice cultivation. *Azolla* itself is a fern. N fixation is carried out by the cyanobacterium *Anabaena azollae* in the leaf cavities of *Azolla*. The most common species of *Azolla* are:

- *Azolla pinnata*: This is the most important species. It is widespread in the Eastern Hemisphere, tropical Africa, Southeast Asia, etc. Of its two forms, *Azolla pinnata* var. *pinnata* and *Azolla pinnata* var. *imbricata*, *pinnata* is more common. Its favourable temperature is 20–30 °C.
- *Azolla caroliniana*: A multitolerant species of *Azolla*, it is pest resistant, shade tolerant and thrives under a wide temperature range.
- *Azolla filiculoides*: It is cold tolerant (-5 °C), and heat sensitive (exceeding 30 °C).
- *Azolla microphylla*: It is heat tolerant but cold sensitive.
- *Azolla nilotica*: Reported to occur in the Nile River in Africa.

On average, dry *Azolla* contains 2.08 percent N, 0.61 percent P_2O_5, 2.05 percent K_2O, and has a C:N ratio of 14:1. It is known to accumulate significant amounts of K. *Azolla* can accumulate 30–40 kg K_2O/ha from irrigation water in the paddy-field. The N-enriched *Azolla* biomass is incorporated into the soil, thus providing the N fixed by the cyanobacteria and all other nutrients absorbed by the fern from the soil and irrigation water. Thus, it is more of a green manure than a conventional biofertilizer. One crop of *Azolla* can provide 20–40 kg N/ha to the rice crop in about 20–25 days.

Azolla requires all the essential plant nutrients for normal growth. Because of its aquatic nature, these elements must be available in the soil water. The deficiency of any one element adversely affects its growth and N fixation. In these respects, *Azolla* behaves like an agricultural crop. P is a key element and its deficiency results in poor growth, pink or red coloration, root curl and reduced N content. Temperature is a key factor that limits the growth of *Azolla* and 25–30 °C is optimal for most species. A pH of 5–8 is optimal although *Azolla* can survive in the pH range of 3.5–10.0. The inoculum for *Azolla* biofertilizer is in the form of dry spores. Application details are provided in Chapter 7.

Phosphate-solubilizing biofertilizers

There has been much research conducted on the use of organisms to increase P availability in soils by "unlocking" P present in otherwise sparingly soluble forms. These microbes help in the solubilization of P from PR and other sparingly-soluble forms of soil P by secreting organic acids, and in the process decreasing their particle size, reducing it to nearly amorphous forms. The earliest known commercial P-solubilizing biofertilizer, Phospho-bacterin, contained *Bacillus megatherium* var. *phosphaticum*. Phosphate-solubilizing organisms include:

- bacteria: *Bacillus megatherium* var. *phosphaticum*, *Bacillus polymyxa*, *Bacillus subtilis*, *Pseudomonas striata*, *Agrobacterium* sp.; *Acetobacter diazotrophicus*, etc.;
- fungi: *Aspergillus awamori*, *Penicillium digitatum*, and *Penicillium belaji*;

- yeast: *Saccharomyces* sp., etc.
- actinomycetes: *Streptomyces* sp., *Nocardia* sp.

In addition to bacteria, the fungus *Penicillium belaji* has been shown to increase P availability from native soil and PR sources in calcareous soils. The responses to soil inoculation of such biofertilizers have been reported, but they are low, averaging about 10 percent, and extremely variable. Based on present evidence, it seems unlikely that inoculation with micro-organisms will contribute significantly to plant P nutrition in the foreseeable future. However, in some countries such as India, the P-solubilizing biofertilizers are becoming popular, ranking next in importance only to the N-fixing *Rhizobium* inoculants. Usually, more than one type of organism is used while preparing a P-solubilizing biofertilizer.

Nutrient-mobilizing biofertilizers

The most prominent among nutrient mobilizers in the soil are the soil fungi mycorrhizae. These form symbiotic relationships with the roots of host plants. These are of two types:

- Ectomycorrhizae: These form a compact sheath of hyphae over the surface of roots of a limited number of plant such as *Pinus* and *Eucalyptus*.
- Endomycorrhizae: These penetrate the roots and grow between the cortical cells. They produce storage "vesicles" ("saclike" structures) between the cells and multibranched "arbuscules" within the cells. Hence, the name vesicular-arbuscular mycorrhizae (VAM). They also produce thin hyphae that grow out up to 2 cm from the root surface.

VAM are ubiquitous in most soils and naturally infect most plants. Responses to field inoculation with VAM are rare except in crops such as onions that have no root hairs to facilitate P uptake and require a rapid supply of P. Responses to soil inoculation do not occur where there is ample P in the soil. Because mycorrhizae cannot be cultured in the same way as rhizobia, commercial inoculation is not possible at this stage. Where inoculation is required, soil from infected plants is used. Application of organic manures stimulates VAM.

The relationship between mycorrhizae and plant roots is useful in improving the capability of plants for soil exploration and nutrient uptake. VAM have been associated with increased plant growth and with enhanced accumulation of plant nutrients, mainly P, Zn, Cu and S, primarily through greater soil exploration by the mycorrhizal hyphae. Out of their special structures, the arbuscules help in the transfer of nutrients from the fungus to the root system and the vesicles store P as phospholipids. Thus, the exploratory capacity of the root system is improved far beyond the zones of nutrient-depleted soil that may surround the root.

Being an obligate symbiont, mycorrhiza inoculum can be supplied in the form of infective soil, infected roots and soil sievings. However, infective roots and growth medium from pot cultures open to the atmosphere can become contaminated with pathogens (fungi, bacteria and nematodes). Mycorrhizae have to be cultured using a particular host. Onions, sorghum and other grasses are suitable hosts. Such cultures are used as inoculum in the form of seed pellets, granules or as such in plastic bags and can be stored at 4 °C for 2–3 months.

SOIL AMENDMENTS

Only very few soils are "by nature" ideal substrates for plant growth. Much effort has been devoted to improving "problem" soils. Generally, the chemical properties of soils are easier to improve than are the physical ones. With increasing intensity of cropping, many methods of soil improvement have become available and proved profitable.

Of the chemical soil properties, the soil reaction (pH) of many soils must be optimized in order to create favourable conditions for plant growth, nutrient availability and to eliminate the harmful toxic substances. Optimizing soil pH is a precondition for the success of nutrient management for crop production. It entails either raising the pH of acid soils or lowering the pH of alkaline soils. Among the soil physical properties, the improvement of soil structure is of great concern to farmers. The texture of sandy, clayey or stony soils may also be improved but to a very limited extent.

Amendments for raising the soil reaction (liming)

Soil acidity is reflected primarily in an increase in H^+ ions and a corresponding decrease in the basic cations. Carbonates (lime), hydroxides and some other basic acting substances are able to neutralize soil acids. The purpose of liming is primarily the neutralization of the cause of soil acidity (H^+ ions and Al^{3+} in very acid soils), thus raising the pH value.

Ca and Mg compounds are mainly used for the amelioration of acid soils. Most liming materials are obtained from limestone deposits that were formed in seas of earlier geological periods. The resulting limestone may be from inorganic precipitates or from carbonate shells. It can range from physically very soft material to very hard rock. Limestone reserves are immense in the form of calcitic and dolomitic mountains. However, there may be regional deficiencies of liming materials as many tropical regions that need them are distant from such deposits.

Liming materials

Common liming materials are:

- Calcium carbonate. It generally contains 75–95 percent $CaCO_3$, corresponding to 42–53 percent CaO (the reference basis for lime effect). A magnesium carbonate ($MgCO_3$) concentration of more than 5 percent is useful. The particle size of hard limestone must be less than 1 mm and that of soft material (chalk) less than 4 mm.
- Calcium magnesium carbonate (dolomite). Its different types contain 15–40 percent $MgCO_3$ and 60–80 percent $CaCO_3$. These products are suitable for acid soils that are also Mg deficient.
- Quicklime (CaO) and slaked lime $Ca(OH)_2$. These are quick-acting amendments for the neutralization of soil acidity, but they are generally

more expensive than natural limes. They have a special role in certain applications, e.g. creating a well-structured soil surface layer for sowing sugar-beet seeds.

The most common liming material is ground natural limestone ($CaCO_3$) with a definite fineness, depending on the hardness of the rock. The harder the rock is, the finer is the grinding needed to obtain equal efficiency. Carbonate limes act slowly because they are only slightly soluble in water and must be dissolved into neutralizing forms. Their solubility in dilute hydrochloric acid (HCl) has recently been accepted as a measure of their reactivity for evaluation purposes. Some have substantial amounts of Mg (an advantage for Mg-deficient soils), whereas others contain small amounts of Mn. Lime amendments not only decrease soil acidity but also have other positive effects (Figure 27).

FIGURE 27
Effects of liming on soil properties

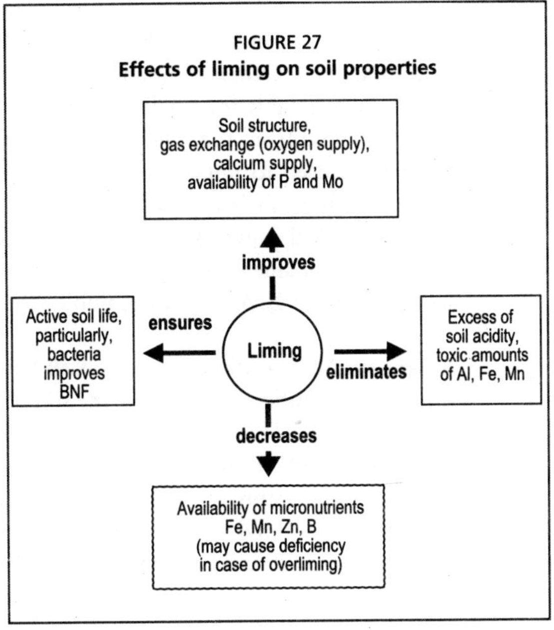

A special kind of lime amendment is marl ("lime earth"). This was used in ancient Greek and Roman agriculture. Marl is a mixture of soil material with 10–30 percent calcium carbonate and it is found in the top few metres of soils of glacial origin. It was rediscovered in Europe in the eighteenth century and used extensively for amelioration of the then acid soils. The mining and distribution of marl requires high labour costs. Lime formed from red marine algae is particularly soft and also contains some B.

Selection of liming materials

In principle, all liming materials can be applied on all soils, but the choice of a material depends mainly on soil texture, local availability and cost. Medium to heavy soils (texture of loam and clay) can be neutralized rapidly with quicklime. However, to maintain the optimal reaction, slow-acting carbonates are more suitable. In coarse-textured soils (sand and loamy sand), carbonate lime is preferable because of the lower risk of overliming where an excessive amount is applied or where the distribution is not uniform. Another aspect of the choice is the presence of by-products. Some limes also contain nutrients other than Ca, some clay minerals, organic matter or micronutrients, which makes them more valuable for sandy soils. The most important of such products is Mg. Application details of liming materials are discussed Chapter 7.

Many industrial by-products have a neutralizing effect on soil acidity and can be used as amendments. Some are easily mobilizable, such as silicates mixed with quicklime. Others contain a certain amount of phosphate and Mg, which makes them suitable for amelioration of acid soils that are also deficient in P and Mg. Press mud from sugar factories using the carbonation process is rich in lime and can be used to improve acid soils. Several PRs also have acid-neutralizing properties. Fly ash is a powdery residue remaining after coal has been burned (as in a thermal power station). It has received considerable attention as a soil amendment for ameliorating acid soils. However, caution is needed to avoid undue accumulation of B, Mo, Se and soluble salts in fly-ash-treated soils.

Amendments for alkaline and alkali soils

Intentional acidification to lower the soil pH may be required on alkaline soils for various reasons. These include removal of negative factors such as micronutrients deficiencies, and removal of excess Na. Soils that have been overlimed may require acidification to improve the availability of Fe, Mn and Zn. Other situations may require an acidic environment for certain crops such as tea. As already mentioned, a certain degree of acidification can be obtained by using N fertilizers that produce an acidic effect where these are cost-effective. However, on soils with a high buffering capacity, this effect may be small.

Amendments for effective acidification are either acids or those that produce acids after decomposition in soil. The most effective substance is diluted sulphuric acid, but its use is technically difficult, costly and inconvenient. In alkali (sodic) soils, the objective is to remove excess exchangeable sodium ions (Na^+) from the rootzone and the undesirable soil dispersion in order to create a favourable environment for plant growth. Common amendments are:

- ferrous sulphate ($FeSO_4$), which yields acid after hydrolysis with water;
- elemental S, which yields acid after oxidation by bacteria to sulphate;
- iron pyrite (FeS_2), which yields sulphuric acid after decomposition (also used for alkali soils);
- calcium sulphate or gypsum ($CaSO_4.2H_2O$), for alkali soils.

The amount of acidifying amendments required depends on the lime content of the soil and other properties. One tonne of S decomposes about 3 tonnes of calcium carbonate. A special test for acidity requirement is recommended in order to avoid unwanted damage. The amount of amendments required for reclaiming alkali soils depend on soil pH and soil texture, with higher amounts needed in soils with very high pH (10 and above) and a high clay content. It is now known that reclamation of only the top 10–15 cm of alkali soils is sufficient. This results in considerable savings in terms of the cost of amendments, water and labour.

Amendments for improving soil texture and structure

In addition to adequate nutrient supplies, a precondition for optimal plant growth is an optimal water supply, adequate aeration of the soil and root penetrability, both in the topsoil and subsoil. Soil physical properties can be improved by creating better soil structure as a precondition for optimal water supply and aeration, and a more favourable soil texture for water retention, root growth and proliferation.

Amendments for soil texture improvement

Light sandy soils lack adequate fine clay particles, whereas heavy clay soils lack enough coarse particles. The consequences of extremely coarse or fine particle sizes are a low potential for natural structure formation. The obvious measures for altering the particle size composition of soils are to supply clay particles to light soils, and sand particles to heavy soils. The key issue is the quantity to be applied and its practical feasibility. The addition of 1 percent of a mineral component is equivalent to adding 30 tonnes/ha of material to the 0–20-cm layer of a topsoil weighing 3 000 tonnes/ha. Thus, increasing the clay content of a sandy soil from 4 to 10 percent in order to convert it into a loamy sand requires 6 × 30 or 180 tonnes/ha of clay material. This would involve substantial transportation costs even if such material were available free of charge in the vicinity of the field. Where suitable sand or clay material is present in the subsoil, it may potentially be brought to the surface through deep cultivation in order to reduce this problem. The disadvantages associated with extremes of soil texture can to some extent be overcome by the use of all available organic material and crop residues.

Amendments for soil structure improvement

An important measure for improving the structure and opening up the subsoil is correct tillage. However, this results in only temporary improvement, and it should be supplemented by creating favourable conditions for the structure-forming processes in the soil. Several amendments have been developed specifically to improve soil structure. These are usually called soil conditioners and are applied to increase the WHC and resistance to erosion of soils. In fine-textured heavy soils, these are used for creating a crumb structure, chiefly for better aeration.

Many commonly used materials, such as lime and organic manures, improve soil structure indirectly. The following substances contribute to the bonding of the soil particles (which creates good crumb structure):

- inorganic or mineral matter: oxides, lime, silicate coatings, and gypsum;
- organic materials: slimy "glues" (polysaccharides, especially polyuronides) produced by microbes, the hyphae of fungi and humic substances derived from the formation of clay humus complexes (the conditions for which are especially favourable in the intestines of small soil animals, particularly earthworms).

In some soils, it may be necessary to improve or supplement natural crumb formation. This can best be achieved by increasing the saturation of the exchange complexes with Ca through the application of liming materials or gypsum (where liming is not possible). The addition of gypsum may be more beneficial for heavy-textured soils but the quantities required are considerable (2–10 tonnes/ha).

Organic soil conditioners imitate the natural bonding among particles and their effect may be sustained for several years. Various polymer dispersions and powders of polymers with long-chain and filamentary molecules are used. One of the first of these soil conditioners was Krilium (which is based on polyacrylic acid) in the United States of America. Other products developed were derived from polyvinyl acid, e.g. VAMA (polyvinyl acetate and maleic acid anhydride) or polyvinyl propionate. These substances are sprayed on the mechanically loosened soil or spread as powders and "rained in" with water. The quantities applied vary between 0.1 and 2 tonnes/ha and the effect is sustained for several years. However, the considerable cost per unit area restricts their application to horticultural and other high-value crops.

Substances that loosen the soil can improve fine-textured heavy soils. One such product is Styromull, which consists of flocs of polystyrene foam. The foamed material is chemically inert and does not react with the soil. It resists rot and does not become internally moist as it consists of cells filled with air. The addition of these 4–12-mm flocs increases permeability to water and aeration considerably. The amount required is about 10 percent by volume or 1–2 m^3/100 m^2 area. This is an expensive procedure and the risk of polystyrine washing into waterways has to be considered. Improved soil aeration can also be achieved by adding coarse rock powder and crop residues. Special soil conditioners are used to loosen fine-textured heavy soils and for stabilizing coarse-textured soils. The mineral soil conditioners used are ammonium iron sulphate and sodium hydrosilicate colloids. These are sprayed onto the soil surface and worked into the topsoil at the rate of 1–1.5 tonnes/ha.

Sandy soils often dry out easily. However, this can be prevented by adding water-absorbing/storing substances. For example, Hygromull consists of flocs of foamed plastic urea formaldehyde resin. This has fine open pores in the 4–12-mm flocs where water is stored up to 60–70 percent of the volume. Only about 5 percent is decomposed annually with a corresponding part of the N component (30 percent) being mineralized. The quantities applied are 2–4 m^3/100 m^2. Again, this is expensive.

Various plant nutrients and their sources can be utilized for optimizing nutrient supplies and managing them for higher efficiency. Chapter 6 deals with strategies for optimizing plant nutrition and Chapter 7 provides some guidelines for nutrient management, including application techniques.

Chapter 6
Optimizing plant nutrition

GENERAL ASPECTS

The goal of optimal plant nutrition is to ensure that crop plants have access to adequate amounts of all plant nutrients required for high yields. The nutrients have to be present in the soil or provided through suitable sources in adequate amounts and forms usable by plants. The soil water should be able to deliver these nutrients to the roots at sufficiently high rates that can support the rate of absorption, keeping in view the differential demand at various stages of plant growth. Optimal plant nutrition must ensure that there are no nutrient deficiencies or toxicities and that the maximum possible synergism takes place between the nutrients and other production inputs.

The ideal state of optimal plant nutrition may not be easy to achieve in open fields. However, it is possible to come close to it by basing nutrient application on the soil fertility status (soil test), plant analysis, crop characteristics, production potentials and, finally, the practicality and economics of the approach. Proper selection of nutrient sources and their timing as well as method of application are equally important. In the end, farmers should be able to maximize their net returns from investment in all production inputs including nutrient sources. In many countries, farmers do not have the financial resources or access to credit for fully implementing the constraint-free package of recommended inputs. Thus, for optimal plant nutrition to be of value to most farmers, it should also aim to optimize the benefit at different levels of investment.

In spite of all theoretical and practical progress towards efficient crop production, it still depends on some uncontrollable and unforeseeable factors, and on interactions among nutrients and inputs. Decisions on fertilization are normally based on certain assumptions of future events, e.g. weather conditions, that may be assumed to be normal but may not turn out to be so. Because of this general uncertainty, many essential data can only be estimated approximately. Thus, some misjudgements can hardly be avoided – neither by farmers toiling at a low yield level nor by those striving for high yields, and not even in scientific experiments, observations and advice.

From the farmers' point of view, optimization of nutrient supply appears difficult considering the many aspects of nutrient supply, uptake, requirements and use efficiency. This is facilitated by improving soil fertility in total, which means, to a large extent, not only offering an optimal uninterrupted nutrient supply but also providing generally favourable preconditions for their effective use. Therefore, extension personnel and farmers are well advised to maintain the fertility of their soils in a good, functioning state and to improve it continuously.

Chapters 4 and 5 contain the background information for optimizing plant nutrition. Chapter 7 provides the principles and guidelines for nutrient management, followed by some examples of general crop recommendations in Chapter 8. Optimal plant nutrition must lead to balanced and efficient use of nutrients and, thus, also to minimal adverse effects on the environment. This is made possible by combining optimal nutrient supplies with best management practices. Towards achieving this goal at field level, farmers must have access to adequate resources, timely and quality advice, and remunerative market prices for their produce.

Balanced crop nutrition

Plants need a proper supply of all macronutrients and micronutrients in a balanced ratio throughout their growth. The basics of balanced fertilization are governed by Liebig's law of the minimum (discussed in Chapter 3). Formerly, it was rightly concluded that, on many soils, the application of N without simultaneous supplies of phosphate and K made little sense. Today, in view of multiple nutrient deficiencies and increasing costs of crop production, fertilization with N or NPK without ensuring adequate supplies of all other limiting nutrients (S, Zn, B, etc.) makes little sense and, in fact, becomes counterproductive by reducing the efficiency of the nutrients that are applied.

Therefore, in view of the widespread occurrence of other nutrient deficiencies, the scope and content of balanced fertilization itself has changed. It now includes the deliberate application of all such nutrients that the soil cannot supply in adequate amounts for optimal crop yield. There is no fixed recipe for balanced fertilization for a given soil or crop. Its content is crop and site specific, hence the growing emphasis on SSNM. The SSNM approach for rice production systems is in various stages of development in several countries, e.g. China, India, Indonesia, Philippines, Senegal, Thailand and Viet Nam. With particular reference to irrigated rice, the SSNM approach involves the following steps (Dobermann and Witt, 2004):

1. Field-specific estimation for the potential indigenous supplies of N (INS), P (IPS) and K (IKS) and diagnosis of other nutritional disorders in the first year.
2. Field-specific recommendations for NPK use and alleviation of other nutritional problems.
3. Optimization of the amount and timing of applied N. Decisions about timing and splitting of N applications are based on: (i) 3–5 split applications following season-specific agronomic rules tailored to specific locations; or (ii) regular monitoring of plant N status up to the flowering stage, using a chlorophyll meter or leaf colour charts.
4. Estimation of actual grain yield, stubble (straw) returned to the field, and amount of fertilizer used. Based on this, a P and K input–output balance

is estimated and used to predict the change in IPS and IKS resulting from the previous crop cycle. The predicted IPS and IKS values are then used to develop fertilizer recommendations in the subsequent crop cycle.

Depending on the situation, some examples of the components of balanced fertilization (nutrients whose application is needed) for different situations are:

- many intensively cropped irrigated areas: N, P, K, Zn and S, or N, P, S and Zn, or N, P and Zn, or N, P, K and Zn;
- coconut in light soils and in root-affected (wilt) areas: N, P, K and Mg;
- immature rubber plantation: N, P, K and Mg;
- mature rubber plantation: N, P and K;
- many areas under oilseeds: N, P, K and S, or N, P and S, or N, P, Zn and S, or N, P, S and B;
- fruit trees in alkaline, calcareous soils: N, P, K, Zn, Mn and Fe;
- cabbage, cauliflower and crucifers in many areas: N, P, K, S and B;
- legumes in acid soils: N, P, K, Ca and Mo;
- newly reclaimed alkali soils in early years: N and Zn;
- high-yielding tea plantation: N, P, K, Mg, S and Zn.

All other factors being optimal, any deficiency of one plant nutrient will severely limit the efficiency of other nutrients (Figure 28). Imbalanced nutrient supply results in mining of the soil nutrient reserves. It can also lead to losses of the nutrients supplied, such as N, by reducing their rate of utilization. Imbalanced availability of nutrients also encourages luxury consumption of nutrients supplied in excess. This decreases the productive efficiency of all applied nutrients. Imbalanced fertilization is inefficient, uneconomic and wasteful, and it should be avoided.

Balanced crop nutrition is not the same as balanced fertilization. The latter should make the former possible. For example, only soils equally poor in available N, P and K should be fertilized with these three nutrients in balanced amounts. This can best be done using soil-test and crop removal data. Where a soil is rich in one nutrient, fertilization should be directed to the deficient nutrients in order to make balanced crop nutrition possible. Thus, the goal is not balanced fertilization as such but balanced crop nutrition through balanced nutrient application in order to supplement those nutrients that are deficient in the soil.

FIGURE 28
Yield response to balanced plant nutrition

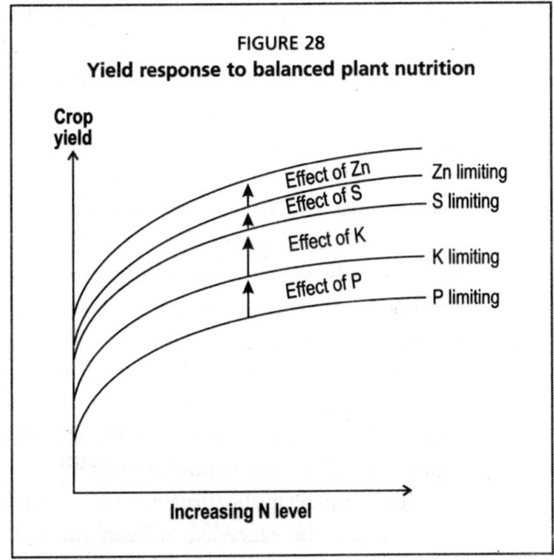

Source: FAO, 1998 (modified to include S and Zn).

TABLE 25
Nutrient-related constraints in relation to increasing yield, example of tea in south India

Productivity (kg/ha of made tea)	Limiting factors
Below 800	None
800–1 000	N and K
1 000–2 000	N, P, K, Zn and lime
2 000–3 000	N, P, K, Zn and liming with materials containing Mg
3 000–4 000	N, P, K, Zn, Mg, Si, B, liming, and transport processes within the soil
More than 4 000	N, P, K, Zn, Mg, Si, Mo, B, liming, and transport processes within the soil

Source: Tandon and Ranganathan, 1988.

Crop nutrition in relation to yield

The requirements for optimal nutrition depend very much on the type of crops grown and the yield level to be attained. The expected yield level largely determines the amount of external nutrient input necessary. It is not so much the yield per se that determines this, but the amount of nutrients removed from the field with the crop produce and the efficiency of applied nutrients. The replacement of nutrients removed at a given yield level is sometimes used to maintain soil fertility on soils that have been built up to a desired level. Here, two sets of fertilizer application norms are used, one for fertility buildup, and the other for fertility maintenance, specifically in case of P.

As the yield goals move up, the "nutrient basket" demanded by the crop also becomes more varied and complex. A soil may have sufficient fertility to support a crop of 2 tonnes/ha but may not be able to support a crop of 5 tonnes/ha on its own. At high yields, it does not remain simply a question of providing N or NPK. This had already been seen in many intensively cropped areas that, in the early 1960s, needed only N. Over a period of time, it became necessary to apply N + P, then N + P + K or N + P + Zn. Now many areas require the application of at least five nutrients (N, P, K, S and Zn) from external sources in order to sustain high yields. This is well illustrated by the example of nutrient needs for increasing levels of tea productivity in south India (Table 25). The principle is the same and holds good for all crops, only the nutrient package differs.

Prevention of excessive fertilization

Overfertilization or excessive fertilization is wasteful and is to be avoided. It goes against the concept of optimizing crop nutrition and also reflects poor application of scientific findings and unprofessional marketing practices. It can also have adverse impacts on the environment. Where high rates of water-soluble fertilizers are applied to crops, transient salt damage to the roots of young sensitive plants should be avoided. Moreover, the excessive or luxury supply of one nutrient can create antagonistic effects that disturb the nutrient balance. For example, high doses of K reduce Mg uptake even where there is a satisfactory Mg supply. Overfertilization not only reduces crop yield and produce quality but also produces suboptimal economic returns.

The optimal application rate of a nutrient can be seen as the cut-off point that is not to be exceeded in most cases. A farmer can continue to benefit from suboptimal rates of application although the benefit is always smaller than at the optimal level. In this respect, fertilizers and other nutrient carriers differ from inputs such as pesticides, which must be applied at a certain critical dosage to be effective. Thus, nutrient application is more flexible, similar to water application, as it enables farmers to operate over a wide range of rates based on their resources and the availability of inputs.

While overfertilization with nutrients such as P can produce significant residual benefits for the following crop, excessive application blocks the farmer's capital unnecessarily. Overfertilization with N invariably leads to lower nitrogen-use efficiency, greater possibility of lodging, pest and disease attack, greater N losses and negative impacts on the environment. Overfertilization with micronutrients can lead to their toxicity, which in many cases is difficult to ameliorate.

From fertilization to integrated nutrient management (INM)

Owing to the widespread use of fertilizers containing N, P and K and their effectiveness in increasing crop yields the world over, the term fertilization has become synonymous with the use of commercial NPK fertilizers. This is a rather narrow outdated concept, which does no justice to the wide field of plant nutrition or to the implications concerning undesirable environmental effects. Although fertilizers have benefited from more systematic and well-defined production and marketing, there are other effective sources of plant nutrients. These include crop residues, organic manures, various recyclable wastes and biofertilizers. Farmers all over the world have been using organic manures for a very long time. Chapter 5 has described various sources of plant nutrients. Diverse nutrient sources can be used in an integrated manner to meet the external nutrient supplies of any cropping system. Towards this end, scientifically, there is no conflict between mineral and organic sources of plant nutrients.

Definition

Although the term fertilization still has a place to describe the actual nutrient supply to crops, it is now gradually being replaced by the wider concept of integrated plant nutrition system (IPNS) or INM. Fertilizers are and will continue to be a major component of INM for producing high yields of good quality on a sustained basis in many parts of the world.

The basic concept underlying IPNS/INM is the maintenance or adjustment of soil fertility/productivity and of optimal plant nutrient supply for sustaining the desired level of crop productivity (FAO, 1995). The objective is to accomplish this through optimization of the benefits from all possible sources of plant nutrients, including locally available ones, in an integrated manner while ensuring environmental quality. This provides a system of crop nutrition in which plant nutrient needs are met through a pre-planned integrated use of: mineral fertilizers; organic manures/fertilizers (e.g. green manures, recyclable wastes, crop residues, and FYM); and biofertilizers. The appropriate combination of different sources of nutrients varies according to the system of land use and the ecological, social and economic conditions at the local level.

The need for INM

The need to adopt a wider concept of nutrient use beyond but not excluding fertilizers results from several changing circumstances and developments. These are:

- ➢ The need for a more rational use of plant nutrients for optimizing crop nutrition by balanced, efficient, yield-targeted, site- and soil-specific nutrient supply.
- ➢ A shift mainly from the use of mineral fertilizers to combinations of mineral and organic fertilizers obtained on and off the farm.
- ➢ A shift from providing nutrition on the basis of individual crops to optimal use of nutrient sources on a cropping-system or crop-rotation basis.
- ➢ A shift from considering mainly direct effects of fertilization (first-year nutrient effects) to long-term direct plus residual effects. To a large extent, this is accomplished also where crop nutrition is on a cropping-system basis rather than on a single-crop basis.
- ➢ A shift from static nutrient balances to nutrient flows in nutrient cycles.
- ➢ A growing emphasis on monitoring and controlling the unwanted side-effects of fertilization and possible adverse consequences for soil health, crop diseases and pollution of water and air.
- ➢ A shift from soil fertility management to total soil productivity management. This includes the amelioration of problem soils (acid, alkali, hardpan, etc.) and taking into account the resistance of crops against stresses such as drought, frost, excess salt concentration, toxicity and pollution.
- ➢ A shift from exploitation of soil fertility to its improvement, or at least maintenance.
- ➢ A shift from the neglect of on-farm and off-farm wastes to their effective utilization through recycling.

These realizations have led to the widening of the concept of fertilization to one of INM, where all aspects of optimal management of plant nutrient sources are integrated into the crop production system. For developing INM practices, the cropping systems rather than an individual crop, and the farming systems rather than the individual field, are the focus of attention. In contrast to organic farming, INM involves a needs-based external input approach, taking into account a holistic view of soil fertility. One of the aims of INM is to obtain high yields and good product quality – in a sustainable agriculture with practically no damaging effects on the environment. INM offers great possibilities for saving resources, protecting the environment and promoting more economical cropping.

Components of INM

The concept of INM is that of a nutrient integrator and not one of nutrient excluder. The major components of INM are the well-known and time-tested sources of plant nutrients with or without organic matter (Chapter 5). These primarily include:

- ➢ mineral fertilizers containing both major nutrients and micronutrients;

- suitable minerals such as PR, pyrites and elemental S;
- crop residues;
- green manures and green leaf manures;
- various organic manures of plant, animal, human and industrial origin;
- recyclable wastes from various sources with or without processing provided these do not contain harmful substances or pathogens above permissible limits;
- animal slurries and biogas plant slurry;
- microbial inoculants (biofertilizers);
- commercial organic fertilizers.

The main features and adoption of INM

The main concern of a farmer is to obtain sustainable high yields under local production conditions. The farmer can profit from the adoption of modern cropping principles, of which sustainability and INM play an important role.

At the farm level, INM aims to optimize the productivity of the nutrient flows through the soil/crop/livestock system during a crop rotation (Figure 29). A balance sheet can be established for every nutrient. However, owing to the complexity involved, only the major nutrients N, P and K are generally considered. The efficiency of a production system depends on the importance of crop uptake versus the total supply of nutrients. High losses of nutrients limit the efficiency. Exploitation of plant nutrient stocks is permissible as long as it does not affect the supply of nutrients and the general status of soil fertility.

Moreover, INM improves the production capacity of a farm through the application of external plant nutrient sources and amendments, and the efficient processing and recycling of crop residues and on-farm organic wastes. It empowers farmers by increasing their technical expertise and decision-making capacity. It also promotes changes in land use, crop rotations, and interactions between forestry, livestock and cropping systems as part of agricultural intensification and diversification. INM involves risk

FIGURE 29
Inputs and outputs of a plant nutrient balance sheet with N as an example

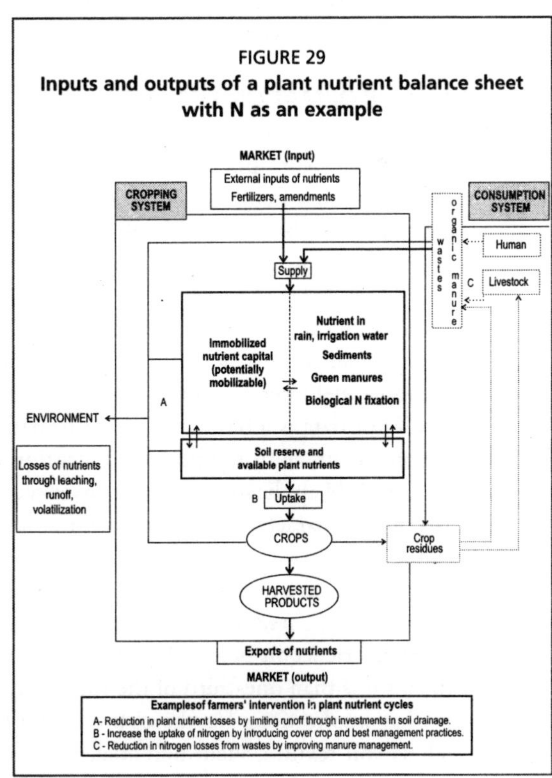

Source: FAO, 1998.

management (risk reduction) and enhances the synergy between crop, water and plant nutrition management.

During the adoption of INM, special attention should be given to sources of nutrients that may be mobilized by the farmers themselves (manures, crop residues, soil reserves, BNF, etc.). Minimization of losses and replenishment of nutrients from both internal and external sources are of major interest. While INM strives for the integrated application of diverse inputs, the use of organic sources cannot replace the use of mineral fertilizers. Although the effects of organic inputs go beyond the nutritional aspects, by contributing to improving soil physical properties and to a better efficiency of fertilizer use, the recycling of organic materials does not suffice to fully replenish the nutrients that are removed by crop harvests. Therefore, an increased and more efficient use of mineral fertilizers in most developing countries is required in the medium term (FAO, 1995).

In countries where a wide concept of crop nutrition beyond fertilization has been recognized, many INM guidelines have already been considered but not adopted on a large scale. In countries with intensive crop production where modern codes of good agricultural practice have been accepted, there is a trend towards better plant nutrient management or integrated crop management systems. This results in a more efficient nutrient use, leading partly to a reduced fertilizer input – even if it means a slightly lower yield level.

BASIC INFORMATION FOR OPTIMIZING CROP NUTRITION

Initial soil fertility status

Balanced nutrient application is a key controllable factor for optimizing crop nutrition on any field. The information on which nutrients to apply and at what rates should be based on a good soil test report. It is assumed that the soil test has already been validated by a high degree of correlation with crop response to the application of the concerned nutrient. The nutrient application rates based on soil tests can be for one optimal yield level or for pre-set yield targets. The optimal yield level is normally the profit-maximizing yield and not the highest achievable yield per se. Thus, the information on soil fertility status as provided by soil test data is a basic piece of information for optimizing crop nutrition for most nutrients, with the possible exception of N. In the absence of reliable soil tests for N, N application in many advanced agricultural areas is optimized on the basis of soil characteristics, growth conditions and crop removal of N at expected yield levels.

Soil testing as a tool for estimating the available nutrient status of soils continues to be a problem area in spite of more than 60 years of intensive research. Analysis of the experience in North America shows that even the best soil test calibration explains less than one-third of the variability in crop response to added nutrients. This has implications for the optimization of nutrient application rates. Factors such as soil texture, yield potential, specific weather conditions and differences between crop cultivars make it difficult to obtain a clear relationship between soil test and crop responses (Bruulsema, 2004). Ideally, the soil test value should be able to capture residual effects of previous nutrient applications. Chapter 4 has discussed evaluation of soil fertility for determining optimal application.

Amelioration of problem soils

Of the many types of problem soils, acid and alkali soils are mentioned here as examples. Amelioration of problem soils is a precondition for optimizing plant nutrition. This is because such soils cannot make the best use of the nutrients applied in the absence of suitable amendments. In fact, soil amendments should precede nutrient application. Once the soils have been amended, the crops grown on them can make efficient use of the nutrients applied and high yields can be obtained on a sustained basis.

Amendment of alkali soils

Alkali soils can be amended with several materials (Chapter 5). Gypsum is the most commonly used amendment. The main purpose of these amendments is to remove excess exchangeable Na from the rootzone, which also results in an improvement in soil physical properties. Once the soil has been amended, near normal rates of N (120–150 kg N/ha) can be applied to rice or wheat. In the initial years after reclamation, optimal productivity can be obtained with the application of N and Zn. Many alkali soils have a high level of soluble P, so that P application is required only after several years (5–10) depending on the crop. Green manuring such soils is useful for optimizing plant nutrition and sustaining productivity (Tyagi, 2000). Without the amelioration of such soils, yields are low and nutrient application is wasteful.

Knowledge of the tolerance of crops to alkalinity can be usefully applied for selecting the most suitable crops for such conditions. Table 26 summarizes the relative tolerance of several crops to exchangeable sodium percentage (ESP). A sound strategy for optimizing plant nutrient use in such soils would be to treat the soil with a suitable amendment and select a salt-tolerant crop cultivar. Selection of a tolerant crop is also beneficial where the soil cannot be amended adequately.

Amendment of acid soils

Acid tropical soils represent a large block of potentially arable soils. Management strategies for them must accomplish the dual task of neutralizing excess acidity (making the soil profile hospitable to plant roots) and correction of nutrient deficiencies. The basis for optimizing plant nutrition in such soils is provided by neutralization

TABLE 26
Relative tolerance of crops to exchangeable sodium percentage in the soil

Range of ESP*	Crops
10–15	Safflower, black gram, peas, lentil, pigeon pea
15–20	Chickpea, soybean, maize
20–25	Groundnut, cowpea, onion, pearl millet, clover
25–30	Linseed, garlic, cluster bean, lemon grass, palmarosa, sugar cane, cotton
30–50	Wheat, rapeseed mustard, sunflower, oats, cotton, tomato
50–60	Barley, beets, *Sesbania*, para grass, Rhodes grass
60–70	Rice, Karnal grass

* Relative crop yields are only 50 percent of the maximum in the alkalinity range indicated.
Source: Tyagi, 2000; Gupta and Abrol, 1990.

TABLE 27

The impact of lime and fertilizer application to maize over 40 years in an acid soil at Ranchi, India

Input applied[1]	Grain Yield	Grain Response	Cost[2] of input	Value[2] of grain	Net returns	Net returns
	(kg/ha)		(Rs/ha)	(Rs/ha)	(Rs/ha)	(BCR)
1956–1969						
0	600	-	-	-	-	-
N	1 500	900	462	4 500	4 038	8.7
NP	2 100	1 500	1 176	7 500	6 324	5.4
NPK	2 400	1 800	1 503	9 000	7 497	5.0
NPK + lime	3 000	2 400	1 943	12 000	10 057	5.2
1970–1979						
0	500	-	-	-	-	-
N	300	-200	1 155	-	Loss	Loss
NP	1 500	1 000	2 339	5 000	2 661	1.1
NPK	2 000	1 500	2 733	7 500	4 767	1.7
NPK + lime	3 600	3 100	3 173	15 500	12 327	3.9
1980–1989						
0	500	-	-	-	-	-
N	30	-470	1 155	-	Loss	Loss
NP	100	-400	2 615	-	Loss	Loss
NPK	300	-200	3 135	-	Loss	Loss
NPK + lime	4 100	3 600	3 595	18 000	14 405	4.0
1990–94						
0	500	-	-	-	-	-
N	20	-480	1 155	-	Loss	Loss
NP	50	-450	2 615	-	Loss	Loss
NPK	100	-400	3 135	-	Loss	Loss
NPK + Lime	4 800	4 300	3 575	21 500	17 925	5.0

[1] Average application rate of N + P_2O_5 + K_2O in kg/ha were 44–44–44 (1956–1968), 104–73–53 (1969–1979) and 110–90–70 from 1980 onwards. Lime applied once in 4 years as per LR.

[2] Economics based on prices in Rs/kg of 10.5 for N, 16.22 for P_2O_5, 7.43 for K_2O, 5.00 for maize grain and Rs440/year for lime (US$1 = Rs44).

Source: Sarkar, 2000.

of soil acidity, improving base status of the subsoil, and planting crop species that can tolerate excess Al.

The amendment of acid soils creates favourable conditions for optimizing plant nutrient use by neutralizing excess acidity and improving the availability of several major nutrients and micronutrients (Figure 17). As a rule, soil amendment, in this case liming, must precede fertilizer application. Without correcting soil acidity, no amount of balanced nutrient application can result in high yields or superior NUE. Thus, plant nutrition is a component of and not a substitute for good management. In many cases, the investment made in costly fertilizers may give very small returns or even result in a loss after a short period of initial success.

Results of a long-term field experiment in the acid red-loam soil at Ranchi in eastern India evidence this clearly (Sarkar, 2000). In this field experiment, which started in the mid-1950s, plots were treated either with N, N + P, N + P + K or N + P + K + liming. The scenario over a period of four decades has been summarized in Table 27 and can be described as follows:

- Stage I (1956–1969): Application of all nutrients (N, P and K) with or without lime increased maize yields and was profitable with the highest profits coming from NPK + lime application.
- Stage II (1970–79): Application of N alone could not increase maize yield any more and investment in N was a total loss. This was partly because in these plots, P and K were being depleted (becoming deficient) and partly because the use of N (as ammonium sulphate) progressively made the soil more acid. Applying NPK raised maize yields and profits. However, the soil acidity was becoming a more dominant constraint than nutrient deficiencies. Not only did the response rates to fertilizer (even "balanced") decline, the difference between NPK and NPK + lime plots widened in terms of yields, response and economic returns.
- Stage III (1980–89): Increasing soil acidity was now deciding the fate of crop growth and no amount of "balanced fertilization" was of help. Application of any nutrient could not even produce as much grain as the unfertilized control plot (500 kg/ha). The limed + NPK treated plots increased maize yield by 3 600 kg/ha as compared with a decrease of 200 kg/ha with NPK. Net returns in the NPK + lime treated plots were nearly Rs18 000/ha (US$410/ha)while NPK application (without lime) resulted in a total loss of money spent on fertilizers.
- Stage IV (1990–94): The same story as in Stage III was repeated with even more unfavourable effects of fertilizer without lime (maize yield 100 kg/ha with optimal NPK application) in contrast to 4 800 kg/ha with the same amount of NPK but applied after liming.

The example in Table 27 is just one out of many examples available to illustrate the crucial role of soil amendments for optimizing crop nutrition.

Nutrient recovery by crops and nutrient removal

An assessment of nutrient additions, removals and balances in the agricultural production system yields useful practical information on whether the nutrient status of a soil (or area) is being maintained, built up or depleted. It also gives insights into the level of fertilizer-use efficiency and the extent to which externally added nutrients have been absorbed by the crop and utilized for yield production. It can also forewarn about nutrient deficiencies that may aggravate in the coming years and need attention.

Figure 30 provides a simplified depiction of nutrient additions and removals. Most of the arrows in this figure also include nutrient recycling to a varying extent. For example, on the input side, part of mineral fertilizers, particularly N, S and K, can leach down but be recycled to the extent the groundwaters are pumped for irrigation. Over a toposequence, the nutrient loss for one field can become the nutrient gain for another field (and farmer). Nutrients from organic manures can enter the plant after mineralization. Atmospheric deposits (N and S) originate from N in the air, gaseous losses and pollution. Similarly, inputs through sedimentation have often been brought in by erosion from higher levels (output) and, in many

FIGURE 30
A simplified depiction of nutrient additions and removals

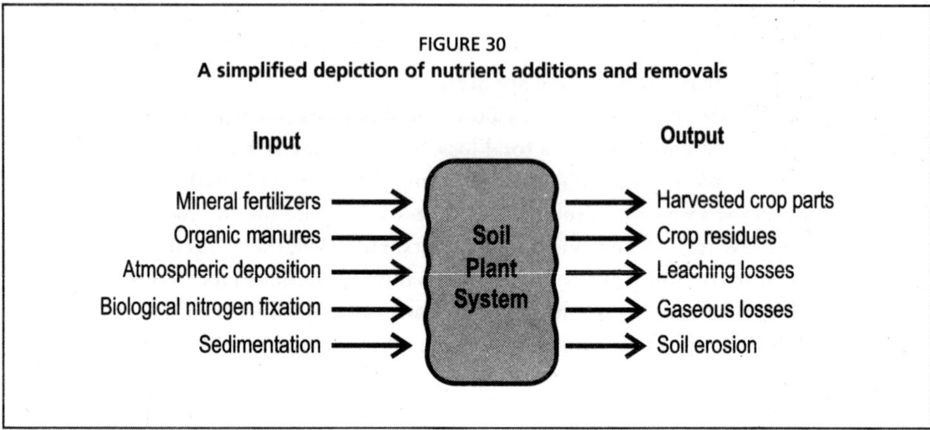

Source: Smaling, 1993.

cases, are actually intersite transfers (30 percent of the soil and nutrients moved by water erosion end up in the sea, the remaining 70 percent stay on the land).

On the output side of Figure 30, harvested crop parts and crop residues both yield valuable organic manures. Most estimates of nutrient removal by crops (from the soil) are overestimates because nutrient removal is often equated with nutrient uptake. This is not the case in many situations. The proportion of nutrients taken up that constitutes nutrient removal can vary from less than 10 percent (as in cardamom) to about one-third (as in coffee) to as much as 90 percent as in several field crops when only stubbles and roots are left behind.

Estimates of nutrient input and output allow the calculation of nutrient balance sheets both for individual fields and for geographical regions. It is a bookkeeping exercise, similar in many ways to keeping a bank account. The extent of nutrient removals from the soil system can provide useful information for optimizing crop nutrition.

Nutrient uptake and removal

At harvest time, plants contain considerable amounts of nutrients in plant parts such as grain, straw, stalks, beets, tubers and fruits, but only a small portion is contained in the roots. Depending on which plant parts are harvested and removed, the nutrients contained in them are removed from the field. In many developing countries where grain crops are harvested manually, the entire nutrients present in grain and straw or stover may be removed from the field. In the case of green manure crops, all plant nutrients in the biomass are returned to the soil and no nutrients are removed, except in situations where legume pods are removed for consumption. In fact, net soil enrichment takes place because of the contribution from BNF in case of leguminous green manures.

Knowledge of nutrient removal from the field is essential for calculating the amounts of nutrients taken away through harvested crops and for establishing a nutrient balance sheet. The nutrient removal data are more useful where computed on the basis of one basic unit of a harvest, e.g. 1 tonne of grain or 1 tonne of straw, so that the total removal at a certain yield level can be calculated easily. Average removal data are useful where nutrients have not been absorbed in excess. Where there is luxury consumption of nutrients, the corresponding removal data can be misleading. In intensive agriculture, N and K data tend to be biased upwards because of this factor. Therefore, larger than necessary amounts may be determined for the replacement of nutrient removals.

Nutrient uptake

Nutrient removal data quoted in the literature for the same crop can vary over a wide range. Table 28 provides some average nutrient removal data. These are based primarily on North American conditions. Nutrient removal data for Indian conditions, representing the tropical and subtropical areas, are presented in Tables 29 and 30 for general and comparative information. These data pertain to uptake per tonne of main produce and include the nutrients present in the by-produce as well. A substantial proportion of N in legumes (pulses, soybean, groundnut, forages, etc) originates from BNF, assuming a satisfactory level of nodulation and N fixation.

Nutrient uptake by crops can vary from less than 50 kg/ha to more than 1 000 kg/ha depending on the crop, variety, the nutrient, its availability, growth conditions and the biomass produced. Major nutrients constitute the bulk of the nutrients taken up. For example, the total amount of nutrients absorbed by wheat and rice (paddy) per tonne of grain production is about 82 kg and 74 kg, respectively. Out of this, N and K_2O alone account for about 75 percent. On an element basis, S uptake is generally similar to P uptake. The six micronutrients taken together add up to about 1 kg/ha (Tandon, 1999).

Higher production through higher cropping intensity also results in substantially higher nutrient uptake, which can range from 400 to 1 000 kg N + P_2O_5 + K_2O/ha/year. The share of N, P_2O_5 and K_2O in nutrient uptake is generally 35 percent N, 17 percent P_2O_5 and 48 percent K_2O, in the ratio 1.0:0.5: 1.4. Thus, every tonne of N removed is accompanied by the removal of 0.5 tonnes P_2O_5 and 1.4 tonnes K_2O on average.

In addition to major nutrients, a grain production level of 10 tonnes/ha through a rice–wheat rotation (6 tonnes paddy + 4 tonnes wheat) can absorb about 3–4 kg of Fe or Mn, 0.5 kg Zn, 200–300 g of Cu or B but only 20 g Mo. Thus, at the same production level, the uptake among nutrients by a crop can vary by more than 10 000 times (260 kg K vs 20 g Mo). Within the group of micronutrients itself, the uptake of Fe and Mn can be 200 times that of Mo. For successful crop production, the crop must be able to access and absorb the indicated nutrients whether these are 150–200 kg of N or K_2O or 15–20 g of Mo.

Nutrient uptake by a crop depends on a large number of factors, both controllable and otherwise. This is why large variations are encountered for a given nutrient or for a given crop even under similar conditions. Nutrient uptake

TABLE 28
Nutrient content of some major crop products and residues

	Nutrient content								
	N	P_2O_5	K_2O	Ca	Mg	S	Cu	Mn	Zn
	(kg/tonne)								
Grains									
Barley (grain)	18.2	7.8	5.2	0.5	1.0	1.6	0.016	0.016	0.031
Barley (straw)	6.7	2.2	13.4	3.6	0.9	1.8	0.004	0.143	0.022
Corn (grain)	16.1	6.3	4.8	0.2	1.0	1.2	0.007	0.011	0.018
Corn (stover)	9.9	3.7	14.4	2.6	2.0	1.4	0.005	0.149	0.030
Oats (grain)	19.5	7.8	5.9	0.8	1.2	2.0	0.012	0.047	0.020
Oats (straw)	5.6	3.4	17.9	1.8	1.8	2.0	0.007		0.065
Rye (grain)	20.9	6.0	6.0	1.2	1.8	4.2	0.012	0.131	0.018
Rye (straw)	4.5	2.4	7.4	2.4	0.6	0.9	0.003	0.042	0.021
Sorghum (grain)	14.9	7.4	4.5	1.2	1.5	1.5	0.003	0.012	0.012
Sorghum (stover)	9.7	3.0	14.2	4.3	2.7				
Wheat (grain)	20.8	10.4	6.3	0.4	2.5	1.3	0.013	0.038	0.058
Wheat (straw)	6.0	1.5	10.4	1.8	0.9	1.5	0.003	0.048	0.015
Hay									
Alfalfa	20.1	4.5	20.1	12.5	2.3	2.1	0.007	0.049	0.047
Bluegrass	13.4	4.5	13.4	3.6	1.6	1.1	0.004	0.067	0.018
Coastal Bermuda	22.3	5.1	19.3	2.7	1.8	1.8	0.001	0.036	0.027
Cowpea	26.8	5.6	17.9	12.3	3.4	2.9		0.145	
Fescue	17.2	8.3	23.6	5.0	1.7	2.6			
Orchard grass	22.3	7.4	27.9	5.0	1.9	2.6			
Red clover	17.9	4.5	17.9	12.3	3.0	1.3	0.007	0.097	0.064
Ryegrass	19.2	7.6	21.4	5.0	3.6	2.5			
Sorghum Sudan	17.8	6.8	26.1	3.5	2.6	2.2			
Soybean	20.1	4.5	11.2	8.9	4.0	2.2	0.009	0.103	0.034
Timothy	10.7	4.5	17.0	3.2	1.1	0.9	0.005	0.055	0.036
Fruits and vegetables									
Apples	1.3	0.4	1.9	0.3	0.2	0.4	0.001	0.001	0.001
Bean, dry	41.7	13.9	13.9	1.1	1.1	2.8	0.011	0.017	0.033
Bell peppers	6.8	2.6	10.8		2.1				
Cabbage	2.9	0.8	2.9	0.4	0.2	1.0	0.001	0.002	0.002
Onions	2.7	1.2	2.4	0.7	0.1	1.1	0.002	0.005	0.018
Peaches	1.2	0.7	2.3	0.1	0.3	0.1			0.000
Peas	58.6	12.5	37.5	10.0	6.4	3.6			
Potatoes (white, vine)	3.0	1.6	5.3	0.2	0.2	0.2	0.002	0.005	0.003
Potatoes (sweet, vine)	2.4	1.1	5.8	0.2	0.2	0.4	0.001	0.004	0.002
Snap beans	15.4	3.7	18.2		1.9				
Spinach	4.5	1.3	2.7	1.1	0.4	0.4	0.002	0.009	0.009
Sweet corn	13.9	4.7	13.5		2.0	1.1			
Tomatoes	2.7	0.9	3.6	0.2	0.2	0.3	0.002	0.003	0.004
Turnips	20.1	8.9	40.2	5.4	2.7				
Other crops									
Cotton (seed & lint)	24.2	9.6	11.9	1.5	2.7	1.9	0.069	0.127	0.369
Cotton (trash)	19.0	5.3	24.0	18.7	5.3	5.0	0.017	0.020	0.250
Peanuts (nuts)	35.0	5.5	8.8	1.5	1.3	2.5	0.010	0.075	0.063
Peanuts (vines)	20.0	3.4	30.0	17.6	4.0	2.2	0.024	0.030	
Soybeans	95.6	20.8	37.6	9.7	5.1	11.7	0.025	0.031	0.025
Soybeans (crop residue)	14.6	2.6	12.1	4.9	1.5	2.0			
Tobacco, flue-cured (leaves)	28.3	5.0	51.7	25.0	5.0	4.0	0.010	0.183	0.023
Tobacco, flue-cured (stalks)	11.4	3.1	28.3		2.5	1.9			
Tobacco, burley (leaves)	36.3	3.5	37.5		4.5	6.0			

Source: Adapted from Zublina, 1991 (updated 1997).

TABLE 29
Total uptake of major nutrients by crops

Group/crop (main produce)	Total uptake of main produce					
	N	P_2O_5	K_2O	S	Ca	Mg
	(kg/tonne)					
Cereals						
Rice (paddy)	20.0	11.0	30.0	3.0	7.0	3.0
Wheat (grain)	25.0	9.0	33.0	4.7	5.3	4.7
Maize (grain)	29.9	13.5	32.8			
Sorghum (grain)	16.4	7.7	25.5			
Pearl millet (grain)	31.8	17.4	61.3			
Finger millet (grain)	24.2	9.5	30.6			
Pulses						
Chickpea (grain)	60.7	9.2	39.2	8.7	18.7	7.3
Pigeon pea (grain)	70.8	15.3	16.0	7.5	19.2	12.5
Lentil (grain)	57.0	14.9	21.6	3.0	7.5	2.0
Green gram (grain)	106.0	48.1	73.2	12.0	71.0	43.0
Black gram (grain)	78.9	14.4	65.6	5.6		
Oilseeds						
Groundnut (seed)	58.1	19.6	30.1	7.9	20.5	13.3
Brown mustard (seed)	64.5	20.6	53.4	16.0	56.5	9.5
Rocket salad (seed)	70.0	26.0	61.1	20.7	19.3	9.3
Soybean (seed)	70.7	30.9	57.7	6.7	14.0	7.6
Safflower (seed)	38.8	8.4	22.0	12.6		
Sesame (seed)	51.7	22.9	64.0	11.7	37.5	15.8
Sunflower (seed)	63.3	19.1	126.0	11.7	68.3	26.7
Linseed (seed)	60.0	18.6	54.0	5.6	31.2	13.1
Castor (seed)	40.0	9.0	16.0			
Tubers						
Potato (tuber)	3.3	0.9	6.2	0.4	1.0	1.8
Cassava (tuber)	5.0	2.3	6.8	0.4	2.7	1.0
Sugar crops						
Sugar cane (cane)	2.1	1.2	3.4	0.3		
Fibres						
Cotton (seed cotton)	43.2	29.3	53.3			
Jute (dry fibre)	35.2	20.3	63.2		39.7	8.0
Fruits						
Mango (fruit)	6.7	1.7	6.7			
Banana (fruit)	5.6	1.3	20.3			
Citrus (fruit)	9.0	2.0	11.7			
Apple (fruit)	3.3	1.5	6.0			
Guava (fruit)	6.0	2.5	7.5			
Pineapple (fruit)	1.8	0.5	6.2			
Sapota (fruit)	1.6	0.6	2.1			
Papaya (fruit)	2.8	0.8	2.2			
Grapes (fruit)	3.9	0.6	6.2			
Zyziphus (fruit)	4.0	1.8	6.3			

Continued

Blank spaces indicate data not available.
Source: Published Indian data summarized in Tandon, 2004.

TABLE 29
Total uptake of major nutrients by crops (continued)

Group/crop (main produce)	Total uptake of main produce					
	N	P_2O_5	K_2O	S	Ca	Mg
	(kg/tonne)					
Vegetables						
Tomato (fruit)	2.8	1.3	3.8			
Cauliflower (curd)	4.0	2.0	4.0			
Cabbage (head)	3.5	1.3	4.2			
Beet root (root)	4.4	2.0	6.7			
Carrot (root)	3.9	1.7	6.6			
Onion (root)	2.7	1.3	3.9			
Plantations						
Coconut (1 000 nuts)	8.1	3.9	12.1	0.0	4.9	1.8
Oil-palm (fruit bunches)	3.7	1.0	4.4			
Cocoa (dry beans)	22.7	10.2	53.3			
Tea (marketable)	178.3	3.5	115.1	10.0	41.7	11.5
Coffee (green beans)	129.0	27.0	174.0	5.0		
Rubber (latex)	30.0	9.0	72.0			
Cashew (nuts)	88.0	25.0	42.0			
Cardamom (dry capsules)	260.0	40.0	520.0			
Forages						
Hybrid Napier (dm[1])	8.5	5.1	17.8	1.9	4.7	2.8
Grasses*						
Mean of 7 crops (dm[1])	9.4	3.4	17.0	2.0	4.6	2.7
Medicinal						
Japanese mint (dm[1])	12.9	7.5	18.5			
Aromatic plants						
Pyrethrum (dm[1])	15.0	12.0	84.0			

[1] dm = dry matter.
Blank spaces indicate data not available.
Source: Published Indian data summarized in Tandon, 2004.

can differ owing to the differences among crops, genetic character of a variety, environment where they grow, fertility level of the field, yield level, luxury consumption, nutrient imbalances and post-absorption events such as lodging and leaf fall. Thus, in order to produce 1 tonne of grain, the uptake by a given crop can vary 1.7-fold in the case of N, 2.3-fold in the case of P and 3.6-fold in the case of K among locations (Tandon, 2004).

Fate of nutrients absorbed by crops

The nutrients taken up by a crop are distributed in different parts of the plant during its life span. In the case of grain crops, 70–75 percent of N and P, 25–30 percent of K and 40–60 percent of S absorbed ends up in the grain, the rest stays in straw/stover. In rice, more than 70 percent of the N absorbed is transferred to the grain while a greater proportion of K, Ca, Mg, Fe, Mn and B remains in the straw. The absorbed S, Zn and Cu are distributed about equally in grain and straw (Yoshida, 1981). In groundnut, out of the nutrients absorbed, the kernels contain

TABLE 30
Average uptake of micronutrients by crops

Crop	Economic yield[1] (tonnes/ha)	Total uptake					
		Zn	Fe	Mn	Cu	B	Mo
		(g)					
Rice	1.0	40	153	675	18	15	2
Wheat	1.0	56	624	70	24	48	2
Maize	1.0	130	1 200	320	130	-	-
Sorghum	1.0	72	720	54	6	54	2
Pearl millet	1.0	40	170	20	8	-	-
Cassava	1.0	45	120	45	5	15	-
Potato	1.0	9	160	12	12	50	< 1
Chickpea	1.5	57	1 302	105	17	-	-
Pigeon pea	1.2	38	1 440	128	31	-	-
Soybean	2.5	192	866	208	74	-	-
Groundnut	1.9	208	4 340	176	68	-	-
Mustard	1.5	150	1 684	143	25	-	-
Sunflower	0.6	28	645	109	23	-	-
Sesamum	1.2	202	952	138	140	-	-
Linseed	1.6	73	1 062	283	48	-	-
Jute (olitorious)	1.0	214	784	251	27	-	-
Jute (capsularis)	1.0	139	368	119	18	-	-
Coffee (arabica)	1.0	35	83	62	82	-	-
Tea	1.0	276	2007	1 933	632	101	-
Guinea grass	269.0	558	2 940	1 880	443	-	-
Berseem	112.0	980	650	580	95	-	-

[1] Data for crops 1–7 are on per tonne yield basis; rest for indicated yield levels.
Source: Published Indian data summarized in Tandon, 2004.

41 percent of N, 52 percent of P, 28 percent of K, 11 percent of Mg and 1 percent of Ca. The leaves and stalks contain 45–50 percent of total NPK absorbed and also the bulk of Ca and Mg (Kanwar, 1983). In potato, harvested tubers account for 80, 83–88 and 70–78 percent of total N, P and K absorbed, respectively. In cassava, the proportion of absorbed nutrients present in tubers is 23 percent of N, 32 percent of P, 38 percent of K, 12 percent of S, 11 percent of Ca and 29 percent of Mg (Howeler, 1978). In jute, the proportion of absorbed nutrients that is returned to the soil before harvest through leaf fall is particularly high.

In tea, 50–65 percent of the N, P, K and Mg absorbed are removed from the field. The figure is about 35 percent for Ca, 25 percent for Mn and 25–50 percent for all the others. In coffee, the nutrient removal follows the order: K > N > P > Ca > Mg > S. The beans take away one-third of the nutrients that the plant absorbs and the remaining amount is retained in the plant biomass. Significant differences in nutrient uptake are observed between the arabica and robusta varieties of coffee. In coconut, the bulk of the nutrients absorbed ends up in nuts, leaves and stipules. Nuts alone account for 51 percent of N, 50 percent of P, 78 percent of K, 23 percent of Ca and 41 percent of Mg absorbed by the cultivar West Coast Tall (Pillai and Davis, 1963). In rubber, 25 percent of the N, 33 percent of the P_2O_5 and 8 percent of the K_2O absorbed is removed through latex. A considerable proportion of the absorbed nutrients is returned back to the soil through leaf litter. In cardamom, less than 10 percent of nutrients absorbed are carried in the capsules. In tree crops, considerable amounts of absorbed nutrients are retained in the trunk and branches. For practical purposes, these can be considered as nutrients removed from the soil.

These and similar data underscore the point that nutrient removals cannot be equated with nutrient uptake, as is very often done particularly for estimating nutrient removals and calculating balance sheets. Although the final economic produce contains only a part of what the crop absorbs, it is the total need of the crop that has to be met by the soil and through external additions for optimizing plant nutrition.

Where crop residues are left on the field, the nutrient content of residues (although a part of uptake) does not constitute removal. Where crop residues are removed, they may be lost forever or returned back in the form of animal dung/ FYM where they are used to feed farm animals. The very heavy losses through erosion highlight the need for large-scale measures in soil and water conservation in order to reduce the depletion of soil nutrients. However, in many cases, these could be intersite nutrient transfers.

Crop recovery of added nutrients and their implications

The amounts of nutrients added through fertilizers and other sources are only partly utilized by the crop (Figure 31). There are four possibilities for what may happen to the added nutrients:

- They enter the pool of available forms and are absorbed by the fertilized plants (recovered portion).
- They are not absorbed but remain available and are partly utilized by the next crop (residual).
- They are "fixed" and thus removed from nutrient cycling for longer periods.
- They are lost from the soil (through ammonia volatilization, leaching, and denitrification in the case of N).

The recovery or utilization rate of an applied nutrient is the portion of the added nutrient that is taken up by the plants. It is expressed as a percentage of the nutrient amount supplied. A recovery of 50 percent means that half of the fertilizer nutrients applied has been utilized by the fertilized crop. The recovery rate for applied

FIGURE 31
An illustration of the partial recovery of applied nutrients by crops

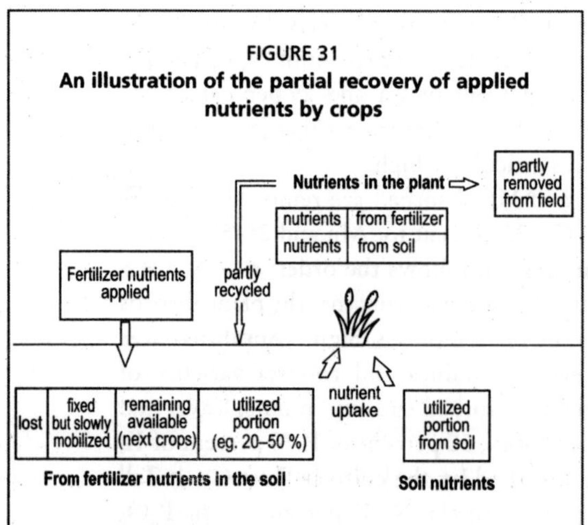

Source: Finck, 2006.

nutrient is often high for K (up to 70 percent), medium for N (35–70 percent), comparatively low for P and S (15–30 percent), and very low (less than 10 percent) for micronutrients.

The nutrient recovery rate is an important indicator of the fertilizer-use efficiency although it may at times include luxury consumption. Existing data on the subject are variable because recovery is affected by the soil, crop growth, root characteristics and production conditions. Nutrient recovery data are approximations with inherent variations. Moreover, the recovery rate of applied nutrients may be seen with reference to different time intervals, such as a specific growth period of a crop, single-crop basis, crop-rotation basis or for several years, as in case of P and some micronutrients. The recovery rate also depends on the extent to which the soil is supplied with nutrients, i.e. whether the soil is deficient or well supplied. Moreover, true recovery must be distinguished from apparent recovery.

The two methods for determining the recovery rate of applied nutrients are:

- Difference method (indirect measurement): The difference between nutrient uptake from fertilized (total uptake) and unfertilized plots is measured as in a fertilization experiment and related to the fertilizer quantities applied. The utilization or recovery rate is then given by the formula:

$$\text{Recovery rate (in \%)} = \frac{\text{total uptake - uptake from soil x 100}}{\text{nutrient amount added in fertilizer}}$$

Example for N (amounts in kg/ha):
N added through fertilizer = 120 kg N
Total uptake from fertilized soil = 100 kg N
Uptake from soil (without fertilization) = 40 kg
Recovery rate of applied N = 100 - 40 = 60 / 120 = 0.5 × 100 = 50 percent.

- Isotopic method (direct measurement): It also requires the conduct of an experiment, but the recovery is determined only on one plot by labelling the fertilizer nutrient with isotopes in order to distinguish fertilizer nutrients from soil nutrients. (For phosphate, the specific activity is the ratio of $^{32}P/^{31}P$ isotopes.)

The utilization rate is derived in three steps (e.g. for phosphate per hectare):

$$\text{1) percent fertilizer P in plants} = \frac{\text{specific P activity in plants x 100}}{\text{specific acivity in fertilizer}}$$

$$\text{2) kg fertilizer P in plants} = \frac{\text{kg P in plants x \% fertilizer P in plants}}{100}$$

$$\text{3) utilization rate (in \%)} = \frac{\text{kg fertilizer P in plants x 100}}{\text{kg fertilizer P added}}$$

The isotopic method is based on the assumption that fertilization does not affect the uptake of nutrients from the soil. However, this may not be completely correct. The fraction of nutrients absorbed from the soil may be reduced by fertilization in many cases and increased in other cases because of the so-called "priming effect",

TABLE 31
Average utilization rate of fertilizer nutrients by the first crop

Nutrient and source	Utilization rate (% recovery)
Nitrogen, mineral	50–70
Nitrogen, slurry	30–50
Nitrogen, manure	20–40
Phosphate, mineral	10–20
Potassium, mineral	50–60
Micronutrients, mineral	0.5–5

so that the method may indicate a higher or lower value than the actual value. To date, no method has been developed to establish "true" values for the recovery of applied nutrients by crops.

Both the difference method and the method using isotopes can be subject to errors. Errors may arise in the difference method because plants respond to nutrient deficiencies by changing root growth and their capacity to absorb nutrients. The recovery estimates using tracers (isotopes) may be affected by internal cycling of nutrients in the soil, such as the mineralization–immobilization turnover in the case of N (Bruulsema, Fixen and Snyder, 2004).

Table 31 presents some average ranges of the recovery of applied nutrients by crops (based mainly on cereals but including some other crops as well). It is possible to achieve even higher values in greenhouse trials, but recovery rates of up to 80 percent are rarely obtainable in soils. The utilization rate can often be increased by careful fertilizer placement, but only on deficient soils. In the first year with intensive cropping, the utilization rates for mineral N fertilizers can be 50–70 percent, e.g. for cereals grown under good conditions. However, recovery of applied N for paddy rice is estimated to range from less than 30 up to 70 percent in Asia (1995–97). (Figure 32).

An understanding of the relationships between crop yields, N use and N recovery can provide important clues to close the existing rice yield gaps. A categorization of selected Asian countries based on rice yield, N use and N recovery presents an interesting picture (Table 32). Most countries, with the exception of the Republic of Korea, Japan and China, fall within the medium- and low-yield groups, indicating considerable scope for raising yields. Although the level of N use in the Republic of Korea (178 kg/ha) is double that of Japan (88 kg/ha), the yield difference is small (0.2 tonnes/ha) as a result of the efficiency factor (recovery). Enhanced nitrogen-use efficiency (recovery of applied N) in the Republic of Korea may lead to optimized N use while maintaining yield levels similar to those in Japan. For countries such as China,

FIGURE 32
Nitrogen-use efficiency in selected Asian countries, 1995–97

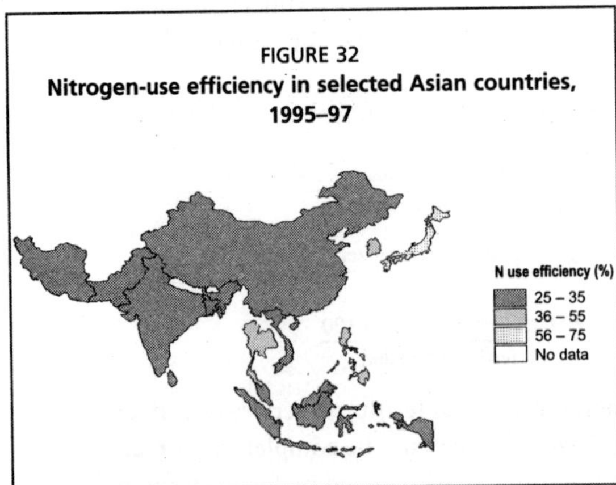

Source: FAO, 2003c.

Indonesia and the Islamic Republic of Iran, improved nitrogen-use efficiency accompanied by optimization of nitrogen-use levels would be a suitable approach for closing the yield gap. The possibilities for raising yields in Viet Nam, Sri Lanka, Malaysia, India and Pakistan remain high, provided that the prevailing low N recovery rates can be improved. The enhancement of N use to medium levels, coupled with efficiency improvement measures, is important for the Philippines, Bangladesh and Thailand (FAO. 2003c).

TABLE 32

Categorization of selected countries based on rice yield, nitrogen use and nitrogen-use efficiency, 1995–97

Country	Yield level[a]	Nitrogen use[b]	Nitrogen-use efficiency[c]
Republic of Korea	H	H	M
Japan	H	M	H
China	H	H	L
Indonesia	M	H	L
Iran	M	H	L
Viet Nam	M	M	L
Sri Lanka	L	M	L
Malaysia	L	M	L
Philippines	L	L	M
India	L	M	L
Pakistan	L	M	L
Bangladesh	L	L	L
Thailand	L	L	M

[a] Yield: H (high) = 5.5 tonnes/ha and higher; M (medium) = 3.6–5.5 tonnes/ha; L (low) = 3.5 tonnes/ha and lower.

[b] N use: H (high) = 120 kg/ha and higher; M (medium) = 81–20 kg/ha; L (low) = 80 kg/ha and lower.

[c] nitrogen-use efficiency: H (high) = 55% and higher; M (medium) = 36–55%; L (low) = 35% and lower.

Among organic sources, the recovery rate of N provided through leguminous green manure can be much higher than the N input from FYM or compost. Part of the unused residual N remains in the soil and can be used for the next crop, and part of it may be lost. For the organic manure slurry applied on the soil surface, the utilization rate of N is about 30–50 percent but this can be improved by injecting it into the soil.

The utilization rate for P fertilizers in the first year can be up to 25 percent, especially with row placement for wide-row crops, but only 10 percent or less with PR applied under unfavourable soil conditions or with broadcast application. The utilization rate for P increases over the longer term as residual effects are considered. Where the utilization rate of fertilizer P is 15 percent in the first year, the residual effect in the second year is about 1–2 percent, and about 1 percent in the following years. Cumulative values for longer periods are: about 25 percent for 10 years; and about 45 percent for 30 years. For very long periods, the recovery may approach 100 percent. Most farmers are not willing to wait that long to adjust their nutrient application rates although it does result in a long-term buildup of the nutrient capital of the soil. With K fertilizers, the first-year utilization rate is about 50–60 percent but long-term rates are higher. The recovery rate of soil-applied micronutrients is extremely low, and for nutrients such as Cu and Zn, a single application can last for several crops.

The assessment of the recovery rate over very long periods is only meaningful with respect to the apparent utilization (discussed below). Fertilizer utilization on well-supplied soils is generally lower than on deficient soils, at least in the first year. This is because the soil already contains sufficient nutrients for the plants, and fertilization serves primarily to replenish reserves.

Fertilizer amounts required according to nutrient removal and recovery

Optimal fertilization should be based on crop removal data in the case of nutrients such as N for which reliable soil test methods are not available. There should be a provision to deduct for luxury consumption from the nutrient removal data, and the effort should be to strive for high recovery of added nutrients. Luxury consumption is particularly relevant for N and K but less so for P, S, Mg, etc. The best way to optimize the application of nutrients that leave a substantial residual effect is to manage them on a crop-rotation basis. Fertilization on the basis of micronutrient removal is not advisable on deficient soils because of their very low utilization rate. Application of these nutrients should be based on available nutrient status of the soil and the period over which a single application can leave significant residual effects (so that micronutrient applications are not repeated each year). For nutrients for which soil application is not very effective (e.g. Fe and Mn), the amounts required can be calculated for foliar applications or in terms of chelates.

Nutrient accounting via input/output balances

Sustainable cropping should not exhaust the soil nutrient supply but improve it to the extent possible. The extent to which this advice is followed depends on the farmer's perception of sustainability and available resources for purchasing fertilizers. This is also an area where INM can play a role by enabling the farmer to recycle all available on-farm and off-farm organic wastes.

A quantitative knowledge of the depletion of plant nutrients from soils may be helpful in devising nutrient management strategies. Nutrient balance exercises serve as instruments to provide indicators for the sustainability of agricultural systems. Nutrient budget and nutrient balance methodologies using various approaches for different situations have been applied widely in recent years at a variety of levels: plot, farm, regional, national and continental (FAO, 2003b).

In agriculturally advanced countries, a farmer can check whether the input by fertilization corresponds to the nutrient removal in order to maintain soil fertility. At the farm level, the amounts of nutrients leaving the farmgate can be used as a criterion for adequate nutrient management. The input of both plant nutrient sources and plant nutrients in animal feed must correspond to the nutrient removal by the crop and in exported animal products. Figure 33 shows the input/output fluxes of plant nutrients (N, P and K) on a farm measured at the farmgate for balance calculation purposes. In this case, the nutrient losses and BNF are not shown. A farmer can carry out such calculations with the aid of standard tables containing nutrient concentrations of fertilizers and feedstuff. Such a calculation also provides information about unaccounted losses, which is required by some fertilizer laws in view of environmental pollution. The problem with this calculation is that unaccounted differences may not only be caused by losses but also by enrichment of soil fertility. Such exercises can be conducted by educated, well-informed farmers who maintain an accurate bookkeeping of various inputs and outputs. Even then, they can benefit from consulting their local farm adviser or extension specialist.

FIGURE 33
Plant nutrient (N, P and K) input/output fluxes on a farm for balance calculation

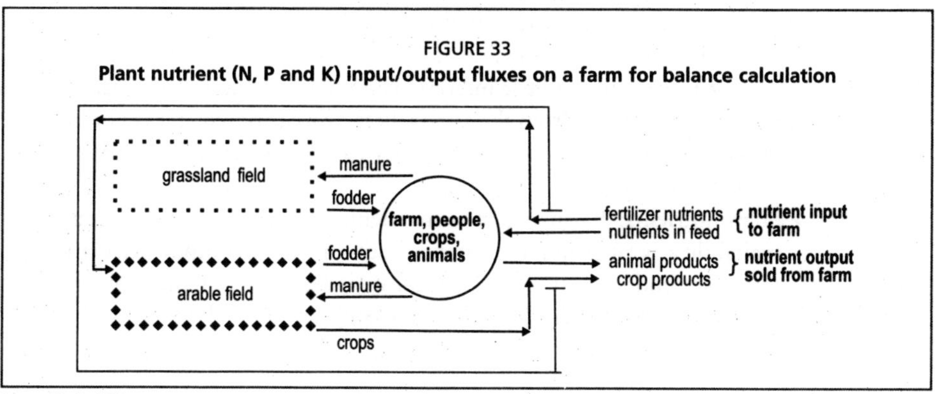

Source: Finck, 2006.

This type of exercise may not be possible for the vast majority of smallholders in most developing countries. In most such cases, farmers have access only to general fertilizer recommendations, supplemented by whatever quantities of organic manures are available in their village. To minimize the depletion of soil fertility, they can base nutrient application rates on soil test results wherever these are available. The farmers can also be encouraged to recycle as much crop residue as possible and, instead of using cattle dung as a source of domestic fuel, recycle it through biogas plants in order to obtain energy as well as manure.

At the regional or national level, an input/output balance of plant nutrients can reveal significant nutrient losses with the sale or export of agricultural products that are not compensated by external nutrient additions. This is a kind of interregional nutrient transfer in which the importing area is enriched with nutrients and the exporting area can be depleted of nutrients (mining of soil nutrients) by exhaustive cropping. While calculating nutrient balances, several nutrient-specific features may be observed. Some possible explanations for these are:

- Nitrogen: Where the N input is much greater than the N output, this indicates a low level of nitrogen-use efficiency, which could be either the result of large losses or of small losses combined with enrichment of soil N reserves. Where the output exceeds the input, there must be a substantial gain from BNF or from depletion of soil N reserves.
- Phosphorus: In intensive cropping, the optimal input of P is usually greater that the P output owing to low P-use efficiency as a result of the enrichment of mineral and organic soil P fractions. This should be considered as a positive long-term effect. This enrichment or buildup of P can contribute to the P nutrition of several crops in succession. This has implications also for the economics of P application (Chapter 9).
- Potassium: The K balance depends largely on the rate of N and K application, any luxury consumption of K, utilization of soil K reserves (particularly from the non-exchangeable fraction) and K losses. K losses are a possibility in coarse-textured soils under high rainfall.

- Calcium, magnesium and sulphur: The Mg balance is similar to that of K, except in neutral and alkaline soils where Mg may be abundant. The Ca balance is generally of little interest. The S balance tends to be negative if the addition of sulphate from the atmosphere or irrigation water is not included, or S-free fertilizers are used particularly for high S-demanding crops, such as oilseeds and fodders.
- Micronutrients: Balancing micronutrients makes little sense because their availability is of major importance (not any input/output calculation). In any case, under most situations, nutrient balances for micronutrients are positive owing to the low use efficiency of applied nutrients by crops (similar to P).

STRATEGIES FOR OPTIMIZING NUTRIENT MANAGEMENT

Nutrient management can be considered from different aspects, such as with the emphasis on soil nutrient status, on crop productivity, on nutrient balances or in terms of the nutrient–water relations.

The ultimate aim of all aspects is to: optimize crop production, maximize positive interactions, maximize net returns, minimize the depletion of soil nutrients, and minimize nutrient losses or negative impact on the environment. Achieving this aim is difficult but not impossible. It requires the application of best available knowledge and inputs as part of a medium- to long-term strategy. For most situations, the required knowledge and inputs are already available. The key is the intelligent management of the various resources.

From soil nutrient exploitation to enrichment

Different strategies of soil nutrient management in cropping systems have evolved over time. These are related to different systems of fertilization. Different strategies may find application simultaneously in the same region, and sometimes on the same farm, and thus be largely responsible for differences in fertilizer input per unit area. The four different strategies concerning soil nutrients are:

- exploitation: exhaustion of soil reserves, no fertilization, decreasing yields;
- utilization: moderate withdrawals from soil reserves, no fertilization, stable yields;
- replacement: maintenance of soil supplies, fertilization to offset removals, stable yields;
- enrichment: enhancement of soil supplies, supplementary fertilization, increasing (high) yields.

Exploitation of soil nutrients

Cropping based on the exploitation (unwise utilization) of nutrients stored in the soil is the oldest strategy of agricultural production. Exploitation cropping uses the natural nutrient capital of the soil. It still plays an important role in crop production in many regions. A common feature of all exploitation systems is that hardly any fertilization or nutrient replenishment is undertaken apart from recycling harvested residues and waste products. This results in nutrient depletion through mining the soil reserves. As a result, the yields decrease from year to year. The available nutrients are consumed until they are exhausted, either because the mobilization rate of organic and mineral reserves is very low or there are only small soil nutrient reserves left to be mobilized. The original fertility of the soil, which had improved over long periods, is thus depleted.

Typical examples of rapidly decreasing soil fertility are found with shifting cultivation in humid forest areas. On newly developed lands with high soil fertility, soil nutrient exploitation may permit highly profitable cropping for several years without fertilizer input. Even outside shifting cultivation, a large number of farmers in many developing countries continue to raise crops drawing primarily on soil nutrient reserves.

Despite all the objections to exploitation cropping as such, controlled exploitation cropping may be useful economically and may even be ecologically acceptable as a stable form of land use, provided that the arable cropping period is limited and that a fallow period is included for regeneration of soil fertility. This may not always be possible in intensively farmed, overpopulated countries, particularly where irrigation or adequate rainfall is available to raise an additional crop. It is a feature of subsistence agriculture in which very little marketable surplus is generated.

Long-term exploitation cropping can cause considerable damage to soil fertility as serious soil degradation may occur. Such serious damage is not completely irreparable, but the cost of regeneration exceeds the short-term gain achieved. Exploitation cropping accompanied by irreparable damage represents destruction of a naturally available potential that humanity, with its continuously shrinking living space, cannot afford. Such an approach is not sustainable for improving crop yields.

Utilization of soil nutrients

This is a less severe version of the exploitation (mining) of soil nutrient reserves discussed above. Similar to exploitation, utilization of soil nutrients involves a certain reduction in the nutrient capital of the soil without a significant decline in the fertility taking place. This may create the impression of a sustainable system of agricultural production without external nutrient input. Such nutrient supply systems can only be practised where the nutrient removals are small and the pool of available nutrients is large and also backed by sufficient a rate of nutrient mobilization from the soil reserves.

In this system, the soil is not impoverished significantly and yields remain constant in spite of annual nutrient removal. However, the fact that yields remain low in such a system makes it unsuitable whenever the farmer wants to improve his yield levels. Then, this strategy will come closer to the exploitation strategy and will have to be replaced by a more balanced output/input regime. No soil, even the most fertile one, can continue to support nutrient removals indefinitely. Again, this system is not sustainable for producing high yields.

Replacement of soil nutrients

The concept of replacing nutrients that are removed or lost from the field permits stable cropping and was practised in ancient civilizations. Examples of this are the natural replacement of nutrients by Nile mud in Egypt, regular use of animal dung as manure in ancient India, and careful compost management in ancient China. Today, especially on most soils with only average fertility, the replacement of all losses is essential for sustaining optimal levels of crop productivity with minimum depletion of the soil reserves.

Maintenance of soil fertility can be partly achieved by using soil-improving crop management practices. These include using nutrient-accumulating plants such as legumes for the accumulation of N or by following crop rotations with different nutrient demands and different rooting depths. Both organic and mineral nutrient sources are suitable for the replacement of soil nutrients. Farm waste products and mineral sources such as silt and marl can also be used as supplements to fertilizers for obtaining moderate to high yields.

The strategy of nutrient replacement is valid only in cases of good initial soil fertility or soils in which the fertility has been built up to an adequate level through repeated fertilization. It is not applicable on naturally poor or depleted soils because fertilization on the basis of removals only can further deplete such soils. The root cause of soil fertility depletion here is that only a part of the nutrients absorbed by the crop are provided by external input and the remaining crop needs are met from soil reserves.

Cropping systems based on the replacement strategy are only rarely used to the full extent. They are very common in a modified form in which the replacement of some nutrients (especially N, P and K) occurs but others are utilized from the soil reserves. This is most common where balanced nutrient application is restricted to the narrow meaning of NPK application. This strategy can allow yields to be kept at medium or even at high levels as long as nutrients other than N, P and K are not limiting.

Enrichment of soil nutrients

Natural soil fertility is often insufficient for sustaining high yields and may further decline after a few years of intensive cropping. Because of this, the level of some nutrients must be increased beyond the amounts needed to replace the removals in order to achieve high yields. Enrichment of soils with nutrients should primarily extend to those nutrients that can be built up and not necessarily to all nutrients. This strategy comprises three approaches: (i) increasing the supply of deficient nutrients beyond the amounts removed; (ii) replacement of removals in the case of nutrients present in sufficient amounts; and (iii) utilization of nutrients from soils endowed with good reserves and nutrient replenishment capacity.

Improvement in soil fertility by nutrient enrichment manifests itself historically by the fact that, in parts of Europe, sugar beet and wheat now produce high yields on soils formerly considered as far too poor for these nutrient-demanding crops. Better nutrient supply over the years and the resulting improvement in soil fertility in general has raised the yield potential of these crops substantially to an upper limit imposed only by climate or other limiting factors that are difficult to correct. Enrichment of the relevant nutrients can be very profitable because of the much higher yield level achieved, provided economic resources are not a constraint.

TABLE 33

Buildup and maintenance approach for making fertilizer recommendations for maize[1]

Bray and Kurtz P_1 – test	P_2O_5 recommended		
	For buildup to optimum	For replacing crop removal	Total P_2O_5
(mg P/kg soil)	(kg/ha)		
4	92	64	156
8	83	64	147
16	65	64	129
24	47	64	111
32	29	64	93
40	11	64	75
45	0	64	64

[1]At a grain yield level of 9 400 kg/ha on a soil with medium P-supplying power in Illinois, the United States of America

Source: University of Illinois, 1994.

The strategy of fertilizing for soil fertility buildup is practised, for example, by farmers in the advanced maize production state of Illinois in the United States of America. Based on the available P status of the soil, phosphate application is recommended with the twin objectives of building up soil available P to an optimal level and replacing P removals by the crop at expected yield levels (Table 33). Once the available soil P status has reached the optimal level, only replacement of P removal is recommended (University of Illinois, 1994). This is a case study of an approach for sustaining high yields.

The concept of enrichment of the limiting nutrients does not mean a perpetual increase in the soil P status, but only an increase up to an optimal supply level that is sufficient for high yields, and certainly not up to luxury supply, which would be both unnecessary and detrimental in view of nutrient losses and imbalances. The enrichment phase is usually a transient one that is followed by a permanent replacement phase, generally at a high yield level. A large number of farmers in many developing countries may not be able to adopt this approach primarily owing to inadequate financial resources, high cost of purchased inputs, and a lack of perception concerning the need for enriching soil nutrient reserves. Many such farmers operate on a season-to-season or at most on a crop-rotation basis. Their weak financial base forces them to look for short-term gains.

INTEGRATED NUTRIENT–WATER MANAGEMENT FOR OPTIMIZING PLANT NUTRITION

Plant needs for water and nutrients are interdependent. Water is not only required for the growth of plants but is also the medium through which nutrients are transported to the roots and absorbed by them. A good water supply improves the nutritional status of crops, and an adequate nutrient supply saves water. With properly coordinated management of nutrients and water, the farmer can increase crop productivity substantially through their efficient use. This holds true both for irrigated and rainfed situations. Application of optimal nutrients without access to adequate water results in poor utilization of the applied nutrients. Similarly, application of low doses of nutrients under conditions of adequate

FIGURE 34
The influence of soil water status on plant nutrition

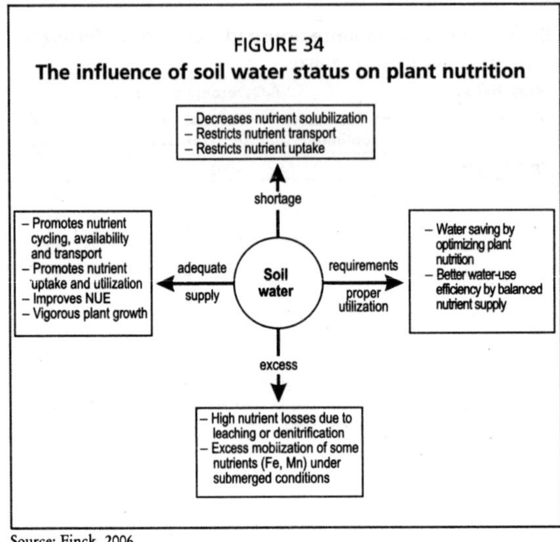

Source: Finck, 2006.

water supply results in a waste of the valuable water resource. Water management is inseparable from good nutrient management practices and vice versa.

Influence of soil water on crop nutrition

Soil moisture conditions have major effects on productive processes such as the accessibility, availability, uptake and use of soil nutrients for crop growth and also on negative processes such as creating anaerobic conditions, and losses of nutrients from the soil (Figure 34).

Water supply

The content of available soil water has a marked influence on several aspects of nutrient supply. Every soil has a certain WHC. This is the upper limit of available water and depends on profile depth, soil texture and soil organic matter content. Irrigation/rainfall above WHC is a waste as excess water is lost by runoff or drainage. Available water lies between field capacity and the wilting point. As adequate (but not excess) soil moisture results in profuse and deeper root growth, both water and nutrients become accessible to plants from deeper soil layers where moisture is adequate.

Where dry conditions restrict water uptake, e.g. during drought, the rate of root extension is reduced in soils of low fertility and the plant is unable to access deeper moist horizons in the soil. In most soils, the nutrient content is highest in the topsoil and this horizon dries out first. Although the plant is able to absorb some water from the subsoil, this may not be sufficient to obtain adequate nutrients for active growth. Phosphate plays a key role in the growth and proliferation of the root system. Where the soil is well supplied with phosphates before planting, the plant can develop a vigorous and deep root system before the onset of mid-season drought. Even when the surface soil becomes dry, such roots are capable of absorbing water and nutrients from deeper layers. In such cases, phosphate application can be considered as an insurance against drought. It not only increases crop growth but also enables a more efficient use of stored soil water that would otherwise have been out of reach of poorly developed roots.

Water and nutrient availability

Soil moisture affects the solubility and, hence, availability of all nutrients. Biological activity in the soil is particularly restricted under conditions that are too wet (owing to lack of oxygen) or too dry. Under very dry conditions, the breakdown of organic matter, and with it the mineralization of organic forms of N and other nutrients into plant available mineral forms, slows down. This may lead to a temporary shortage of N in the soil. Thus, in very dry periods, little accumulation of mineral N occurs. When the rains come, there can be a considerable flush of mineralization, providing available N and other nutrients for plant growth, provided the subsequent heavy rains (as received during the monsoons) do not leach the mineralized N beyond the rootzone.

The use of irrigation can minimize fluctuations in soil biological activity during crop growth. One of the significant effects of irrigation or moderate rainfall is to increase soil nutrient supply from organic sources. However, such increases are seldom sufficient to meet the additional demand for nutrients resulting from greater plant growth. Mineralization of other nutrients such as P and S also increases with adequate soil moisture.

The availability of mineral potassium (K^+) and other cations is also improved by a satisfactory soil moisture status. In dry soil conditions, the cations in general are more tightly bound to soil colloids, not easily exchangeable and, therefore, are less available, or rather less accessible, to plants. In addition, as the volume of soil solution is smaller, the amount of sparingly soluble nutrients, such as P, is reduced and plants are unable to absorb them in required quantities.

In waterlogged soils, the concentrations of ammonium ions, P, Fe and Mn increase, but the content of nitrate-N decreases because of leaching and denitrification. The uptake of many nutrients by rice such as N, P, Mn and Fe increases under waterlogged conditions but the uptake of other cations may be reduced. In this respect, the upland rice system is closer to most other cereals and quite different from the flooded-rice system.

Water and nutrient mobility

As nutrients need to move only a short distance, adequate soil moisture favours the mass flow of nutrients, especially N, with the soil solution to the root surface. Movement by diffusion within the soil solution is important for several nutrients including P and K and it is aided by adequate soil moisture. Moreover, the uptake of nutrients by crops is also enhanced where the plants have an adequate water status. Efficient use of nutrients within the plant for growth and metabolism also depends on a satisfactory uninterrupted supply of water. Where sufficient water is not available, transport of absorbed nutrients within the plant is restricted. This also restricts their use for metabolic activities and plant biomass production, which can ultimately have an adverse effect on the yield and nutrient content of the economic produce.

Water and crop response to nutrients

The growth and yield response of a crop to fertilizer application is very much influenced by the level of water supplied. Crop response is a synthesis of the various factors affecting crop growth, nutrient availability and nutrient uptake. The

FIGURE 35
Effect of rainfall on crop yield

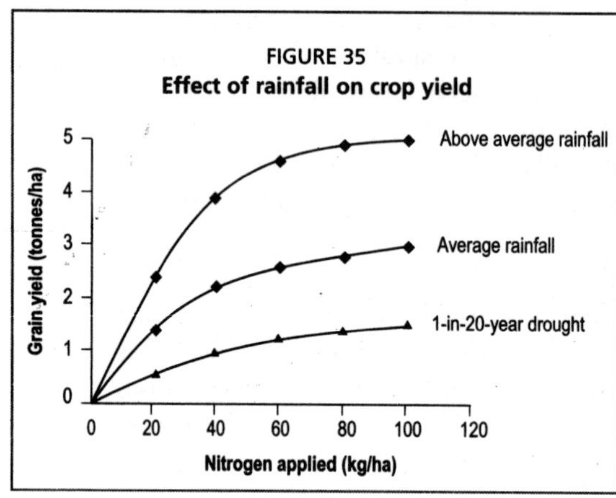

greater response to N, as well as a higher yield level, with increasing rainfall is shown in Figure 35. Such variations in rainfall greatly affect the optimal rate of nutrient application. For crops raised largely on stored soil moisture, an estimate of the moisture in the soil profile before planting is as valuable as an estimate of the available nutrient status of the soil. Consequently, more nutrient input is required to make use of a better water supply, and the economically optimal rate of nutrient application also rises. Where plants have access to adequate water but not to adequate nutrients, this amounts to an underutilization of the valuable water resource.

Water and nutrient-use efficiency

In agronomic terms, NUE means the increase in yield obtained per unit of applied nutrient. It is the same as rate of response and can be calculated as: NUE = (yield of fertilized plot – yield of control plot)/amount of nutrient applied.

Many aspects of crop management influence the actual yield level and the response to applied nutrients. In relation to water supply and management, NUE may be improved by minimizing the fertilizer losses from the soil that are caused by poor water management, for example leaching or denitrification. The NUE can also be improved by ensuring that lack of water does not at any stage retard crop growth or nutrient uptake appreciably. Excess water can be a cause of nutrient losses, and insufficient water at a critical stage can limit growth and yield. It is also important that all other production inputs and management factors be adequate.

The timing of water application influences NUE considerably through its effect on crop yield, which can be reduced substantially where water supply through irrigation or otherwise is deficient at the most critical stages of crop growth. In most crops, the active vegetative growth stage and the reproductive growth stage have been found to be most critically affected by moisture deficiency as summarized below:

- rice: head development and flowering > vegetative period (active tillering) > ripening;
- wheat: flowering > yield formation > vegetative period (crown root initiation);
- sorghum: flowering and yield formation > vegetative period;
- maize: flowering > grain filling > vegetative period;
- peas: flowering and yield formation > vegetative period;

- potato: stolonization and tuber initiation > yield formation > early vegetative growth;
- groundnut: flowering and yield formation, particularly pod setting;
- safflower: seed filling and flowering > vegetative period;
- cotton: flowering and boll formation;
- sugar cane: period of tillering and stem elongation > yield formation.

Water and nutrient losses

There are three main ways in which water status and water management can influence loss of nutrients from the soil–plant system.

Excessive rainfall, or excessive irrigation, resulting in the passage of water through the soil profile through deep percolation will carry with it soluble nutrients, particularly nitrate, sulphate and B. In temperate climates with moderate or high rainfall, the amount of rainfall during winter can cause appreciable loss by leaching of these nutrients. This is particularly the case where high amounts of such nutrients may be present in the soil at the beginning of winter (owing to breakdown of crop residues at the end of the growing season). The amount of loss depends on how much water moves through the soil profile and the stock of soluble nutrients. The extent of nutrient losses must be considered when determining nutrient application rates.

Leached nitrate can also enter water bodies or become denitrified under anaerobic conditions within the soil profile. Such conditions can exist within pockets or compact zones within an otherwise aerated soil. Waterlogging causes loss of N through denitrification of nitrate. In flooded-rice soils, nitrate levels can be kept low by placing ammonium or amide source of N, such as urea supergranules (USGs) in the reduced soil zone and by proper water management. However, in upland soils, nitrate levels are often quite high, such that periodic waterlogging by heavy rainfall as in a monsoon-type climate or excess irrigation can result in a large loss. As free-draining soils become waterlogged less readily, this risk is greatest on the high clay fine-textured soils.

Ammonia volatilization from urea and some ammonium-containing fertilizers is influenced by temperature, soil reaction and soil water status. Under very dry conditions, little loss occurs, and in stable wet soil conditions, ammonium remains in solution. However, where soil moisture status is intermediate, or where the soil or floodwater loses water rapidly by evaporation, volatilization of ammonia can be appreciable. This is particularly observed where urea is surface broadcast without incorporation on alkaline soils with inadequate moisture during periods of high temperature. Chapter 11 examines various routes of N loss from soils and the means to minimize them.

Crop nutrition influencing water demand

Water requirement of crops

Effective water management requires careful planning of crop production at farm level. Water requirement means the quantity of water needed for transpiration from the green plants, evaporation from the soil and other water losses during application. Crops require 300–800 litres of water for transpiration in order to produce 1 kg dry matter. The amount of water consumed is both plant specific and climate dependent. It is also determined largely by the nutrient supply and the size of crop canopy or leaf surface. To minimize water requirements, various losses such as those during conveyance of irrigation water, runoff, seepage by deep percolation, leaching and waterlogging should be avoided. Water requirement (WR) must be met from water stored in the soil profile (Sw) plus rainfall (Rw) plus irrigation (Iw). Therefore, the irrigation water requirement (IR) = WR - (Sw + Rw). Even where the total amount of water is sufficient, this may not ensure high yields if there is a water deficit in critical growth stages (listed above).

Crop nutrition and water demand

A good nutrient supply also creates higher osmotic pressure in plant cells, which results in a better resistance to drought. Potassium ions (K^+) play an important role in regulating the functioning of stomata in the leaves that control water loss. Thus, a good supply of K can conserve water. Phosphate promotes early root growth, which allows better access to water from deeper soil layers and also shortens the growth period. This leads to early ripening, which reduces water demand. To a certain extent, a shortage of water can be compensated for by optimizing plant nutrition. Under low rainfall, nutrient input, especially of N, should be adjusted to the amount of stored soil water (Figure 36).

Water-use efficiency

As in the case of any production input, the efficient use of water is also of practical interest. Water use in crop production is not confined to transpiration from plants. Additional water losses such as evaporation must be considered in calculations of water-use efficiency (WUE). WUE is defined as the economic crop yield (Y) per unit of water used by the crop for evapotranspiration (ET). It is expressed in kilograms of crop per millimetre of water used:

$$\text{WUE} = \frac{Y}{ET} \text{ kg/mm}$$

In recent years, WUE has increased considerably owing to substantial yield increases as a result of improved nutrient supply, especially of N, P and K. As water supply is often a limiting factor in crop production and irrigation is both expensive and finite in quantity, any practice that increases yield per unit of water used is important. Good nutrient supply must complement irrigation or else part of the additional water will be wasted, leading to a drop in WUE. Once full crop cover is achieved, water use (ET) from the field is controlled mainly by incoming solar energy, nutritional status, etc. In these circumstances, any input factor that increases economic yield improves WUE.

Optimizing plant nutrition should aim to maximize both NUE and WUE. The best way to achieve this will depend on the soil fertility status, the water regime in a given production system, and moisture conservation practices such as mulching.

FIGURE 36
Response of rainfed wheat to nitrogen in soils with different stored moisture levels

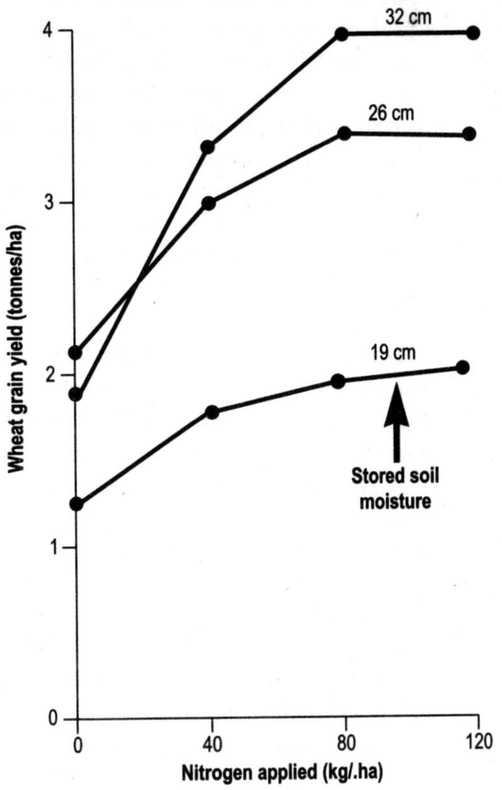

Source: Meelu, 1976.

For rainfed dryland crops, the plants often have to face moisture stress at some stage of growth. Whatever the level of fertilizer used, the factor most often limiting production is water supply. Fertilizer rates must be decided in relation to the level of water supply from stored soil moisture and the anticipated rainfall, which determine the yield (Figure 36). It is advisable to apply N in more than one split in order to take advantage of rainfall expected during crop growth. Under very "dry" conditions, too much fertilizer applied before planting or very early on during crop growth may affect crop yield and WUE adversely by stimulating excessive vegetative growth, which uses up the limited water supplies leaving very little water for the reproductive and grain-filling stages of growth. This is a case where a luxuriant crop stand can be counterproductive.

For irrigated upland crops, the fertilizer requirement is normally high and the amount to be applied can be decided in relation to soil fertility level, expected yield and local management practices. Both NUE and WUE will be maximized by providing adequate amounts of both water and nutrient inputs for full growth and yield. Their applications should be timed so that crop nutrient and water needs are always met.

In wetland rice, provided water management is good, yields are determined by climate, season, variety, management and the nutrients applied. The amount of fertilizer, method of application and timing are all important. Generally, the NUE and WUE are lower in such systems compared with upland crops because of the large volume of water required and high N losses. The efficiency of both the inputs can be improved by applying N in 2–3 splits during crop growth and by using efficient N carriers. There is scope for economizing on water in flooded-rice culture because, if the soil can be kept saturated, waterlogging or deep submergence may not be required.

PLANT NUTRITION AND RESISTANCE TO STRESS

A crop can suffer from several types of stresses during its growth. These may be caused by soil, moisture, temperature, salinity, nutrient deficiencies or toxicities, pests and diseases. The response of crops to various stresses is often affected by their nutrient status. Optimizing plant nutrition can enable the crop to withstand such stresses and emerge with minimum loss of yield. The role of some plant nutrients such as K in this regard has been investigated in considerable detail. The subject of plant nutrition and resistance to various climate and other stresses is discussed in brief here. Vlek and Vielhauer (1994) provide a detailed review of the subject with special reference to N, P and K.

Tolerance of plants to water stress

Water stress to varying degrees is often experienced by plants at some stage even under irrigated conditions. However, it is more frequent in dryland farming and areas where irrigation is not assured.

A crop receiving balanced nutrition is able to explore a larger volume of soil in order to access water and nutrients. Plants facing moisture stress can also suffer from nutrient stress owing to the very close association between water and nutrient availability. According to Vlek and Vielhauer (1994), the main stress in relation to N management is probably the uncertainty of rainfall where irrigation is not available. Where rainfall is excessive or very intense, N is subjected to leaching or denitrification, while with drought it has a tendency to remain in the soil, unutilized by the crop.

P has a marked effect on root growth. Hence, crops deficient in P are not able to access water from deeper soil layers owing to poor root development. Therefore, such crops are more susceptible to drought than crops with adequate P and, hence, a well-developed root system. In contrast, crops overfertilized with N develop too much vegetative growth relative to the root size. This results in rapid water loss from the plant canopy, which depletes soil water faster than does a crop receiving balanced fertilization. Such crops are very susceptible to drought. Where the situation is not remedied by irrigation or timely rains, the net result is a large drop in yields. In legumes, moisture stress retards nitrate reductase activity, protein synthesis and N fixation severely.

K has an osmotic role in the plant that enables the plant tissue to hold on to its water. The movement of K in and out of the guard cells that surround the stomata on plant leaves is responsible for the opening and closing of these cells, which greatly assists in reducing moisture loss when the plant encounters moisture stress. Where plants are deficient in K, the stomata cannot function properly and the water loss from plants can be very high. Application of K has been shown to enhance the drought resistance of plant under moisture stress. During recovery from moisture stress, K can help the plant to maintain higher growth rates.

Tolerance of plants to lodging

Lodging or displacement and breaking of the stem from its upright position are common in several crops, especially cereals and grasses. Depending on the severity of lodging, the effect may be permanent or reversible to a certain extent. Crucial growth stages in cereals that are associated with yield loss as a result of lodging are heading and early grain-formation periods. Lodging in the case of traditional tall varieties of rice and wheat under N fertilization and their low genetic yield potential were some of the major reasons for the development of dwarf, stiff straw HYVs. These HYVs had higher yield potential that could be realized because these could also respond to higher rates of N application without lodging.

Lodging is particularly severe on windy days where plants with weak stems contain high levels of N. It is an interactive effect of plant type, environmental conditions, soil texture and nutrient management. Plants low in K are susceptible to lodging because they have thinner stems as a result of insufficient K. Lignification of the vascular bundles in stems is impaired under K deficiency. Such plants generally have weak stems. Plants well supplied with K have thicker stems and greater stem stability. Resistance to lodging is basically governed genetically, but adequate K supply decreases the tendency to lodge. The role of K in enhancing plant resistance to lodging has been well documented in several crops such as maize, rice, wheat and oilseed rape (Kant and Kafkafi, 2002).

Tolerance of plants to salinity and alkalinity

In saline and alkaline soils, exchangeable Na is present in very large amounts compared with exchangeable Ca and K. Na is not an essential plant nutrient. There are indications of an association between the tolerance of a crop or a crop variety to salinity and its K status. Salt-tolerant crops are generally found to contain more K than crops susceptible to salinity. It has been shown that crop varieties that can absorb K in preference over Na are relatively more tolerant to salinity and alkalinity (Rana, 1986).

In a comparison between a salt-tolerant wheat variety (Kharchia) and a salt-sensitive variety (HD 4530), it was observed that both the varieties produced similar yields at an ESP of 7 percent. However, at an ESP of 43 percent, Kharchia still produced 2.5 tonnes of grain per hectare whereas HD 4530 yielded 0.75 tonnes/ha. The ratio of Na/K absorbed at 43 ESP was 0.43 in Kharchia and 2.59 in HD 4530. This indicates that Kharchia was capable of absorbing more K and excluding Na, but that HD 4530 was unable to restrict Na uptake (Joshi, 1980). In tomatoes, the K^+/Na^+ selectivity ratio was also higher in the salt-tolerant variety than in a non-tolerant variety (Kant and Kafkafi, 2002). These results suggest that maintaining adequate levels of K and K^+/Na^+ ratios in plant cells is essential for normal growth under saline conditions.

Tolerance of plants to cold

Nutrients can have both positive and negative effects on cold tolerance. Plants that have been overfertilized or those receiving imbalanced nutrition produce soft leaf

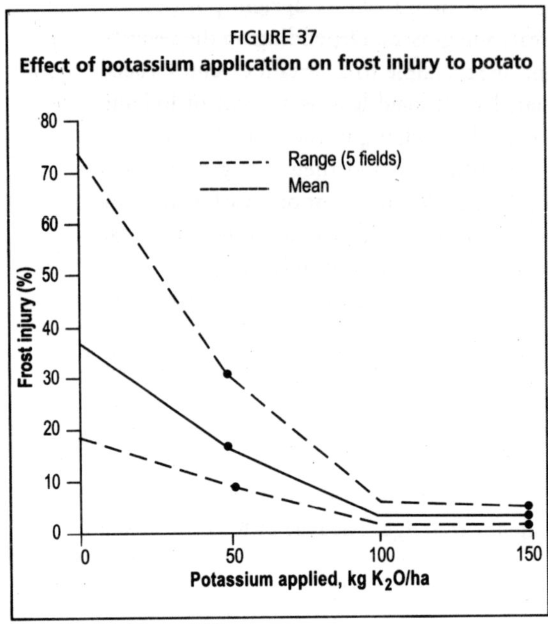

Source: Grewal and Sharma, 1978.

tissues that are susceptible to cold and frost damage. K has a key role in regulating cell sap concentration and this helps plants tolerate cold stress caused by very low temperatures. Potato plants well supplied with K have been found to withstand frost better than plants low in K. In the northern plains of India, the frost injury rate was 36 percent in potatoes grown without K application, 16 percent at an application rate of 50 kg K_2O/ha and 2 percent at an application rate of 100 kg K_2O/ha (Figure 37). The higher K content of plants lowered the freezing point of the cell sap, enabling them to survive spells of frost. For a given crop, the susceptibility to frost also varies with the variety. K application can increase the frost resistance of the frost-sensitive varieties.

B supply is sometimes associated with reduced frost damage. The best evidence for this has come from eucalyptus and pine trees although some indications are also available for apples and grapes (Shorrocks, 1984).

Resistance of plants to pests and diseases

Of several nutrients whose role has been studied, N and K have been investigated in considerable detail. A summary of the effects of nutrients on disease and insect resistance is presented below:

- Nitrogen: Excess N results in luxuriant plant growth, which makes them more attractive to insects and susceptible to disease and leaf-feeding insects.
- Phosphorus: A good supply helps plants resist disease, particularly bacterial leaf blight in rice, possibly by balancing the adverse effect of excess N. A good P supply also provides tolerance against infections with some bacterial or fungal crop diseases (e.g. phytopthora of potatoes).
- Potassium: K improves disease resistance by maintaining tightly closed stomata, which prevents the entry of pathogens into leaves. It also improves stem strength, which reduces lodging, which in turn reduces insect and disease damage and crop quality.
- Calcium: Adequate Ca is reported to reduce the incidence of club root in Brassica crops.

- Boron: B-deficient plants are more susceptible to powdery mildew. Adequate B in plants reduces the incidence of club root in Brassicas.
- Manganese: Mn deficiency causes increased incidence of blast and black spot diseases.
- Copper: Cu-deficient plants are considered to be susceptible to airborne fungal pathogens.
- Chloride: Application of Cl-containing fertilizers may reduce incidence of "take-all" (root and crown rot) in wheat by inhibiting nitrate production and reducing pH at the root surface.
- Silicon: High N and low K uptake reduce Si uptake, which makes rice more susceptible to blast disease. A low silica content in leaves makes them softer and more succulent, making them susceptible to attack by leaf-feeding/ sucking pests.

N and K are known to exert a profound influence on the susceptibility or resistance of plants towards many types of pests and diseases. A high N content of the leaf tissue is known to make plants susceptible to a number of diseases and attack by pests. The adverse effect of N can be neutralized to a considerable extent by providing balanced crop nutrition, particularly optimal N:K ratios. In contrast, plants deficient in K are more susceptible to disease than those that have been adequately fertilized with K. The subject has been reviewed in detail by Perrenoud (1990).

Rice plants deficient in K or with a poor N:K balance are particularly susceptible to brown spot disease, stem rot and bacterial leaf blight. The incidence of the disease may also be affected by the amount of vegetative growth. Experiments with rice have shown that the incidence of brown spot increased with N supply at all K rates. The problem was most severe where N was applied in the absence of K because the growth stimulation brought about by N resulted in an internal dilution of K and an increase in infection potential. Adequate supply of B is associated with reduced incidence of ergot disease on barley. Seed treatment with B has also been reported to provide resistance to tomato, capsicum and cabbage against damping off fungi (Shorrocks, 1984).

NUTRIENT MANAGEMENT IN DIFFERENT CROPPING SYSTEMS

Plant nutrition problems are rare where a small population utilizes a large area of fertile soil. In contrast, almost any nutrient input is justified in cases of low production levels in relation to the food and fibre demands of the population. There is a great variety of cropping systems between these two extremes, each of which requires different system of nutrient management. All cropping systems have limitations imposed by natural and economic conditions. The objective of optimizing nutrient management is to make the best use of soil and applied nutrients within the characteristics and demands of specific farming systems for optimal production with minimal depletion of soil nutrient status. The topics in this section are interrelated with those in the earlier section on strategies for optimizing nutrient management.

Exploitive cropping at low yield level

Historically, cropping without external nutrient application has been common in many parts of the world. Exploitation of soil nutrients basically means cultivating crops until available soil nutrients have been exhausted (mined) and the yields have declined markedly. In the end, such fields must be abandoned and left to return to natural vegetation for regeneration. A typical example of exploitation cropping is shifting cultivation used by subsistence farming in certain tropical forest areas (discussed above).

This system is exploitive because nutrient losses are not compensated for by input. Nevertheless, it is stable to a certain extent as long as there has been no serious soil deterioration during the cropping period and there is sufficient land available for long regenerative phases under natural vegetation. For this to happen, there needs to be about seven times more land available than is actually needed to support the population. The poor reputation of shifting cultivation as a misuse of soil resources is mainly a consequence of the deviation from the original concept by shortening the forest fallow period and, thus, not allowing the soil enough time for regeneration. This mostly occurs as a result of an increased population pressure. With increasing populations, such systems need to be replaced by more stable and productive types of farming systems.

Sustainable agriculture at low to medium yield level

The concept of sustainable agriculture has gained a high priority. Sustainable agriculture has already been defined and described in Chapter 2. It involves the successful management of resources for agriculture to satisfy human needs while maintaining or enhancing the quality of the environment and conserving natural resources. Systems of this kind involve complex interactions and require integration of all production factors.

A prominent concept for sustainable agriculture is low-input sustainable agriculture (LISA). LISA is supposed to optimize the management and use of internal production inputs (mainly on-farm nutrient resources) in order to obtain satisfactory and sustainable crops yields and profitable returns. LISA is a subtype of organic farming. It is a production at the lower end of the crop response curve and not expected to meet the food and fibre need of heavily populated countries where most of the available arable land is already being farmed. With continuous growth in population and a near stable agricultural area, LISA would hardly be capable of providing adequate food and fibre for the expanding population.

Low-input agriculture and its associated low to medium productivity may be required for compelling natural and economic reasons. Extensive sustainable agriculture (low input, low output) in vast areas of developing countries is an example. It may also be deliberately promoted and practised for ideological reasons such as biofarming or ecofarming in developed countries. It is certainly more suitable for subsistence agriculture, for the production of high-value produce demanded by a section of the population, and for products with a "niche" market rather than for meeting the food needs of the population as a whole.

In areas with severely yield-limiting factors as in dryland farming areas, extensive farming with low input and low to medium yields still has its place. The main emphasis in this type of system lies in the use of mobilized soil nutrients and internal nutrient cycling via organic substances. However, complete cycling is difficult to achieve because of unavoidable losses. Typical examples of this approach are small subsistence farms with no or little means for nutrient input. In other systems, fertilizer input is deliberately kept low as its efficiency is known to be low under stress conditions induced by water shortage and periods of drought. Harvesting and recycling of rainwater on or off the farm holds the key to optimizing crop nutrition and increasing crop yields.

Intensive sustainable agriculture at high yield level

Sustainable agriculture cannot be equated with subsistence agriculture for the vast majority of cropland in the world. Sustainability is by no means confined to low-input conditions but can be achieved at any level of production where inputs and outputs are in balance and the best land-use practices are followed. Such systems could be called adequate-input sustainable agriculture (AISA). As demonstrated in Western Europe and elsewhere, high but adequate rates of nutrient application result in sustainable production with high yields without significant adverse effects on soil fertility or the environment. Farming systems of this kind are rather diverse, ranging from rainfed to irrigated areas, but they have many similarities in terms of nutrient management.

Research results from many parts of the world show that high crop yields are sustainable through balanced and integrated nutrient management supported by suitable amendments to address problems such as excess acidity or alkalinity. There is hardly any challenge or role for modern science and technology if sustainable agriculture is to be restricted to low-productivity subsistence farming.

The long-term experiments at Rothamsted in the United Kingdom have been in existence for more than 150 years. Results of continuous cropping for more than 100 years (1952–1967) show an average wheat yield of only 1 tonne/ha in an untreated plot and about 2.5 tonnes/ha in plots receiving either 35 tonnes FYM/ha or only fertilizers at the rate of 146 kg N + 75 kg P_2O_5 + 100 K_2O/ha.

In the United States of America, the oldest experimental plots, known as Morrow Plots, have been in existence since 1876 at the University of Illinois. Based on results obtained over a period of more than 100 years from these plots, Darmody and Peck (1993) concluded that well-treated soils could provide food and fibre continuously at high levels. Average maize grain yield in the best rotation coupled with optimal fertility management was 8.6 tonnes/ha compared with 2.2 tonnes/ha in untreated plots under continuous corm. These results contain a significant message for countries that are continuously striving to meet the food and fibre needs of an expanding population from a resource base that is expanding either slowly or not at all.

In a long-term experiment at Aiza, Fukushima Prefecture, Japan, a set of fertilizer treatments with and without organic manures and amendment were initiated in 1920. Even in the 1980s, the untreated control plot was able to sustain paddy yields of about 4 tonnes/ha, but plots receiving only NPK through fertilizers produced twice as much. Nearly 70 years of continuous fertilizer use have not had a negative effect on the physical, chemical and biological properties of this paddy soil (von Uexkull and Mutert, 1993).

In a study to evaluate changes in the properties of agricultural soils over a 60-year period, researchers in California, the United States of America, analysed 125 soil samples collected in 2001 for which reference samples taken around 1945 were also available. By comparing the analytical values obtained from the two reference years, their overall conclusions were that while increased clay percentage may indicate accelerated soil erosion, the soils of California have maintained their chemical quality over the past 50–60 years (DeClerck and Singer, 2003).

Results from a number of long-term field experiments were started in India in the early 1970s using high-intensity crop rotations involving 2–3 crops in succession per year under irrigated conditions. On the whole, these experiments have shown that high levels of crop productivity (8–12 tonnes grain/ha/year) can be sustained by integrating optimal and balanced fertilizer application rates with 10–15 tonnes FYM/ha/year. These experiments have established that fertilizer is the key input for increasing crop productivity, but also that the integrated use of fertilizers and FYM or lime where needed give higher and more sustainable yields as it could also correct some micronutrient deficiencies and improve soil physical and biological properties (Swarup, 2000).

Even under rainfed dryland conditions, medium to high crop yields can be sustained through an integrated use of fertilizers and organic manures. Results of a nine-year field trial with dryland finger millet in the red soils at Bangalore, India, show that the best yields were obtained when recommended rates of fertilizer were applied in combination with 10 tonnes FYM/ha. It was only at this input level that grain yields of 3 tonnes/ha and above could be harvested in eight out of the nine years (Table 34). A considerable portion of the yield potential would have been lost if either of these inputs had been omitted.

The goal of intensive sustainable agriculture at high yields is to utilize, as far as possible, the yield potential of high-yielding crops by eliminating all nutritional constraints through INM including fertilization and maintaining high soil fertility,

TABLE 34
Effect of fertilizers and FYM on the productivity and stability of dryland finger millet over nine years at Bangalore, India

Annual treatment	Mean grain yield	Number of years in which grain yield (tonnes/ha) was			
	(kg/ha)	< 2	2–3	3–4	4–5
Control	1 510	9	0	0	0
FYM (10 tonnes/ha)	2 550	1	6	2	0
Fertilizer 50–50–25 (kg/ha N–P_2O_5–K_2O)	2 940	0	5	4	0
FYM (10 tonnes/ha) + 25–25–12.5 (kg/ha N–P_2O_5–K_2O)	2 900	0	6	3	0
FYM (10 tonnes/ha) + 50–50–25 (kg/ha N–P_2O_5–K_2O)	3 570	0	1	5	3

while simultaneously protecting the crop against disease and insect damage. However, there is a negative aspect of nutrient management under such systems. This happens where there is heavy reliance on the fertilizer input while neglecting the soil nutrient reserves and those available in various organic sources. This tends to occur where cheap chemical fertilizers are readily available. This has led to the public misconception that intensive cropping is essentially a "nutrient-wasting" system.

Sustaining crop productivity at a high yield level has proved possible in many progressive agricultural areas, even in parts of so-called developing countries such as Punjab State in India. The dependence on fertilizers for adequate food and fibre production continues to remain because of continuous growth in human population and little expansion in the net cropped area. Food production can be enhanced by better nutrient cycling and prevention of losses. However, the food demands of an increasing population cannot be met only from organic sources or from fertilizers alone. They require an active pre-planned INM approach. As part of integrated crop production, INM will be a decisive factor in attaining the goal of sustainable high yields and profitable crop production without negative effects on the environment.

Harnessing BNF is an important component of INM and this is not confined to a particular cropping system or productivity level. Although considerable amounts of N can be fixed by legumes, whether or not this results in a buildup of soil N or the N nutrition of the following non-legume crop, depends on the amount of N fixed, the amount of N removed in the crop products and the residues. In many cases, growing a legume in a rotation contributes significantly to the N nutrition of the following crop. Where crop yields are high and a large amount of N is removed in the harvested product, the effect may be small or even negative. In grass–legume pastures, the transfer of N from the legume to the pasture is small, and the N passes from the legume to the grass primarily in the manure and urine from the grazing animal or after the decomposition of legume residues.

Biofarming and ecofarming

Biofarming and ecofarming are forms of organic farming. They refer to special farming systems that exclude the application of manufactured mineral fertilizers or pesticides, but use natural minerals such as PR, animal manures, compost and legumes as nutrient sources. Such systems place considerable emphasis on nutrient cycling. It is claimed that with this production system a better food quality is produced and that the environment is better protected against unwanted pollution from agricultural chemicals. The system is workable because of the higher produce prices realized, which compensate for the generally lower yields obtained.

The general term biofarming denotes a group of similar and yet different systems of nutrient supply. Biological dynamic agriculture (the oldest, orthodox type of system initiated by Steiner in 1924) excludes all kinds of commercial mineral fertilizers. In contrast, major groups (e.g. Bioland) exclude mainly water-soluble mineral fertilizers, especially N fertilizers, but permit other major nutrient sources if they are natural products such as PR, crude salts of K and lime. Micronutrients are allowed only where there is an obvious deficiency. The rejection of water-soluble N fertilizers, whether nitrate-containing ones or urea, has no scientific basis. It is an ideological concept based on the philosophy of going back to nature.

The general features of permitted practices under organic farming as set by the International Federation of Organic Agriculture Movements (IFOAM, 1998) are:

- Inputs manufactured by chemical processes should not be used.
- Water-soluble N and P fertilizers are avoided as a matter of principle.
- Soluble potassium sulphate and micronutrients are permitted provided a threatening deficiency is documented through analysis.
- PR and other natural minerals with a low solubility can be used.
- Weeds are removed or damaged by mechanical soil treatment or the use of flame.
- Extensive crop rotation and intercropping are adopted, while monocultures are avoided.
- Herbicides and synthetic pesticides are prohibited and genetic engineering is not accepted practice.

Although the claims for superior quality food by avoiding chemical fertilizers and chemical crop protection have not been substantiated, a limited number of consumers support this production of so-called "natural" food by paying premium prices. The further claim that these types of biofarming and ecofarming systems cause less pollution of water bodies because they do not use any chemical fertilizer input should be questioned. Although a lower amount of N leaching is often achieved per unit of land, it rarely holds true per unit of crop produced, especially because almost twice the area of land is required for biocropping and ecocropping than with conventional farming.

However, organic farming does have a place as one of the many farming systems. It is more of a class enterprise rather than a mass enterprise. It is best suited for producing organically grown produce for which consumers are prepared to pay the higher price demanded. Based more on belief than on fact, it automatically favours the exclusion of certain technologies and inputs because these go against the belief. This approach conventionally ignores the existence and operation of nutrient cycles in soils through which mineral and organic nutrient forms are interconvertible (and beneficially so because plant roots feed only on mineral nutrient forms regardless of whether these are derived from mineral or organic sources). Such compartmentalization of nutrients into organic (natural) and mineral (artificial) overlooks the basic fact that these two forms not only coexist but are interchangeable in soils.

Organic agriculture faces the same environmental and sustainability problems with crop nutrient management as does mainstream agriculture: emissions of ammonia and nitrous oxide, nitrate leaching, energy use, and depletion of PR resources (Laegreid, Bockman, and Kaarstad, 1999).

Optimizing nutrient management in diverse cropping systems

There is a multitude of cropping systems in use throughout the world. These range in intensity from raising one crop per year (as happens in many rainfed dryland areas) to 3–4 crops per year in irrigated/assured rainfall areas on the same piece of land. Wherever adequate rainfall or irrigation is available and the climate permits, raising two grain crops in succession within a year is possible. In many areas, the whole cropping system or rotation is completed within one year. In other areas, a given system may be rotated after 2–3 or more years. Only some of the nutritional features of the main types of cropping systems are discussed here.

Annual crops in different rotations

Short rotations that include crops such as rice, wheat, maize, oilseed rape, barley, vegetables and fodders are highly nutrient demanding and, therefore, rely mainly on high external nutrient input. Except for N, especially where no legumes are involved, nutrient management is more concerned with the whole rotation than with individual crops. Fertilizers are applied to maintain a high nutrient supply utilizing both the direct (fertilized crop) and the residual effects. This is sometimes referred to as "rotation fertilization". For example, in temperate climates, substantial amounts of mineral N often remain in the soil after oilseed rape, which is usually followed by winter wheat. The wheat crop utilizes the residual nutrients in autumn before the main leaching period. Longer rotations, which include crops such as sugar beet, potatoes or even legumes with their extra gain of N, often have more soil tillage, soil cover and, thus, nutrient mobilization than cereals.

One of the most intensive and nutrient-demanding rotations in parts of South Asia is the rice–wheat rotation. In India, this rotation is practised on more than 10 million ha, primarily in the northern alluvial plains. Under optimal management, grain yields of 8–12 tonnes/ha/year can be harvested. Optimizing nutrient management in this system includes the application of NPK and other required nutrients such as S and Zn. The wheat crop must receive its optimal rate of P application while rice can benefit to a considerable extent from the residual effect of P applied to wheat. On highly P-deficient soils, P must be applied to both crops. Incorporation of green gram residues after picking the pods before planting rice is an effective green manuring practice in this system. In general, research recommendations provide for application of the full recommended rates of fertilizer to the wheat crop, while 25–50 percent of the recommended fertilizer to rice can be saved through the use of 10 tonnes/ha FYM, *Sesbania* green manure and crop residues (Yadav *et al.*, 2000). Information is also becoming available on INM in this highly intensive system (Table 35).

Annual crops in monoculture

In several tropical and subtropical areas, high-intensity monoculture is practised wherever the rainfall is well distributed or where adequate irrigation is available.

Wetland rice has its special problems of nutrient management owing to the strong reducing conditions of the submerged soil in which several mobilization

TABLE 35
Examples of INM packages and their comparison with fertilizer recommendations for rice–wheat cropping in different agroclimate regions of India

Region	Mineral fertilizer recommendation (kg/ha)	Integrated nutrient management recommendation (kg/ha)
Trans Gangetic Plain	Rice: 120 N + 60 P_2O_5 + 60 K_2O + 20 zinc sulphate	Rice: 60 N + 30 K_2O + 10 tonnes/ha FYM or poultry manure
	Wheat: 180 N + 60 P_2O_5 + 30 K_2O	Wheat: 150 N + 30 P_2O_5 (through SSP) + 30 K_2O + *Azotobacter* or *Azospirillum* + PSB
Upper Gangetic Plain	Rice: 120 N + 60 P_2O_5 + 40 K_2O + 20 zinc sulphate	Rice: 90 N + 30 K_2O + 10 tonnes/ha FYM or green manuring with *Sesbania/Leucaena* lopping
	Wheat: 120 N + 60 P_2O_5 + 40 K_2O + 40 S	Wheat: 90 N + 60 P_2O_5 (through SSP) + 30 K_2O
Middle Gangetic Plain	Rice: 100 N + 60 P_2O_5 + 40 K_2O	Rice: 50 N + 30 P_2O_5 + 20 K_2O + green manure (green gram stover) + 20 zinc sulphate in calcareous soils
	Wheat: 120 N + 80 P_2O_5 + 40 K_2O	Wheat: 90 N + 60 P_2O_5 + 30 K_2O + 10 tonnes/ha FYM
		or
		Rice: 75 N + 45 P_2O_5 + 30 K_2O + 15 kg/ha BGA + 10 tonnes/ha FYM + 20 zinc sulphate in calcareous soils
		Wheat: 100 N + 65 P_2O_5 + 30 K_2O
Lower Gangetic Plain	Rice: 80 N + 60 P_2O_5 + 40 K_2O	Rice: 40 N + 45 P_2O_5 + 30 K_2O + 10 tonnes/ha FYM or green manure + 10 tonnes/ha *Azolla* or 10 kg/ha BGA + 20 zinc sulphate
	Wheat: 120 N + 60 P_2O_5 + 60 K_2O	Wheat: 90 N + 45 P_2O_5 (through SSP) + 45 K_2O

Source: Sharma and Biswas, 2004.

and fixation processes take place (Chapter 5). A major unresolved problem is the low recovery of fertilizer N, which is mainly applied through urea in these systems. Usually, only 30–50 percent of the added N is taken up by the crop compared with about 70 percent in intensive well-managed wheat cropping. The low N efficiency is a consequence of N losses by various routes.

Extensive on-farm trials suggest that the adoption of appropriate crop and nutrition management practices can minimize the effects of diminishing returns at increasing N application rates mainly on account of N losses. In order of importance, the limiting factors that smallholder rice farmers using prill (or granular) urea can address are: (i) too few split applications, resulting in substantial N losses and consequent inadequate N supply to meet crop requirements at various growth stages; (ii) cultivars that may be insufficiently N responsive; and (iii) inadequate initial plant population. A multilocation on-farm trial/demonstration project on irrigated rice (1995–98), funded by Japan and implemented by FAO in Indonesia, the Philippines and Malaysia, demonstrated that deep-placed USG enables a 21-percent N saving in comparison with 70 kg/ha N applied as prill urea in three splits (FAO, 2003c). Urea coated with Nimin, a commercial extract from neem (*Azadirechta indica*) seed, has been widely tested, especially in India. This reasonably inexpensive biological product shows great promise for resource-poor farmers, with an average yield increase of 5–10 percent over uncoated prill urea. Supergranules made with Nimin-coated urea and placed deep show further improvement over the USG technology.

In many rice-growing areas, wherever the climate permits, 2–3 rice crops can be raised in succession within a year. For example, in India, rice–rice annual rotation is practised on almost 6 million ha. Supply of N through BGA and *Azolla/Anabaena* symbiotic systems has some promise and could potentially replace a portion of N fertilizer.

Annual crops with short-term fallow

Fallowing may be required for weed control in humid climates or for water storage in the soil in dryland farming. In the absence of crop removal, fallowing also conserves mobilized soil nutrients, thus providing an extra nutrient supply for the next crop. Fallows can be bare or with a plant cover, depending on the main purpose. Bare fallow is a period of nutrient and water accumulation. In overpopulated, land-scarce countries, land is rarely left fallow by choice. It is more a consequence of the farmer's inability to raise an additional crop under low rainfall or inadequate stored soil moisture. The vegetation cover during the fallow period can be used effectively as a mulch or even as a green manure.

Multiple-cropping systems

Multiple cropping refers to the cultivation of two, or often more than two, crops on the same field in a year. The concept of multiple cropping includes cropping practices where sole or mixed crops are grown in sequence, simultaneously one after another, or with an overlapping period. A distinction is made between sequential cropping and intercropping. Sequential cropping can involve growing two, three or four crops a year in sequence or ratoon cropping. Intercropping involves mixed/row/strip intercropping (simultaneously) or relay intercropping (overlapping).

Optimizing plant nutrition in multiple-cropping systems revolves around:

- adjusting for residual effects of nutrients such as P, S and micronutrients (e.g. applying P on priority to wheat and green manure to rice in a rice–wheat rotation, and FYM on priority to maize in the maize–wheat rotation);
- prioritizing the application of fertilizers to those crops in the system that have a poor root system and are poor users of applied nutrients (e.g. potato in a potato–maize system);
- planning for a short-duration catch crop that can feed on residual fertility in between two main crops (e.g. green gram in a maize–wheat–green gram annual rotation);
- practising INM keeping in view crop characteristics (e.g. green manuring where possible before planting rice or inoculation of the rice field with BGA/*Azolla* in rice-based cropping systems);
- phasing of fertilizer application among crops in a rotation so that maximum direct plus residual gains are obtained (e.g. P application on priority to wheat in rice–wheat, maize–wheat or sorghum/millet–wheat rotations, S application to an oilseed crop in an oilseed–cereal rotation);

- in mixed cropping, such as with cereals and legumes, the fertilizer application is primarily determined by the cereal, and the legume seed can be inoculated with *Rhizobium* culture;
- nutrient management in multiple-cropping systems should be finally decided by the economics of the yield response to various nutrient applications, particularly where the component crops fetch different market prices (e.g. a yield response of 1 tonne oilseed is more valuable than a yield response of 1 tonne cereal).

Depending on the strategy of nutrient management used, the gains from multiple cropping can vary considerably. Results from several long-term experiments employing multiple-cropping rotations for example have shown that: (i) intensive cropping with only N input is a short-lived phenomenon; (ii) sites that were initially well supplied with P, K or S became deficient over a period of time when continuously cropped using N alone or S-free fertilizers; (iii) in most situations, optimal fertilizer application + 10–15 tonnes FYM/ha/year was required in order to sustain crop yields; (iv) soil fertility status was improved or depleted depending on input–output balances as well as by soil properties; and (v) fertilizer rates considered as optimal still resulted in nutrient depletion from the soils at high productivity levels and in the process themselves became suboptimal application rates.

These experiments demonstrated that the same field that produced 1 300 kg grain/ha from two crops grown without fertilizer application could give 7 424 kg grain/ha when the crops received optimal application of the nutrients required (Nambiar, 1994).

OPTIMIZING NUTRIENT MANAGEMENT IN DRYLAND AND IRRIGATED FARMING

The following sections discuss some aspects of nutrient management under varying regimes of water availability. These range from dryland farming, to conventional irrigated farming and, finally, to flooded soils used for wetland rice production. The aspects discussed are general and applicable to various types of cropping systems described above. These all point to the need for integrated management of nutrients and water in order to optimize the efficiency of and returns to nutrient application.

Nutrient management in dryland farming

In rainfed dryland farming systems, the yield is usually limited by a shortage of water, rainfall being not only scarce but also variable and, thus, unreliable. The main nutritional problem is the shortage of total and available N owing to the low SOM content. In order to make the best use of the scarce soil N resource at sowing time, the N requirement of the crop should be adjusted for the nitrate flush occurring from rapid mineralization at the onset of the rainy season. In practice, this is not easy because of the uncertain onset of the rainy season. There can also be some upward movement of nitrate from the subsoil by evaporation.

The natural N supply may be sufficient for low yields, e.g. 1–3 tonnes grain/ha. However, for medium yields, additional N sources such as farm waste materials or even mineral N should be added where there is sufficient moisture. Grain yields of 3–4 tonnes/ha are sustainable under dryland farming where the system is managed properly, as shown in Table 33. The growing of grain and fodder legumes is widely practised in such areas. In order to derive maximum benefit, adequate phosphate application should be ensured and the legume should be inoculated with an appropriate *Rhizobium* strain in order to maximize the gains from BNF.

Mulching is difficult in these environments because of a shortage of organic matter. However, where available, it can be used for soil protection or mixed into the topsoil as a nutrient source. In very hot climates, mulching can also reduce water loss from the soil and reduce soil temperature. An increase in the very low SOM level is desirable, but the possibilities are limited because of high mineralization rates. The application of organic substances is often limited by competitive use of crop residues, etc. for fodder, fuel and roofing. Another possibility to conserve the natural nutrient supply and plant available water is the use of a bare fallow. However, this may reduce SOM and risk soil losses from erosion.

In addition to N, the P supply is often insufficient either because to low available P in the soil or slow mobility towards plant roots. As P is especially required for root growth and as deep rooting may be decisive for crop survival during dry spells, a good P supply is important beyond its actual role as a nutrient. A good K supply is also essential to reduce transpiration losses from crops. However, for dryland farming on many arid soils, there is generally sufficient available K for at least low to medium yield levels. The same holds true for Mg and S.

Poor availability of micronutrients in neutral to alkaline soils results in a frequent deficiency of Fe and/or Zn. Some improvement in their availability can be made by using strongly acidifying N fertilizers such as ammonium sulphate and, to a lesser extent, urea. However, ammonia volatilization under such systems should be minimized.

Considerable production potential still exists in dryland areas but it can only be realized by combining moisture conservation and the recycling of rainwater with optimal nutrient supply. Special climate and biotic stress factors must be taken into account while managing such soils. However, cropping systems in semi-arid regions that use common agricultural practices may not always be sustainable. They can potentially be made so by the application of the existing research knowledge for INM and the harvesting of the rainwater in combination with farmers' accumulated experience.

Nutrient management in irrigated farming

Irrigation supplies a vital input (water) for crop production and also brings some nutrients with it. It also stimulates the mineralization of SOM and the solubilization and transport of nutrients from sparingly soluble to available inorganic forms.

Irrigation results in considerable dilution of the soil solution. This has the advantage of lowering the osmotic pressure, but the disadvantage of lowering the concentration of nutrients, which cannot be replenished rapidly. There is a relative increase in the concentration of monovalent cations such as K^+ in the soil solution caused by cation exchange. The resulting increase in K supply may temporarily reduce the supply of Mg. The Ca concentration also decreases, but this has no detrimental effects in view of its large total supply.

When a soil is saturated, the pore space occupied by air also becomes filled up with water, creating anaerobic conditions. Where the saturation is temporary and followed by deep percolation, this leads to leaching of soluble nutrients. Where it is prolonged or results in waterlogging, chemically reduced conditions set in. This results in more intensive mobilization and re-supply from mineral nutrient reserves, especially at high temperatures. Nutrients such as Fe and Mn are converted from unavailable to available forms because of the reduced conditions. As the intensity of the reduction varies, so does the availability of these nutrients, resulting in the appearance and disappearance of Fe-deficiency symptoms during the irrigation cycle. Where the redox potential is lowered permanently, iron oxides can be reduced to such an extent that Fe toxicity can occur.

Apart from the flooded-rice soils, there are dry periods in between wet periods in most irrigated soils. These could be caused by a high rate of deep percolation, high evapotranspiration loss or inadequate supply of irrigation water. The drying out of the soil during the dry phase between irrigation periods increases the soil solution concentration by evapotranspiration but reduces the rate at which these nutrients can be transported to the roots. The concentration of divalent cations such as Ca^{2+} increases relative to the monovalent cation K^+.

More severe drying finally results in immobilization of mobile nutrients, i.e. conversion from the soluble and mobile forms to the reserve fraction. Phosphates precipitate, Fe and Mn are oxidized and, thus, are less available (reverse of what happens during flooding). K is adsorbed more strongly, the degree of which depends on the content of clay minerals in the soil. However, these temporary deficiencies at the end of the dry phase may be compensated for by mineralization of plant nutrient reserves. These features of irrigated soils must be taken into account when determining optimal nutrient application rates as the relatively high production level must be supported by more intensive fertilization. Fertilizer can also be supplied with the irrigation water via fertigation (Chapter 7). Many aspects covered in the above section on integrated nutrient–water management are also applicable to this section.

Grasslands or permanent pastures and meadows

The growing of either grassland or arable fodder crops for animals results in a special internal farm nutrient cycle that benefits arable crops. In these systems, the export of plant nutrients in meat or milk is lower than with harvested plant products. Fertilization of grassland has two main goals: a high yield of palatable fodder for substantial production of milk, meat and wool; and good health (including good fertility) of the domestic animals. The fertilization required depends on the production target (e.g. amounts of milk and meat), on the soil nutrient supply and on the grassland utilization system, such as grazed or fodder cut for conservation.

Principles of grassland nutrition

For proper animal nutrition, grassland fodder should contain large amounts of protein, carbohydrates (energy carriers), vitamins and flavouring substances. It should also have optimal amounts of mineral nutrients but no toxic organic substances or excess inorganic nutrients.

Two different aspects must be considered for optimal nutrient supply to plants and animals. First, an optimal mineral composition of the plant not only increases the content of valuable organic substances, such as amino acids, proteins, carbohydrates and vitamins, but also the supply of minerals. Only a limited amount of essential minerals can be given to the animals directly. Second, the mineral requirements of plants and animals differ in some respects. These are:

- similar requirements for plants and animals: P, S, Ca and Mg;
- larger requirements by plants than animals: K, B and Mo;
- larger requirements by animals than plants: Na, Cl, Ca, Mg and some micronutrients;
- required only by animals: I, Co, Se and Cr.

A knowledge of the fodder composition (protein and mineral nutrients) at the time of pasturing or haymaking is an essential precondition for the efficient production of valuable fodder. Milk production requires large amounts of energy and protein as well as a high mineral content. Meat production initially requires fodder that is very rich in protein, but later more energy is required. Fertilization also serves to control the botanical composition of the pasture. The proportion of grass in the pasture increases with increasing amounts of N and K, while the proportion of legumes decreases.

Soil reaction can and should be slightly lower than on arable fields of the same soil texture. In fact, slight to moderate acidity is often useful. Where liming is required, the reaction should stay below neutral.

Thus, the target for nutrient application of grassland consists of supplementing the natural concentrations until the optimal supply range is reached (Table 36). Luxury supplies, or even excess, may lead to problems such as reduced feed intake of other nutrients or decreased absorption of minerals in the animal. The concentration of minerals in the fodder generally decreases with age owing to dilution and maturity effects. Therefore, data on concentrations must refer to a definite growth stage. For grassland, a suitable reference stage is shortly before the beginning of flowering.

Some aspects of nutrient supply in grassland

Most intensively managed grasslands are short in N supply, and N fertilization is almost always required for high yields. The amount of N needed depends on:

TABLE 36
Optimal mineral concentrations of grassland fodder on a dry-matter basis

Major nutrients			Micronutrients			Beneficial nutrients for animals	
Name	A[1] (%)	B[2] (%)	Name	A[1] (µg/g)	B[2] (µg/g)	Name	(µg/g)
P	0.3	0.4	Fe	50	60	I	0.3
Ca	0.5	0.7	Mn	40	60	Co	0.1
Mg	0.15	0.25	Zn	20	30	Se	0.1
K	2.0	2.0	Cu	5	8		
Na	-	0.2	Mo	0.3	0.3		

[1] For high grass yield and medium milk production.
[2] Fodder for highly productive cows, i.e. 20 litres milk/day, intake of 12 kg of dry matter.
Source: Finck, 1992 (data from various sources).

growth conditions, the desired yield level and the protein content of the fodder. For 20-percent protein, 3 percent N must be in the dry matter, which results in a requirement of 30 kg N/tonne of dry matter. On average, 1 kg of N produces 25 kg of dry matter. In many areas, legumes supply the N to the system and grazing management is required to maintain them in the sward.

The P concentration in grass should be 0.3–0.4 percent. Where P is a yield-limiting nutrient, considerable improvement can be achieved by P application. This is because it encourages the growth of legumes and, thereby, the N supply to grasses. The choice of the P form is of minor importance, especially on moist grassland with a good mobilization capacity. On strongly acid P-sorbing soils, PR is recommended.

The natural supply of K should suffice for high fodder yields in many situations. However, where the forage is cut and removed, K may need to be applied. Large amounts of K can be supplied with animal slurry, but excess K can decrease the supply of Mg. Potassium chloride is the preferred source of K.

The large Ca concentration required cannot be attained easily by grasses, which often contain only 0.4 percent Ca. Many herbs and especially legumes contain more than 1 percent Ca. The Ca:P ratio should be 1.5–2:1. The Ca concentration can be increased by liming, but this should only be done up to the optimal pH value, which is somewhat lower than seven.

Mg is often a limiting factor for grass growth on acid soils. Animals can suffer from grass tetany (hypomagnasaemia) where the Mg concentration of the grass is very low or Mg absorption from the fodder is inhibited. The critical Mg concentration in the fodder for high-performance dairy cows is about 0.25 percent. Moreover, the ratio K:(Ca + Mg) should be less than 2.2:1 (expressed in equivalents per kilogram). Magnesium sulphate or any other Mg source can be used.

A deficiency of Cu causes poor growth of cattle and "lick disease". Cattle require 1 µg/litre Cu in their blood and for high milk yields; this is achieved with about 8 µg/g Cu in the fodder. Animals often prefer plants or plant parts with higher Cu concentrations. For proper Cu utilization by the animals, the Ca concentration of the fodder should be below 0.8 percent, Mo should be less than 3 µg/g, and S concentration in the range required for optimal plant growth. Cu deficiency in grassland can usually be corrected for several years by adding 3–5 kg Cu/ha through any Cu-containing fertilizer.

Sufficient Mn, even for high requirements, is generally supplied where the pH value of grassland remains in the slightly acid range. However, on neutral soils the high Mn concentrations required for high milk yield and animal fertility may not be reached. A simple way to increase Mn supply is through soil acidification by using acid-forming N fertilizers. Zn requirements for high milk yields are significantly greater than the Zn needs of plants. However, many soils supply sufficient Zn. Zn application is required only where the optimal Zn status is not reached. Fe, B and Mo are usually present in sufficient amounts in the fodder, but Mo may need to be applied to acid soils for better N fixation by legumes.

Some grasses absorb only small amounts of Na and contain less than 0.01 percent Na whereas some herbs, e.g. white clover, have Na concentrations of more than 0.4 percent. It does not seem necessary to cover all the Na requirements of animals via grass, but a relatively high Na concentration is desirable. Deficiencies of I and Co are rare but a shortage of Co on acid sandy soils, often together with Cu deficiency, can occur. Se deficiencies are more widespread than formerly assumed. However, care should be taken with general application of Se on all grasslands as its optimal range is narrow and high concentrations are toxic. Cr seems to be required only in extremely small amounts.

Beneficial elements, such as V, Ni, Si and bromine, which are required only in very small amounts, are generally supplied by the soils. The silicic acid in many grasses occurs in the form of needles, which may cause injury to the digestive tract of the animals.

Chapter 8 provides recommendations for the fertilization of intensively used grasslands.

Chapter 7

Guidelines for the management of plant nutrients and their sources

PRECONDITIONS FOR SUCCESSFUL NUTRIENT MANAGEMENT

Improvement in the nutrient status of soils and crops is successful with respect to yield increase as well as environmental acceptance when it is integrated into the crop production systems considering the many interactions involved. Plant nutrients should not just be added to the soil, but management practices should ensure their maximum uptake by plants. The total nutrient supply from external sources including fertilizers plus available soil nutrients should be balanced, the soil nutrient supply should be utilized without exhaustion, and external inputs should be used to the extent required. In short, the application of nutrients should be balanced, efficient and economic on a sustainable basis. Simultaneous application of all 16 essential plant nutrients is not called for except in solution cultures. Nutrients and their combinations to be applied can be indicated best through soil and plant diagnostic techniques.

Before applying nutrients, whether through organics or mineral fertilizers, it is advisable to consider the following guidelines as basic requirements for nutrient use. In addition to these, available diagnostic techniques should be fully utilized in decision-making. Plant nutrients, their role and deficiency symptoms have been discussed in Chapter 3. Chapter 4 has examined the dynamics of plant nutrients in soils along with diagnostic techniques for the nutrient status of soils and plant. Chapter 5 has described the materials that supply these nutrients. This chapter provides information on principles and practical guidelines on nutrient management, application techniques of fertilizers and other sources of nutrients such as organic manures and biofertilizers. Chapter 8 provides some illustrative nutrient recommendations for a number of field crops and grassland.

The general agronomic preconditions for successful nutrient management include: (i) selection of a high-yielding and locally adapted crop variety; (ii) proper seed-bed preparation and cultivation practices; (iii) proper sowing or transplanting to ensure optimal plant density; (iv) good soil and water management practices under both irrigated and rainfed conditions; and (v) sufficient plant protection against possible yield losses.

Basic requirements of good soil fertility

The basic requirements of good soil fertility include:

- optimal soil reaction within a practical range;
- sufficient organic matter by applying organic manures for improved soil structure, water storage capacity, nutrient supply and satisfactory activity of soil organisms;
- a stable porous soil structure with no compact layer (which restricts root growth);
- good drainage;
- water availability, especially during periods of water stress and long dry spells;
- removal or neutralization of toxic substances, e.g. in strongly acid (Al), polluted (toxic heavy metals) or saline/alkali soils (excess chloride, Na, etc.).

Soils that are very rich in a nutrient and are able to release it at an acceptable rate in relation to crop demand would generally need its application only to the extent of crop removal replacement. This calls for periodic monitoring of the soil nutrient status because the "very rich" condition does not last indefinitely, particularly under intensive cropping. At the same time, it is necessary to differentiate between nutrients that are mainly applied on a crop-to-crop basis, such as N, and nutrients that leave a significant residual effect. The latter are not to be applied to each crop but on a cropping-system basis (P, S, Mg and micronutrients such as Zn and Cu). Large applications of Mg resulting from the use of dolomitic limestone can last for several years. In deciding the frequency with which such nutrients need to be applied, the degree of their fixation by soil constituents needs to be taken into account. The system is a dynamic one and it should be managed accordingly.

Basic issues for timing nutrient supply

The application of organic manures, fertilizers and liming materials should be timed when these are most effective. Organic manures and liming materials should be applied several weeks before sowing. The same holds true for materials that need to be converted into soluble and plant available forms in the soil before they can contribute to crop nutrition. Such materials include ground PR, elemental S products and pyrites. However, leguminous green manures grown before rice can be incorporated into the puddled soil a few days before transplanting rice as their rate of decomposition is quite fast.

Fertilizers can be applied both at or before planting and during crop growth. The decision about when and how much to apply depends on: crop duration; total amount of a nutrient to be applied; nature of the nutrient, especially with regard to its transformation and mobility; availability of water; and anticipated outbreak of pests and diseases.

In general, the total amount of N is applied in 2–4 instalments starting from a basal dressing. Where the crop is raised largely on stored soil moisture, the entire N is to be applied pre-planting, preferably below the soil surface. For winter crops, N is to be applied partly in autumn but mainly in spring in 2–3 dressings. In the case of N-deficiency symptoms in the standing crops, immediate N application via leaves or soils is suggested. Phosphate and potash fertilizers are mixed into the top layer in moderately fertile soils, especially in narrow-row crops. They are placed strategically or drilled below the seed in wide-row crops, especially in low-fertility soils and soils with high P-fixing capacity. The K needs of several fruit and vegetable crops are very high and must be met from the early stages of crop growth. S is also normally applied before planting.

Special emphasis is needed on certain nutrients for specific soils and crops. For example, legume crops generally need only a small starter dose of N in spite of their high N requirement. This is because these crops are able to procure much of their N through N fixation where conditions favour adequate nodulation and N fixation. In many grain legumes, *Rhizobium* inoculation is a standard recommended input and is given through seed-coating before planting. For nutrients such as Fe and Mn, foliar application is far superior to soil application and their application needs to be timed with crop growth.

Common mistakes in nutrient management

The implementation of optimal plant nutrition is more difficult than generally assumed. As a result, deviations from the optimal supply frequently occur. In practical agriculture, owing to many uncontrollable variables, perfect implementation of scientific findings is rarely possible. Efficient nutrient management should start by avoiding common mistakes. Some suggestions for avoiding common mistakes in nutrient management are provided below:

- Maintain the soil in good condition as the basis for high NUE. Common mistakes include: overlooking too high or too low soil pH, inadequate organic matter, and poor soil structure.
- Apply adequate nutrients in order to achieve a realistic yield level. A common mistake is to strive for an unrealistic yield level. Where excess N is given for an unrealistic yield, a part of the N remains unutilized and may be lost.
- High yield levels are rarely reached on the basis of own practical experience alone. A common mistake is make insufficient use of available diagnostic techniques.
- Ensure a balanced supply of nutrients taking into account available soil nutrients. A common mistake is the overapplication or underapplication of some nutrients, e.g. part of NPK remains ineffective where there is S or Zn deficiency, and part of N remains unused where there is P deficiency.
- Check whether nutrients other than NPK, such as Mg, S and micronutrients, should be applied to a crop with high requirements. A common mistake is to overlook hidden hunger, which can limit growth and yield.
- Select the right kind of fertilizer material. A common mistake is the failure to consider the secondary effects of fertilizers, e.g. the S component for increasing the oil content in oil crops and protein content in legumes. In addition, acid-forming fertilizers can be used in high pH soils to bring the pH towards optimum and help in mobilizing deficient nutrients such as Mn and Zn.

- Use fertilizers with a low cost per unit of nutrients where they are equally effective. For example, per unit of P, TSP is cheaper than SSP (where S is not a limiting factor) although, TSP is more expensive than SSP on a per-bag basis. A common mistake is to cost fertilizers on a per-tonne or per-bag basis.
- Nutrients that benefit more than one crop through residual effects should be evaluated and costed differently to nutrients that do not leave a significant residual effect. A common mistake is to equate N and P in a similar manner in terms of their agro-economic response.
- Fertilizer use should give maximum net returns with a minimum benefit–cost ratio (BCR) of 2:1 – the higher the ratio, the better. A common mistake is consider only the BCR, disregarding the absolute net return.

The following sections discuss guidelines for nutrient management and application techniques separately for different nutrients and their sources. Chapter 6 has discussed crop recovery of applied nutrients. Here, after a discussion on the management of individual nutrients, guidelines are provided for the application and management of different sources of nutrients (fertilizers, organic manures, and biofertilizers).

GUIDELINES FOR NUTRIENT MANAGEMENT THROUGH FERTILIZERS

Nitrogen

N is a key nutrient in crop production. The action of N fertilizers on crop growth and yields is a summation of the efficiency with which it is utilized for crop production in terms of yield and quality. Because the correct use of N fertilizer is of great importance from both a production and environmental standpoint, important guidelines for efficient N use are provided here.

Selection and effect of different forms of N in fertilizers

For most crops, the N form (NH_4^+ or NO_3^-) is of minor importance although some plants appear to have a specific preference for one or the other. It might be expected that plants would prefer ammonium as it is directly usable for protein synthesis whereas nitrate must first be reduced to ammonium, which requires energy. For practical purposes, the two major N forms can be considered as largely equally effective. However, in view of its side-effect as a soil acidifier, ammonium is slightly superior in neutral soils where there are no gaseous losses of ammonia. The inferiority of nitrate in paddy rice is because of losses through leaching and denitrification. Nitrate can have an edge under moisture stress such as in dryland farming owing to its greater mobility.

A general shortcoming of most N fertilizers is their high solubility in the soil and rapid action compared with the much slower growth rate of crops. The practical solution to this lack of synchrony is repeated N application through splits during the growth season. Differences in the rate at which N is released play an important role in the selection of N fertilizers for soil application. Nitrate is effective immediately and free in the soil solution. Ammonium acts moderately quickly as, after exchange from charged surfaces, it can be taken up by the roots and, within a short time, soil bacteria can also transform it into nitrate. Urea acts somewhat more slowly because of the decomposition required to convert its amide form to ammonium, which is a temperature-sensitive process. Slow-release N fertilizers have a very slow and sustained action, which is useful for turf grasses, intensive gardening, greenhouses, high-value crops and for situations involving high N losses and environmental concerns, e.g. sandy soils, high-rainfall areas. Controlled-release N fertilizers, somewhat similar in action, employ techniques such as creating physical barriers through coating easily soluble granules with polymer films, resins, molten S, gypsum, and lac.

For most crops and cropping systems, the N form is of minor importance under good conditions of nutrient transformation and uptake in the soil. This means that the farmer can generally use the cheapest form of N. However, there are important exceptions. Under cold conditions in early spring, quick-acting nitrate fertilizers are superior to ammonium or urea fertilizers unless there is sufficient available soil N to meet the initial needs of the crop. With high temperatures, even urea is sufficiently quick acting except under dry conditions. In the case of acute N deficiency in growing crops, an instant supply of N is required. In such situations, the best option is foliar spraying with urea or N solutions, or top-dressing with nitrate. Under conditions favouring denitrification, as in rice fields, only ammonium or urea fertilizers should be used. In S-deficient fields, ammonium sulphate would in general be superior to S-free N carriers.

Rate of fertilizer N

The amounts of N to be applied depend on the difference between crop requirements and the supply of available soil N, which depends on mineralization of organic matter and residual N from the previous application. The rate of N is also modified by the inclusion of a legume in the system, and by the purpose for which the legume is grown (as a green manure, as an intercrop or as a grain legume in sequence cropping). Sometimes, a grain legume is raised for harvesting the green pods and its residues are ploughed in, which also contributes to the total N supply. Where insufficient N is applied, the expected yield will not be obtained. Where too much is applied, this will decrease the N-utilization rate, increase the danger of lodging in small cereals and lower the disease resistance of crops. Consequently, especially for intensive cropping, reliable diagnostic procedures are very helpful for supplementing the farmer's own experience. Towards this end, the LCC is finding acceptance as a guide to N applications for rice, maize and some other crops.

Timing of N application

Crops need a continuous supply of available N for high yields, especially during the rapid vegetative growth period. For the supply to be adequate before the periods of peak requirement, N fertilizer should be applied in good time in order to avoid even a temporary deficiency. Where a large single application is made to young plants before or at sowing time, this avoids any deficiency during the early growth stages. However, it may lead to initial oversupply with N lowering plant resistance to diseases, favouring early lodging, causing higher losses during wet periods, and often resulting in a short supply at the yield formation stage. Moreover, the total N requirement of the crop is difficult to assess with only one pre-plant N application. This approach can be used for fertilizing a dryland crop raised primarily on stored soil moisture.

When part of the total N is applied to young plants at the beginning followed by one or two supplementary N applications according to requirements, it results in higher distribution and labour costs. However, the N reserves of the soil are better utilized, transient deficiencies are avoided, and fertilization can be better adjusted to crop needs. The number of portions (splits) in which the total amount of N is to be applied depends on several factors, such as:

- type of crop and its duration;
- total N to be applied;
- soil texture;
- water availability;
- likely outbreak of pests and diseases;
- availability of labour;
- weather conditions.

Depending on the climate, soil moisture status and labour availability, the proportion of total N applied before sowing may range from a small starter dose to the full dose of N. As a general guideline, for irrigated cereals, not more than 30–40 kg/ha should be given at a time. For late N supplies intended to increase grain protein, foliar spraying with urea has proved effective in many situations. Under severe climate conditions, unusual application strategies may be required, such as the application of ammonia-N before winter for the following summer crop in order to facilitate early planting.

Method of N application

Fertilizers applied on the soil surface should reach the main rooting zone without delay and losses. On moist soils or areas receiving frequent rainfall, this is the case with most N fertilizers as they are all water soluble. However, top-dressed fertilizer granules of urea or ammonium-N may remain on the surface during dry periods and lose N as ammonia where exposed to sunshine on neutral to alkaline soils. Fertilizers such as anhydrous ammonia are injected at a certain depth in the soil with special equipment and precautions. For most crops, it is not necessary to place N fertilizers into the rootzone, the exception being crops raised on stored soil moisture. Deep placement of large USGs in the reduced zone of flooded-rice soils is an N-conserving technology that contributes to more efficient N use. Application methods such as foliar spraying or fertigation are covered in a later section.

Minimizing N losses

The purpose of efficient and profitable N application is to obtain a high utilization rate of the applied fertilizer nutrients by the crop in the first year itself by maximizing N uptake and minimizing losses. Losses of N are not only wasted fertilizer costs – they also have unwanted pollution effects. Losses can be kept below a tolerable level through appropriate crop–soil–water–nutrient management. Farmers tend to tolerate higher N losses where fertilizers are cheap or subsidized, which is not desirable.

Losses of N can potentially be reduced and N utilization by the crop increased by treating urea or ammonium-containing fertilizers with a nitrification inhibitor that delays the conversion of ammonium into nitrate, thus releasing less nitrate for leaching and/or denitrification. The first nitrification inhibitor was an organic compound called N-Serve [2-chloro-6(trichloromethyl)pyridine]. Generally, nitrification inhibitors have not proved successful under field conditions for large-scale application. However, favourable results under field conditions have been obtained in India by treating urea with the oil obtained from the seeds of the neem tree (*Azadirachta indica*), which have been shown to possess nitrification-inhibiting properties.

Secondary effects of N fertilizers

In addition to the direct effect of N as a nutrient, the influence of its positive and negative secondary effects should be taken into account. The main secondary effects are: the supply of other nutrients with the N, such as S, Mg, Ca and B; salt damage of young plants following the application of N close to the seedlings; damaging effects of minor constituents of urea, such as biuret during foliar spray; and the herbicidal or fungicidal effects resulting through application of fertilizers such as calcium cyanamide. The application of N fertilizers can bring about changes in soil reaction with associated nutritional effects. The conversion of ammonium into nitrate creates acidity because nitrification is an acid-forming process. At an assumed utilization rate of 50 percent N, the loss of Ca from the system owing to the application of various N sources would be: 0.4 kg CaO/kg N through CAN; 1 kg CaO/kg N through urea; and 3 kg CaO/kg N through to AS. However, there can be a gain of 1 kg CaO/kg N through calcium nitrate application.

Strong soil acidification as a result of N fertilizer application is a disadvantage in acid soils because this acidity must be compensated for by liming in order to maintain an optimal pH range for better nutrient availability and microbial activity. However, in intensive agriculture on high pH soils, the acidifying effect of N fertilizers may result in additional mobilization of nutrients such as Fe, Mn and Zn. This short-term acidification contributes towards a more balanced nutrient supply. Acidification of alkaline soils may be advantageous because it increases P supply by making calcium phosphate more soluble and also increases micronutrient availability.

Phosphorus

Selection of the appropriate P fertilizer

The choice of P fertilizer to be used depends on several soil factors, climate conditions, crop characteristics, economics and secondary effects of fertilizers.

In spite of numerous comparative studies made worldwide, no universally applicable advice can be given. However, some suggestions may be helpful. Water-soluble P fertilizers are best on slightly acid to neutral and alkaline P-deficient soils, particularly for short-duration crops with an immediate need of available phosphate. However, a high degree of water solubility can be a disadvantage in soils with strong P sorption where phosphate ions are transformed rapidly into less available forms. Phosphate forms with only moderate water solubility give the best results on moderate to slightly acid soils.

Slow-acting PRs require sufficient amounts of soil acidity and biological activity for conversion into easily available P forms. Their special advantage is their lower cost and a lower solubility, which decreases the rate at which the P is adsorbed in soils rich in active Fe or Al compounds. The use of very slowly acting PR is restricted to strongly acid soils and on perennial crops such as rubber, tea, and oil-palm. Thus, depending on the soil and crop situation, P fertilizers ranging from fully water soluble to zero water solubility can be utilized effectively.

The form of P is much more important on P-deficient soils than on those well supplied with P. The relative importance of higher water solubility decreases as the soil P status improves and the crop duration increases. Therefore, from a practical point of view, for cropping systems that have received an optimal supply of P for some years and P is needed mainly for the maintenance of an adequate P level, both the quick and somewhat slower-acting P forms can be equally effective. In spite of what is known about the effectiveness of various P sources, many farmers tend to buy the cheapest P source based on the price per unit of P_2O_5 and, sometimes, erroneously, even on the basis of price per bag. They should, for example, not be tempted to buy "cheaper" PR if it will not be effective under their conditions.

Rate of P application

This important aspect has been discussed together with diagnostic methods in Chapter 4. The general guideline is to decide the optimal rate of P based on soil fertility levels, response rates and the cost of P. There are two strategies for deciding the P application rate. First, on P-deficient or strongly P-sorbing soils, sufficient P is applied to meet the plant demand for low and medium yield levels. The second strategy is to raise the P level of the soil up to the optimal range and maintain it there by adding sufficient P to replace the P removed by the crops, a concept that has proved effective in sustaining high yields. Farmers can select the strategy based on whether they are interested in short-term response or long-term soil fertility buildup as well. The resources required for adopting the buildup plus maintenance approach are also an important aspect in decision-making.

Timing of P application

In order to make the best use of a P fertilizer, it should be applied according to its properties. Water-soluble forms must be applied at or before sowing time into the rootzone with as little as possible soil contact (granulated products or "placed" near the roots); top-dressing afterwards will have a delayed effect because of slow penetration into the soil. However, on P-deficient soils, a delayed application (up to one month after sowing) is better than no application at all, particularly where the desired P fertilizer is not available in the market in time. Phosphate fertilizers, such as powdered PR, that must be solubilized in the soil before they can furnish P for crop use, should be applied 3–4 weeks before sowing and well mixed into the topsoil in warm areas. For seasonal crops, incorporation of PR up to 10–15 cm depth following broadcast has been suggested (FAO, 2004a 2004b). Application in autumn is advisable for summer crops in temperate areas.

Methods of P application

The solubility and availability of soluble P fertilizer is better protected where there is restricted contact between soil and fertilizer. This happens where the fertilizer is concentrated locally in small zones near plant roots rather than being distributed evenly within the whole field. Minimizing soil contact means less and slower conversion into moderately available soil-P forms. Placement can improve the utilization of water-soluble P fertilizers by up to 25 percent in the first year, with the residual effect being hardly affected. In contrast, the best approach for insoluble fertilizers such as PRs is to maximize the soil fertilizer contact by spreading and mixing them with the whole topsoil.

Placement increases P uptake especially under: (i) low P supplies in the soil; (ii) dry periods or years; (iii) wide spacing of plants (e.g. maize); (iv) low rates of P application; and (v) plants with short vegetative growth periods (by enabling a rapid start of initial root growth). In contrast, the special efforts and costs of placement are hardly worthwhile with narrow-row crops in soils with good moisture conditions in humid regions. Special machinery can be used to place fertilizer around the seed (contact fertilization), alongside the seed (row fertilization) or underneath the seed (strip fertilization). Where specialized machinery is not available, placement can be achieved by ploughing and applying the fertilizer under the seed row before sowing.

Utilization of P fertilizers

Compared with N and K fertilizers, the recovery rate of P fertilizers by crops is low. About 15 percent of the P added is utilized during the first year, the range being 10–25 percent. The utilization of P by subsequent crops continues through residual effects, which may continue for a long time, reaching a rate of about 50 percent within 20–30 years. However, for economic reasons, only the residual effects of a few years can be considered (Chapters 6 and 9). For a better utilization rate of applied P, the fertilizer should be given directly to the most responsive crop in the rotation. For example, in rice–wheat or maize–wheat rotation, the best direct plus residual responses are obtained where P fertilizer is applied to wheat while the succeeding crop of rice or maize is allowed to feed on soil reserves and residual P. This is also because wheat is a winter-season crop and benefits more from direct P application as the low temperatures are not very favourable for adequate release of soil P.

In contrast to N, phosphate is rarely leached out of the soil. This is the primary reason why residual effects of P are more important than those of N. Where leaching does occur, the amounts are generally less than 1 kg P/ha and insignificant from a pollution point of view.

Secondary effects of P fertilizers

The selection of P fertilizers is not only a matter of the P form, it must also consider secondary effects. Some phosphate fertilizers also supply S, Mg, Mn and Si, while others have an enhanced soil-structure-improving capacity. Some P fertilizers decrease and others increase soil reaction, and some are superior in immobilizing harmful substances. For example, where SSP gives better yields of crops than does TSP, this may be because of the S supplied through SSP. Where Thomas phosphate (basic slag) is superior to SSP, this may be because of the additional liming effect or Mg supply.

Potassium

Selection of K fertilizer

The selection of K fertilizers is relatively simple compared with that of N and P fertilizers. All soluble K fertilizers are more or less similar with respect to their K-use efficiency. The main choice is between potassium chloride and potassium sulphate. For plants that are tolerant to chloride and whose quality is not impaired by high Cl, the cheaper potassium chloride (MOP) is preferred. For plants that are sensitive to high Cl for quality or other reasons, potassium sulphate or potassium nitrate is a better choice. Of the agricultural crops, potatoes and tobacco and many horticultural crops belong to the chloride-sensitive group. However, the Cl component is suitable for "salt-liking" plants, such as sugar beets and palms, and it brings extra beneficial effects. The K component of NPK complexes is similar to the K in straight fertilizers.

Timing and method of K application

It is a standard practice to apply the total amount of K just before sowing or planting by mixing it into the top layer. It is placed when the NPK complexes are drilled. At later growth stages, top-dressing on the soil surface is also effective. Where very high amounts are required, there may be some salt damage to young plant roots during dry periods. In order to avoid this, split applications are preferable. Split application of K together with N can be a useful strategy where leaching losses of K are considerable (as in sandy soils under high rainfall). Losses through leaching occur mainly in periods of high water penetration on sandy or peat soils with a low storage capacity. Placement of K is advisable in cases of single plant fertilization, e.g. trees and tea bushes. On most production sites, K losses are insignificant from both an agricultural and an environmental viewpoint.

Secondary effects of K fertilizers

Several K fertilizers also provide other nutrients that can have a beneficial effect on crop yields and produce quality, excluding Cl. Potassium sulphate also contains S, which can be useful on S-deficient soils and for high S-demanding crops. For crops with a high Mg requirement or on Mg-deficient soils, fertilizers with a combination of K and Mg are recommended, potatoes being a typical example. In such situations, potassium magnesium sulphate can be used. Potassium nitrate also provides readily available N and is a preferred source for several horticultural crops. The chloride component of MOP is particularly useful in the nutrition of sugar beets and palms. On grassland, the Na in K fertilizers can be of benefit to grazing animals. In some countries, Na is considered an impurity and a maximum permissible limit is set.

Sulphur

S can be applied to the soil through any suitable S carrier. The choice depends on: crop, local availability, price and the need for other nutrients. All sulphate sources are generally equally effective as they contain S in the water-soluble, readily available sulphate form. S is applied automatically where sources such as AS, SSP or APS are used to provide N, P or N + P. Rates of S application generally range from 20 to 50 kg S/ha depending on the S status of soil and crop demand. Higher rates are generally needed on sandy soils and for oilseed crops. In most cases, S is applied at or before sowing along with N, P, K or Zn when two nutrient fertilizers are used. Where sulphate salts of micronutrients are used to correct specific micronutrient deficiencies through soil application, the S added through them should be taken into account in deciding the total rate of S to be applied. However, such materials cannot be selected to supply S where their micronutrients are not required.

Where elemental S or pyrites are used, these should be applied 3–4 weeks ahead of planting through surface broadcast on a moist soil followed by mixing. This allows sufficient time for the insoluble S in them to be converted to the plant available sulphate form. The rate of oxidation of elemental S is controlled by: the particle size of the material; temperature; moisture; and the degree of contact with the soil. S in materials of finer particle size oxidizes at a rapid rate. Where S deficiency is noticed in a growing crop, this can be corrected by providing a top-dressing with ammonium sulphate, or a suitable liquid S fertilizer can be given as foliar spray. Where the S application rates are medium to high, a significant residual effect can be expected.

Calcium

Several Ca fertilizers have been described in Chapter 5. Specific fertilization with Ca is not often needed as most soils have a satisfactory status of available Ca. Significant amounts of Ca are applied where acid soils are limed with calcium carbonate or with dolomite. Ca is also delivered wherever gypsum is applied as an amendment or as a source of S, and where N is provided through CAN. In many areas, gypsum application to groundnut is specifically recommended in order to meet the high demand for Ca during pod formation. It should be applied in furrows.

The rate of Ca application may vary from zero for cereals on calcareous soils to 500 kg Ca/ha for bananas under humid tropical conditions. To correct Ca deficiency in standing crops, foliar sprays with water-soluble materials such as calcium chloride or preferably calcium nitrate can be given. In many apple-growing areas, e.g. in South Africa, it is common to use $CaCl_2$ sprays (0.5 percent) or calcium nitrate (0.65 percent) at 40–45 days after flowering to avoid the occurrence of "bitter pit" (FAO, 1992).

Magnesium

Mg application is more widely recommended than that of Ca. Fertilizers containing Mg have been described in Chapter 5. Sufficient Mg is added where acid soils are limed using dolomitic limestone. Most fertilizers containing magnesium sulphate are equally effective as sources of Mg. In very acid soils, especially under plantation crops, the mineral magnesite can also be used to apply Mg. For cereal crops on acid soils, the rate of Mg application can range from 10 to 50 kg Mg/ha depending on the Mg status of the soil and crop needs. Higher rates of 30–120 kg Mg/ha are recommended for grasslands in order to avoid grass tetany in animals. For high-yielding crops in the tropics, some recommended rates are (in kilograms of Mg per hectare): pigeon pea 18; rice, cotton and coffee 20; cassava, maize, potatoes and pineapple 30; yams 34; sugar cane 35; and bananas 50 (FAO, 1992).

Mg fertilizers can be applied to the soil or given as foliar spray. The readily water-soluble Epsom salts ($MgSO_4.7H_2O$), magnesium chloride and magnesium nitrate are used as foliar sprays either to prevent losses in yield and quality caused by to acute Mg deficiency or as part of the regular fertilizer schedule.

Boron

Common sources of B have been described in Chapter 5. Most B fertilizers are soluble borates. Various borates differ in their B content depending on the amount of water in their structure. Slow-release boron frits have a longer-lasting effect than soluble sources. They are particularly suited for sandy soils and high-rainfall areas to reduce leaching losses of B. Because of the small quantities involved and in order to ensure uniform application, B is sometimes applied through boronated fertilizers. A wide range of boronated fertilizers are produced around the world.

In order to avoid any chance of toxicity, B should be applied only where its deficiency has been confirmed. The recommended rates on B-deficient soils for most crops range from 0.5 to 2 kg B/ha. Higher rates of 2–6 kg B/ha are indicated for almonds, grapes and walnuts (Shorrocks, 1984). B can be applied to the soil or through foliar spray. Soil application is generally given before sowing. Higher rates of B application are more appropriate for broadcast application, whereas lower rates would be more suitable for side-dressing. In all cases, direct contact of the fertilizer with the seed should be avoided. The concentration of B in spray solution can range from 0.1 to 0.5 percent but should be decided on the basis of on local conditions.

The application of B fertilizers poses more problems than other micronutrients because of the highly different requirements of crops in a rotation. Crops with a high demand should be well supplied with B especially for high yields but not excessively, because a following crop that has a low B requirement may be damaged instead of being nourished by residual B.

Chlorine

Chloride is rarely applied deliberately although it is delivered wherever chloride-containing fertilizers such as MOP, calcium chloride and MOP based NPK complexes are used. It is a nutrient to be kept in mind where fertilizing palms on sandy soils or sites away from the sea. Practical recommendations for the application of chloride to coconut and oil-palm are available (IFA, 1992). For coconuts under Malaysian conditions, the rate of application ranges from 0.11 kg Cl/tree at an age of 6 months and increasing progressively to 0.9 kg Cl/tree. Oil-palms are considered to be deficient in Cl where their leaves contain less than 0.25 percent Cl in the dry matter.

Copper

Cu can be applied through a variety of inorganic salts and chelates. These have been discussed in Chapter 5. Cu application should normally be based on the available-Cu status of soils. Both soil applications and foliar sprays are suitable. A single pre-plant soil application can be effective for several crops grown in succession, and each crop need not receive Cu fertilizer except on organic soils. For soil application, the rates of Cu applied vary widely from 1 to 23 kg Cu/ha (Shorrocks and Alloway, 1988). Normally, recommended rates are 1.5–4.5 kg Cu/ha where banded, and 3–6 kg Cu/ha where broadcast (FAO, 1983).

Because Cu is complexed strongly by SOM, the amount applied (5–10 kg Cu/ha) is high compared with plant requirements. A single application is sufficient for several crops. Application rates are lower on sandy soils or those with a low organic matter content. Cu fertilizers leave a significant residual effect on the following crops, hence, there is no need for annual applications. Cu fertilizers should be well mixed with the topsoil. On grassland, they penetrate only slowly into the soil.

The commonly advocated concentration for spray application is about 0.025 percent Cu (100 g Cu/ha as copper sulphate, equivalent to 400 g $CuSO_4.5H_2O$). However, some specialists do not advocate the use of copper sulphate for foliar spray because it can be phytotoxic even at low concentration and can also corrode the spraying equipment (Shorrocks and Alloway, 1988). To save on application costs, foliar sprays of Cu can be carried out using chelates and oxychloride of copper, which are compatible with many agrochemicals and can, therefore, be applied with a fungicide or a herbicide. Spray application has the advantage of delivering Cu directly to the plant, which is not the case with soil application if Cu is strongly adsorbed in unavailable forms. In some cases, dusting of maize seed with copper sulphate or soaking of oat and vetch seed in 1-percent solution of copper sulphate has also been found to be effective.

Iron

Iron chlorosis is considered to be one of the most difficult micronutrient deficiencies to correct in the field (Tisdale, Nelson and Beaton, 1985). A number of Fe fertilizers have been described in Chapter 5. The most common fertilizer for soil application is ferrous sulphate. However, the soil application option is generally not preferred owing to the rapid oxidation and immobilization of the ferrous to ferric iron in the soil. Rates of ferrous sulphate applied to the soil range from 20 to 100 kg/ha of $FeSO_4.7H_2O$ (19 percent Fe). The efficacy of soil-applied ferrous sulphate improves where it is mixed with an organic manure and applied.

The commonly recommended method of Fe application is through foliar sprays either as inorganic salts or preferably through chelates of Fe with EDTA, EDDHA, etc. The Fe-EDTA chelate is useful only in slightly acid soil while Fe-EDDHA is unique as its stability remains constant over a wide pH range of 4–9. Where ferrous sulphate is used for foliar spray, its concentration ranges from 0.5 to 2 percent. The sprays have to be repeated several times at 10–15-day intervals. In calcareous soils, Fe availability can be increased by using acidifying materials such as elemental S wherever its use is economic.

Manganese

A number of Mn fertilizers are available (Chapter 5). As with Fe, foliar application of Mn is generally more effective than its soil application. For soil application, manganese sulphate is a superior source of Mn compared with other sources. However, soil application is generally uneconomic owing to the conversion of applied Mn into insoluble forms. In spite of being only slightly water soluble, manganese oxide can be a satisfactory source of Mn. It must be finely ground in order to be effective. Mn deficiency induced by liming or high pH can be corrected by soil acidification, e.g. by the use of elemental S or by applying Mn fertilizer along with AS.

The rate of Mn application varies from 1 to 25 kg/ha. The lowest rates are for foliar spray and the highest rates pertain to soil application by surface broadcast. When Mn fertilizer is banded, usually half the rates for broadcast application are needed. For foliar application, Mn can be applied either through a 0.5–1.0-percent solution of $MnSO_4$ or through a suitable chelated compound. For wheat in Mn-deficient soil, the recommendation is to give one spraying of 0.5-percent $MnSO_4$ solution (at a per-hectare rate of 2.5 kg $MnSO_4$ in 500 litres of water) 2–4 days before irrigation followed by 2–3 additional sprays at weekly intervals on sunny days. The natural organic complexes and chelates of Mn are best suited for spray application.

An alternative to adding Mn to the soil is to improve Mn availability by: using acidifying-N fertilizers such as AS; compacting loose soils; and preventing excessive soil drying. All of these measures favour reducing conditions that produce plant available Mn^{2+} ions. However, such practices may reduce the availability of other nutrients.

Molybdenum

Mo is required by crops in the smallest amounts of all micronutrients. A number of fertilizers containing Mo have been described in Chapter 5. Rates of Mo application are generally very low, ranging from 25 to 150 g Mo/ha. It can be applied to the soil, given through foliar spray or through seed treatment. The optimal rate of Mo depends primarily on the soil, the crop and the method of application. In order to obtain satisfactory distribution of the small amount Mo applied to soil, Mo fertilizers are sometimes combined with multinutrient fertilizers. For example, in Australia, MoO_3 is incorporated into PR pellets (Tisdale, Nelson and Beaton, 1985). Mo can also be applied through SSP fortified with 0.05 percent Mo. In the case of strongly acid soils, the amounts need to be doubled. Mo can also be applied to the seed, to the nurseries or by soaking seeds in a solution of Mo fertilizer. Mo fertilizer may not be required where the soil supply is improved by liming, loosening and better drainage.

Zinc

Among micronutrients, Zn deficiency is perhaps the most widespread. Zn can be applied through a number of inorganic and chelated compounds (discussed in Chapter 5). Zinc sulphate is the most commonly used source of Zn. Soil application rates of Zn are typically in the range 4.5–34 kg Zn/ha in the form of zinc sulphate (broadcast or sprayed in an aqueous solution onto the seed bed). Higher application rates are often used for sensitive crops, such as maize, on alkaline and/or calcareous soils as opposed to for maize on non-calcareous soils (Alloway, 2004). In India, where Zn deficiency is a widespread problem, soil application of 5 kg Zn/ha is advised on coarse-textured soils, and 10 kg Zn/ha on fine-textured soils. One application can last for 3–6 crops.

In the rice–wheat rotation, where Zn availability is low, the application of Zn to rice is more profitable. In Brazil, 5–7 kg Zn/ha through zinc sulphate is generally used to correct Zn deficiency in both lowland (paddy) and upland rice. The amount of Zn required to be applied to a wetland rice soil depends on soil characteristics, source of Zn, severity of Zn deficiency, and variety of rice to be grown. Generally, 10 kg Zn/ha as zinc sulphate or root dipping in 2-percent zinc oxide is adequate for most situations (Neue and Mamaril, 1985). Application of Zn to the floodwater or to the soil surface has been found to be more efficient than its incorporation into the wetland soil.

As with most crops, the normal way of correcting Zn deficiency in wheat soils is to surface broadcast a Zn compound, usually zinc sulphate at 5–20 kg Zn/ha to the seed bed and incorporate into the topsoil. Where the Zn fertilizer is to be banded (placed to one side and below the seed in the row), then a lower rate of 3–5 kg Zn/ha is used. For foliar applications (usually of a chelate such as Zn-EDTA), an even lower rate of 0.015–0.25 kg Zn/ha is used. In order to correct Zn deficiency in a standing crop, the crop can be sprayed with a 0.5-percent solution of zinc sulphate (0.5 kg of zinc sulphate in 100 litres of water). Before spraying, 250 g of unslaked lime (0.25 percent) should be added to the solution in order to neutralize the acidity of the zinc sulphate (Gupta, 1995).

Because of the small amount of Zn required, special procedures have been developed, e.g. dipping roots into zinc oxide slurry/paste, and hammering a zinc nail into a Zn-deficient tree so that the sap may dissolve some of the Zn and take it up. Other alternatives include dipping the roots of rice seedlings in a 1-percent zinc oxide suspension before transplanting or mixing zinc oxide with pre-soaked rice seeds before direct seeding. Dipping potato seed tubers in 2-percent zinc oxide suspension is also effective. The high seed rate (3.0 tonnes/ha) of potato makes it possible to supply the micronutrient needs of potato through soaking.

GUIDELINES FOR FERTILIZER APPLICATION

Basic aspects of fertilizer application

Recommendations for the application of nutrients are generally made on a nutrient basis (Chapter 8). However, these are never applied as nutrients but in the form of specific products such as fertilizers and manures. Various sources of plant nutrients have been described in Chapter 5. The method of application of fertilizers and other nutrient sources is a very important aspect of nutrient management. At the field level, this also means fertilizer management. Fertilizers containing the same nutrient differ markedly not only in their chemical properties and nutrient content but also in their physical characteristics. All of these determine the method of fertilizer application. Thc crop, soil and available equipment and labour are equally important.

The objective is to apply a fertilizer in such a way that the nutrients in it contribute as much as possible towards crop production. This can be accomplished by ensuring that fertilizers remain in the active rootzone, improve the soil fertility and produce minimum negative effects on the environment. A prerequisite of correct fertilizer application is its uniform distribution over all the treated area whether it is surface broadcast or applied in a restricted manner. The method of application should follow the research findings about the most suitable technique for a given soil and crop situation. The following section deals primarily with solid fertilizers. The methods of fertilizer application in general have also been described in FAO/IFA (2000). Liquid materials are discussed in a later section. Guidelines for the application of organic manures and biofertilizers follow this section on mineral fertilizers.

Multinutrient fertilizers vs single-nutrient fertilizers

Farmers want fertilization to be effective, simple and cheap. This can be achieved through the use of straight fertilizers or suitable complexes. In the case of straight (single-nutrient) fertilizers, a separate fertilizer has to be purchased for each nutrient to be applied (urea for N, TSP for P, MOP for K, $ZnSO_4$ for Zn, etc). Where a suitable multinutrient product is available in which the ratio of nutrients is close to or similar to the ratio of nutrients recommended, then one fertilizer can do the job. For example, where agronomically suitable, a 15–15–15 complex can provide any amount of NPK if these are to be applied in a 1:1:1 ratio, or a product of the grade 20–20–0 can deliver N and P if recommended in equal (1:1) amounts.

For SSNM, a multinutrient fertilizer that matches the exact nutrient needs of a field is very often not available. In such cases, either separate single-nutrient fertilizers are selected or a tailor-made mixture or bulk blend is prepared. In many situations, a suitable multinutrient fertilizer can be selected for the basal dressing followed by a straight fertilizer for top-dressing.

Both approaches of whether to prefer single-nutrient carriers or multinutrient products have their advantages and drawbacks. The use of single-nutrient fertilizers often provides flexibility, lower cost per unit of nutrient and the advantage of applying only those nutrients that are needed and will generate an economic benefit. However, this approach involves purchasing, handling and applying several materials and possibly making mistakes in computing the quantities of fertilizers required to deliver the desired nutrient rates. Mistakes can also occur while mixing different fertilizers not only in terms of quantities but also in terms of compatibility.

Multinutrient fertilizers have their special advantages, especially with bulk blending and on-farm mixing. Of the economic arguments, the difference in price per nutrient unit is often decisive. Where single-nutrient fertilizers can be obtained more cheaply, there is a strong incentive to use them and either to distribute them separately, mix them on the farm before application or to make use of cost-effective bulk-blending facilities. Where the farmers are not adequately trained but their soils need the application of several nutrients, they should apply a suitable multinutrient fertilizer rather than deciding and purchasing separate fertilizers for each nutrient needed. Chapter 5 includes some guidelines for the handling, storage and mixing of fertilizers.

In view of the multitude of soils and cropping systems under cultivation, only a few suggestions can be provided here for the application of multinutrient fertilizers. In general, the grade to be selected should come closest to delivering the nutrients in the ratio recommended for the crop. Otherwise, a suitable combination can be sought. For example, application of 40 kg each of N, P_2O_5 and K_2O can be made by: (i) selecting separate fertilizers for each nutrient; (ii) a 1:1 N:P_2O_5 complex plus a straight K source; and (iii) a 1:1:1 N:P_2O_5: K_2O complex or blend.

As N is the component most liable to loss, these fertilizers must be applied with an eye on high nitrogen-use efficiency. PK fertilizers or NPK types with little N are very useful for provide a good initial supply allowing N to be applied later according to the special crop needs. Special soil nutrient supplies will also influence the choice of fertilizers. On a soil especially rich in available K, NP fertilizers will be the choice, whereas NK fertilizers are the right choice on soils rich in phosphate. Most multinutrient fertilizers have an acidifying influence on the soil reaction, similar to N fertilizers.

The choice between solid, liquid and gaseous fertilizers depends on factors such as economics, efficiency and ease of operation, and on whether fertilization and crop protection can be partly combined. These can generally be evaluated according to farm-specific conditions.

Size of fertilizer particles

Theoretically, fine, powdery material mixed thoroughly into the topsoil layer would result in the most uniform distribution within the rootzone. However, this is not always so and it is often too costly. The use of granular, water-soluble fertilizers represents a compromise between uniformity of distribution and ease of application. The granule size of water-soluble fertilizers is generally standardized so that 90 percent of the granules are 2–4 mm in diameter. Large granules have the advantage of a reduced immobilization, which is especially important for phosphates. Very large supergranules of 1–2 g are sometimes used for placement in rice and for trees.

Because water-insoluble fertilizer granules would release nutrients too slowly, they are granulated in such a way that powdery material is only bound loosely. Thus, in moist soils, the granules disintegrate rapidly. In all cases, the granules must be sufficiently stable to withstand transportation and spreading. When the granulated fertilizer disintegrates into powder in the soil, it should have close contact with soil particles in order to achieve the necessary mobilization (Figure 38).

FIGURE 38
Penetration and incorporation of fertilizer nutrients into the rootzone

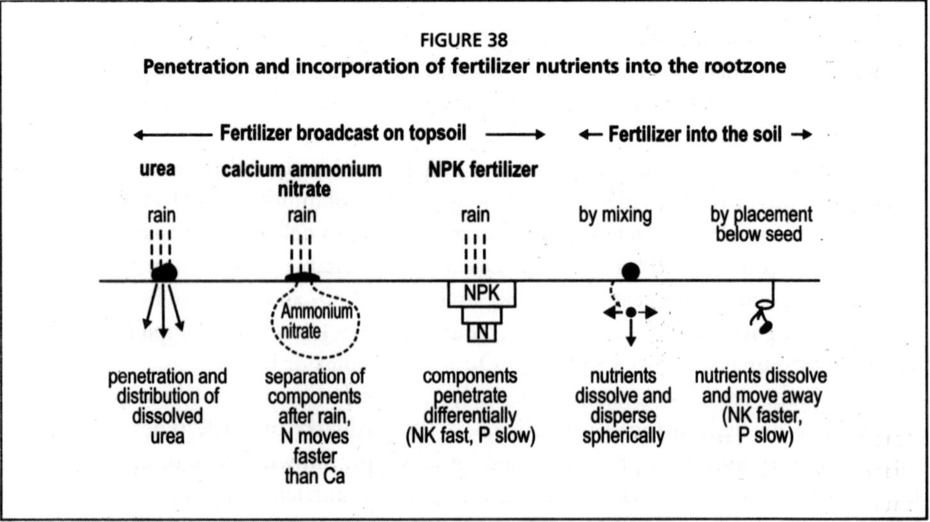

Source: Finck, 1992.

Fertilizer distribution on the soil surface

The application of granulated fertilizers on the soil surface is the easiest and most common procedure. The fertilizer granules should be distributed as uniformly as possible in order to supply each plant with nutrients in more or less equal amounts. This is not an easy task. Experienced farmers are able to spread fertilizers by hand with considerable accuracy but mechanical distribution is superior in most cases. The difficulty of hand spreading uniformly 120 kg N/ha through a standard NPK fertilizer requires the distribution of 24 million granules per hectare (2 400 granules/m^2).

Non-uniform fertilizer distribution is a sign of faulty application. It results in some plants receiving too little or too much nutrient within the same field. The deviation from uniformity should not exceed 10 percent. The principle of homogeneous distribution on the whole field has its limitations where the soil in the field has variable nutrient status. In such cases, precision fertilization is required (discussed below).

Penetration of surface-applied nutrients into the rootzone

Fertilizers spread on the soil surface, whether bare soil or with plant cover, will penetrate slowly into the top layer if they are water soluble and if there is sufficient moisture. Dryness after fertilization results in a delay in fertilizer nutrient uptake because the applied nutrient cannot be transported to the roots owing to inadequate moisture. Water-insoluble fertilizers such as PRs or elemental S products need to be mixed into the rootzone after application on the surface. The incorporation of insoluble fertilizers applied to grassland is generally left to slow mixing by soil fauna. Because this is a slow process, a good supply of nutrients should be given during seed-bed preparation or at sowing.

During the penetration process, fertilizer components of different solubilities in the same product separate. For example, in the case of calcium ammonium nitrate, the $CaCO_3$ remains on the surface much longer than does the easily soluble ammonium nitrate. Once in the soil, the nitrate moves more quickly than does the ammonium. In the case of an NPK complex fertilizer, the N component moves more quickly than the K and much more quickly than the P (Figure 38).

Placement of fertilizers

Placement usually means positioning the fertilizer in a desired region or depth at sowing, either at the side or below the seed. It is normally done where the entire field is not to be treated or where restricted soil fertilizer contact is desired, as in the case of highly water-soluble but relatively immobile nutrients such as water-soluble phosphates. Placement is also the preferred method of fertilizer application for crops planted in widely spaced furrows, e.g. maize, potato, sorghum, sugar cane and pineapple (except for bushes and tree crops). Fertilizer placement generally results in a better rate of nutrient utilization by the crop and, thus, higher NUE compared with a broadcast application. It is an also effective

FIGURE 39
Fertilizer placement by different methods (with seed in rows or banded)

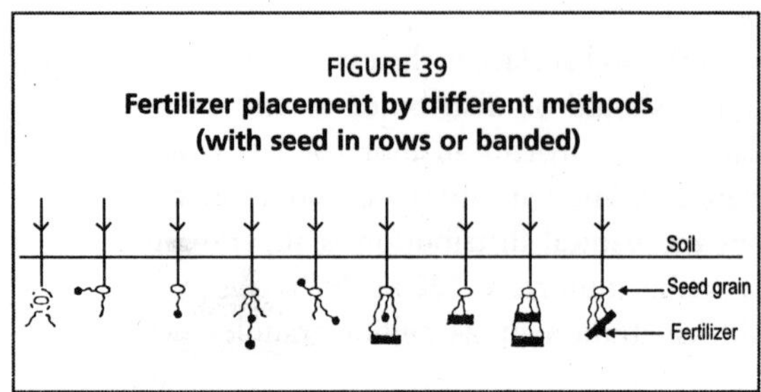

Source: Finck, 1992.

method under resource constraints where small rates are to be applied in soils of low to medium fertility.

Where placed beside the growing plants under wide spacing, the fertilizer is termed a side-dressing. Placement is suitable for all nutrients, but best results are obtained with N and phosphate in fields with wide-row crops. The benefit of placement is greatest at low rates of application and during early growth in periods of dry or cool weather when nutrient uptake is impeded. Its key advantage is that it places the nutrients in the rootzone where they are needed. Its main disadvantage is the higher cost of application.

Placement can be profitable for small cereals in dry areas, but for wheat in humid areas it would hardly justify the extra costs. For micronutrients, placement can take the form of seed treatment, which provides a good initial supply as for example with Mo fertilizers. When roots of rice seedlings or potato seed tubers are dipped or soaked in nutrient solutions before planting, this also results in a kind of placement in the rootzone. Fertilizer placement requires combined sowing and fertilizing machines that place the fertilizer in different ways below or next to the seed (Figure 39).

Fertilizer placement is generally made at sowing time or soon after in a number of ways:

- in a band a few centimetres to the side and below the seed;
- in a band directly below the seed, although this may hinder growth of the tap-root;
- in immediate contact with the seed, termed combine drilling (only in moist soils and mainly with phosphate as close contact with N may damage the seed);
- in one or two bands on one or both sides of plant rows;
- by spot application between plants as in the case of USGs between rice hills or as in the case of ring placement around trees.

Application equipment for solid fertilizers

The main problem with fertilizer application is non-uniform distribution in the field. Compared with the widely used and tedious spreading of fertilizer by hand, mechanical distribution is labour-saving and more precise. However, it should also be cost-effective. Precise and more expensive spreading procedures may be worthwhile for expensive fertilizers used to produce high yields on medium–large farms. The amount of fertilizers to be spread ranges from about 50 kg/ha to more than 1 500 kg/ha. The cost of distribution can range from 10–20 percent of the total fertilizer costs.

The requirements for suitable mechanical distributors are:

- delivery of exact rates;
- uniform distribution of the fertilizer with a deviation of less than 10 percent;
- distribution to be independent of slope and speed;
- ease of handling, operation and maintenance;
- resistance to corrosion;
- energy efficient.

Beyond the simple box-type of distributors, there are ejection distributors and high-precision distributors. They all have their advantages and limitations.

Box distributors

Box distributors with a width of 2–5 m operate with a simple mechanical system of moving chains, rotating plates or a moving lattice. They can be adapted to apply both granular and fine-grained fertilizers. However, they have only a small capacity and can only be operated at slow speed, which limits their use.

Ejection distributors

Ejection distributors (centrifugal spreaders) operate on the principle of ejecting fertilizer granules by using centrifugal force either by spinning discs or by oscillators. The simplest spinning-disc equipment operates with one disc that spreads granular fertilizers with an acceptably uniform distribution. Those with two counter-rotating discs or oscillating-spout distributors provide even better distribution. Such distributors are also suitable for fertilizers with a finer particle size. Spinning-disc types are the most common ones for cheap and relatively uniform fertilizer spreading. The fertilizer is metered from a hopper onto a rapidly spinning disc and flung laterally to a width of about 10 m on each side (Figure 40). They cover wide strips of the field at a reasonable speed and accuracy. About half of the strip receives the full amount of fertilizer whereas

FIGURE 40
Equipment for fertilizer distribution

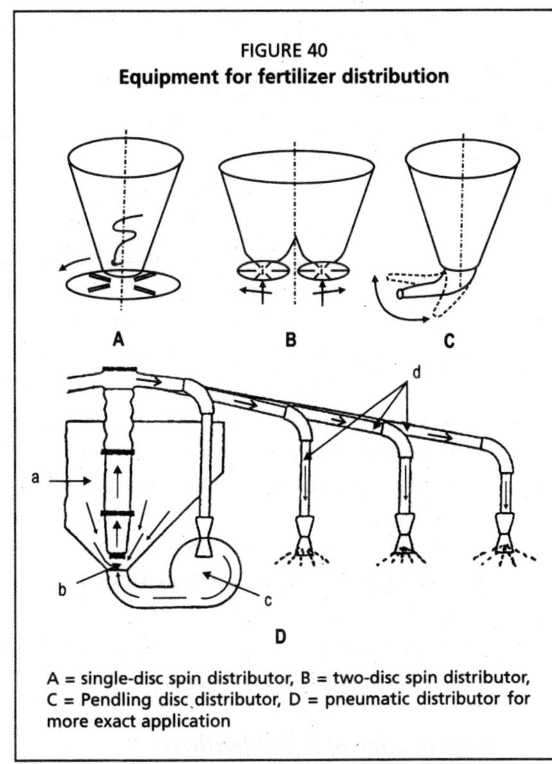

A = single-disc spin distributor, B = two-disc spin distributor, C = Pendling disc distributor, D = pneumatic distributor for more exact application

Note: (a) = injector gate, (b) = air blower, (c) = outlet pipes, (d) = delivery points.
Source: Finck, 1992.

towards both ends the amounts decrease. This gradient is compensated for by a system of overlapping in order to obtain full uniformity for the whole strip.

The distributors are either connected to a tractor with a container volume of 300–1 000 litres or have their own container, which can hold up to 4 000 litres. The rate of fertilizer distributed ranges from 50 to 2 000 kg/ha. The accuracy of distribution is usually about ±10 percent, up to a maximum of ±20 percent. As medium accuracy suffices for most purposes, broadcasting with simple types of spinning-disc distributors is very common.

High-precision distributors

For more accurate and precise distribution of fertilizers with varying physical characteristics, pneumatic types of distributors are preferable. However, they are much more expensive. In these distributors, the granules are transported through tubes by air pressure and finally blown on small plates about 1 m apart. The result is a semi-circle distribution with good overlap. Such machines cover a width of up to 15 m or more and the container volume ranges from 1 000 to 2 000 litres. They can deliver fertilizer at rates ranging from 30 to 2 000 kg/ha and they are suitable for fertilizers of average granule size, for mixtures and also for small granules and urea prills. They provide a sufficiently uniform distribution.

Aerial application of fertilizers

An increasing amount of fertilizer is distributed by aircraft. However, this method is generally more expensive than other methods. Large areas can be fertilized in a short time, especially at low fertilizer rates. The method is applicable in difficult terrain, be it paddy fields or steep mountain areas. One advantage of aerial application over normal soil application is that the wheels of vehicles cause no soil compaction or damage to crops. However, the method has little practical feasibility for smallholders in developing countries. Aerial application requires careful and precise marking of application areas in order to avoid accidental contamination of open waters. The maintenance and marking of buffer areas around watercourses and water bodies (to avoid drift or accidental application of fertilizers directly to surface water) is mandatory in certain countries. Aerial applications have to be done during favourable atmospheric conditions when the likelihood of significant drift is lowest.

Application of liquid and gaseous fertilizers

Several liquid and gaseous fertilizers have been described in Chapter 5. Some of these require special application techniques while others can be sprayed on the leaves with conventional sprayers.

Application of liquid fertilizers

Liquid fertilizers serve two different purposes, either to supply nutrients to the soil or to provide direct nutrient supply to plants through foliar sprays. Fertilizer solutions provide for better soil transport and distribution of nutrients compared with granular fertilizers. Suspensions, which are concentrated solutions with small suspended solid particles, usually have higher nutrient concentrations than do liquid solutions. Both require solid and corrosion-resistant tanks or silos for storage and transport, good safety measures and special application equipment. Liquid fertilizers that can be applied on bare soil or on soils covered with plants include: fertilizer solutions, fertilizer suspensions and organic materials such as animal slurry. These can be materials containing one or more nutrients including macronutrients and micronutrients.

Application of liquid fertilizers on bare soils is made through special nozzles spaced about 50 cm apart and operating at pressures of 100–300 kPa that deliver relatively large drops. Being turbid liquids, suspensions require special nozzles that do not become blocked by the small solid particles. Different travel speeds and discharge rates that can be regulated from 0.5 to 4 litres/minute permit the application of 10 to 300 kg N/ha. Position markings are required in order to avoid overlapping. Concentrated solutions or suspensions cause no osmotic problems on bare soils. This is because they enter the topsoil layer through pores and are diluted by the soil moisture.

The application of concentrated fertilizer solutions through a canopy of young plants can cause serious osmotic damage. Therefore, the solution should be diluted 2–3 times with water so that the leaves can tolerate the osmotic stress. An alternative method is to apply through dropper tubes, which deliver the solution on the soil surface under the crop canopy. Driving on well-defined wheel paths is the best guarantee for properly joining the individual fertilizer strips.

Aqueous ammonia may lose ammonia through evaporation. Therefore, it should be applied into the soil by special machines. The problems encountered are similar to those with slurry application. Liquid fertilizers are very suitable for injection fertilization into deeper layers for trees by using special lances with fixed top and lateral nozzles.

The application of liquid fertilizers to soils has advantages and disadvantages:

- ➢ advantages:
  - application of dissolved and, thus, immediately available nutrients,
  - simple filling procedure of containers by pumps (labour-saving),
  - very precise fertilizer distribution (superior to spreading of solids),
  - large area can be fertilized in a short time (5–10 ha/hour),
  - fertilization can be combined with compatible crop protection sprays;
- ➢ disadvantages:
  - nutrients in soluble forms (liquids) are generally more expensive than those in solid forms,
  - large amount of water must be transported,
  - complete fertilization is rarely possible, hence, application of solids is also needed,
  - transportation and storage requires expensive tanks and safety measures,
  - nozzles must be corrosion-resistant,
  - handling is generally more expensive than with solid fertilizers.

Application of gaseous fertilizers

In practice, fertilization with gases is restricted to anhydrous ammonia. It is a widespread practice in countries with large farms, a low ammonia price and high cost of solid N fertilizers (e.g. the United States of America). Anhydrous ammonia is applied from pressurized tanks. It leaves the distributing device as a gas after the pressure has been released and enters the soil as a gas. The problem with its application is in correctly dosing the liquefied gas at a pressure of about 1 000 kPa from the field tank with the aid of pumps and allowing for the speed of travel, temperature, etc. Pressurized ammonia is subject to special safety regulations concerning the strength of containers and pipelines, corrosion damage, possible injury to the operators, and toxicity of the gas.

Anhydrous ammonia is best applied into bare soil. It must be injected sufficiently deeply into the soil in order to avoid losses by evaporation. This is minimized by devices with special injection prongs that disturb the soil as little as possible, so that no opening at the surface is left. It can also be introduced into the soil by lances as in the case of liquid fertilizers. A precondition for this is to maintain the soil at medium soil moisture level, i.e. the soil must be neither too wet nor too dry.

Foliar fertilization

Leaves absorb nutrients as a natural process by which plants obtain additional nutrients from rainwater. This principle is utilized in agriculture by spraying the foliage with dilute solutions of the desired nutrients. Foliar fertilization is generally recommended for supplying additional N, Mg and micronutrients, but it can also be used to provide P, K and S.

Role of foliar fertilization

In practical farming, foliar fertilization is used as a quick remedy for unexpected deficiencies, for late supply of N during advanced growth stages, as a preventive measure against unsuspected (hidden) deficiencies, and to overcome fixation of nutrients in soils (e.g. Cu, Fe, Mn and Zn). The main advantage of foliar fertilization is the immediate uptake of the nutrients applied. Its shortcoming is the limited amounts that can be supplied. Nutrients present in inorganic salts or in chelated forms can be used for foliar application. The materials suitable for foliar fertilization have been described in Chapter 5.

For foliar application to be effective, a substantial amount of the deficient nutrient must be added, but it should not cause plant damage, leaf scorching, and negative osmotic effects. The solutions must be dilute (1–2 percent), especially if they contain nutrient salts. Foliar fertilization is at the best a supplement to soil application and not a substitute for it. Crops are less sensitive to organic compounds because they have only a slight osmotic action. Therefore, urea is better tolerated by leaves than is nitrate or ammonia and it enables the application of concentrations up to 15 percent with low-volume sprayers. Where urea is used for foliar sprays, it should contain no more than 0.25 percent biuret. The same applies to micronutrient sprays through chelates vs inorganic salts.

With the exception of N, foliar application can supply only very limited amounts of the major nutrients such as P and K compared with their total requirements. The situation is a little better for Ca, Mg and S, but even these can be added only in limited amounts, which are often insufficient in a single application. The best results are obtained with micronutrients because a relatively large portion of the total requirement can be supplied in a single spraying. In cases of marked deficiencies or mobility problems within the leaf, repeated sprayings with micronutrients are essential, as in the case of Fe and Mn. Foliar fertilization can be combined with crop protection spraying, but the mixed components must be compatible.

Practical operation of foliar application

For foliar application, several types of sprayers are employed. A greater volume of solution is required per unit of area in the case of high-volume sprayers. The commonly employed procedures involve: (i) spraying about 400 litres/ha of a solution in fine 0.1–0.2-mm droplets; or (ii) high-pressure, low-volume spraying where the solution is blown at the leaves in very small droplets. Higher nutrient concentrations can be used with low-volume sprayers than with high-volume sprayers. In either case, there should be good adhesion of the solution to the leaves. This can be improved by adding special detergents and stickers.

Spraying is most effective, and the risk of scorch is minimized, where the spray droplets do not dry rapidly. This is best achieved by spraying on cloudy days or in the early morning or late afternoon. Application of N solutions should be avoided during the early growth stages. In the case of multiple deficiencies, combinations of nutrients are applied with special combined fertilizers, containing for example N, Mg and micronutrients. Per-hectare amounts up to 30 kg N, 1 kg Mg and 0.1–0.5 kg micronutrients can be applied in a single foliar spray.

During foliar fertilization, it is important to maintain the proper concentration suitable for the particular crop. This is usually stated on the bags containing special foliar fertilizers. Some general figures for concentrations are given below for foliar fertilization on certain crops at 400 litres/ha using solid fertilizers:

- urea (46 percent N):
  - solution of 8–15 percent = 14–28 kg N/ha for cereals, oilseed rape, etc.,
  - solution of 2.5–5 percent = 5–10 kg N/ha for beets and potatoes,
  - solution of 0.5–1 percent = 1–2 kg N/ha for fruit trees, vegetables;
- magnesium sulphate (10 percent Mg): solution of 2 percent = 0.8 kg Mg/ha for cereals and fruit trees;
- iron chelate (5 percent Fe): solution of 0.2 percent = 0.04 kg Fe/ha for fruit trees;
- manganese sulphate (24 percent Mn): solution of 1 percent = 1 kg Mn/ha for cereals;

- copper chelate (14 percent Cu): solution of 0.13 percent = 0.07 kg Cu/ha for cereals;
- Solubor, Octoborate (20 percent B): solution of 1.5 percent = 1.2 kg B/ha for beets and oilseed rape.

Fertilization through irrigation (fertigation)

In fertigation, fertilization is combined with irrigation, and the nutrients are supplied together with the water. In reality, it is a type of liquid fertilization. In the past, mainly N was added to water in furrows and through sprinkler irrigation. However, with the increasing use of microirrigation, fertigation on a precisely controlled small scale (microfertigation) has been developed. Beyond maximizing yields and quality of crops, the aim of fertigation is improved utilization of nutrients and lower water consumption, while minimizing pollution by surplus nutrients. The saving may be up to 30–50 percent of water and nutrients. Fertigation can improve crop yields in fields and greenhouses substantially. N utilization is higher and there are reduced losses through nitrate leaching. In addition, plants take up more phosphate compared with P placement, and the uptake of other nutrients is also enhanced.

The nutrient application process

Microirrigation distributes the nutrient solution to individual plants via drip or trickle irrigation operating at about 100 kPa pressure, or via minisprinklers operating at about 200 kPa pressure. The advantage is a constant supply of soluble (available) nutrients right into the rooting zone in order to meet the daily crop demand. The goal is to feed the plants in synchronization with their growing nutrient requirements. However, establishment and application costs are generally much higher than for broadcast fertilization combined with sprinkler irrigation.

Drip irrigation produces small zones of wet soil volumes with relatively uniform water content. The distance of nutrient movement from the input point differs from one nutrient to another. Nitrate and sulphate are transported farther than phosphate, which is more liable to immobilization near the site of deposit but less so than with broadcasting or adsorbed cations of K or Mg. As in the case of broadcasting, processes in the rhizosphere may affect nutrient uptake through fertigation as well. Fertigation of a partial soil volume with a confined root system allows a precise control of nutrient supply, thus, avoiding deficiencies or excess, as well as salinity hazards (except on poorly drained clay soils). The size of the root systems can be modified to some extent, but smaller volumes need better control of nutrient supply.

Suitable fertilizers for fertigation

Fertilizers for fertigation must be readily and fully water soluble, and the combined solution should be within the acidic pH range (about pH 5) in order to ensure nutrient mobility and availability. Nitrate and urea are better distributed in soils than is ammonium and, therefore, they are more suitable.

The main difficulties are with the common phosphates and even with polyphosphates because of their potential precipitation as Ca phosphates. Because of this, acidic P fertilizers such as phosphoric acid (e.g. 1 g/litre of pH 2.2), MAP, mono-potassium phosphate (MPP) and the more expensive urea phosphate or glycero-phosphate are often recommended. Of the K fertilizers, potassium nitrate and potassium sulphate are preferred to MOP because they contain no salinity-causing chloride. Recent research in Israel has shown that KCl can partly replace KNO_3 in fertigated tomatoes without adversely affecting growth and yield (Imas, 2004). The mixing of fertilizers must be undertaken carefully to avoid mistakes or compatibility problems. Moreover, an unwanted early precipitation of micronutrients in the soil can be avoided by using chelates such as EDTA, or more stable ones such as DTPA and EDDHA in soils of neutral reaction. Iron chelates are more effective than those of Mn or Zn.

Operational aspects

Fertigation requires corrosion-resistant mixing and pumping equipment and small lateral tubes for distribution of the nutrient solution through special nozzles. The lateral tubes can either be put on top of the soil or installed as subsurface fertigation. With the latter system, the nutrients are delivered into the centre of the root system, the root volume is increased and the rootless topsoil layer is kept dry. This has the advantage of reducing weed growth, but crop germination and establishment must be assured. Early plant growth is stimulated by pre-plant fertilization through broadcasting or placement, and this improves the efficiency of fertigation.

Fertigation requires special management skills as a breakdown in the system can have serious consequences. The composition of the nutrient solution and its uninterrupted flow must be controlled carefully. The nutrient composition is based on the daily consumption rate of the crops in the field. This can be obtained from guidelines for different crops. The crop growth is generally divided into ten segments in order to aid nutrient management. The required nutrient rates (expressed in kilograms per hectare per day) are in the range of 0.3–6 for N, 0.05–0.8 for P and 0.3–10 for K. The nutrient concentration must be high enough to produce high yields but must not cause salinity damage or related problems. A suitable concentration of the irrigation water is about 100 mg/litre (0.01 percent) of N and K.

A special problem with fertigation is clogging of the solution emitters. These can become blocked by precipitation of carbonates and/or phosphates, by suspended particles, by a biofilm of microflora or by fine roots. Special cleaning methods have been developed to prevent and remove the substances causing the blockages. Because of the complex application technology, fertigation is suitable only for advanced farmers and it requires considerable capital investment. Careful and frequent monitoring is required, preferably using simple field methods applicable to the farmer. A more detailed discussion of this topic can be found in Bar-Yosef (1999).

Hydroponics

Hydroponics is a system where the plant roots grow in a nutrient solution instead of the soil. Although soils are the "natural" growth substrate for plants, soil-less crop cultivation has been employed successfully. With intensive hydroponics, very high yields can be produced. However, it requires special equipment such as corrosion-resistant containers and pumps, devices for measuring out solutions of salts and acids, and suitable analytical instruments.

Ingredients required

The major ingredient is a suitable water supply, such as rainwater, which is low in mineral components. The fertilizers used must be water-soluble solids or liquids that can be mixed easily to prepare a concentrated stock solution. The required dilute nutrient solution is prepared from this stock solution by dosage pumps. A reliable monitoring system is essential for a well-functioning hydroponics system in order to maintain the correct composition of nutrients and to keep the salinity within a tolerable range. The solution has to be checked frequently by measuring its electrical conductivity.

Common fertilizers containing major nutrients for making the stock solutions are:

- salts: NH_4NO_3, $Mg(NO_3)_2$, Ca $(NO_3)_2$, K_2SO_4, KH_2PO_4, and $MgSO_4$;
- acids and alkali: HNO_3, H_3PO_4, and H_2SO_4; KOH for pH adjustment.

Micronutrients are added as salts or chelates in the required low concentration. The composition of the nutrient solution depends on crop requirement and growth stage. It generally has a total soluble-salt concentration of 0.2–0.7 mg/litre (1–2 mS electrical conductivity); N and K each at about 0.1–0.2 mg/litre, P about 0.01 mg/litre and a pH of 4–5.

Nutrient supply

The different techniques of nutrient supply are: (i) static solutions that are changed at certain intervals; (ii) flowing or cycling solution where the original concentration is maintained by dosing; and (iii) supply of solution over short intervals alternating with water. Compared with fertigation, the advantage of hydroponics is an even better control of optimal plant nutrition as there are no soil-related complications such as fixation of applied nutrients. On the other hand, the investments and the needs for control are higher. The advantages for plant nutrition via hydroponics are best utilized where other growth factors, such as temperature and CO_2, are controlled. Because soils are not required, hydroponics can be used in locations with poor or no agricultural soil.

As with any production system, hydroponics has its advantages and disadvantages:

- advantages:
  - nutrients are supplied in soluble forms and remain easily available,
  - the nutrient solution contains the whole range of nutrients with optimal ratios,

- the solution can be adapted easily to changing plant requirements during growth,
- no toxic substances are present to disturb plant growth;

➢ disadvantages:

- there is no buffering capacity in the event of deficiency or excess, therefore, good control is required,
- the oxygen supply to the roots is less than in soils, thus, an external air supply is required,
- roots have no solid anchorage, thus, mechanical support or an inert porous material is needed.

Precision fertilization

Variability and uncertainty are dominant features of field crop production. There are differences between nutrients in the type of variation encountered in field situations. For P and K, the variation is mainly spatial and location-related, but for N there is an additional large temporal (time-related) variation. These are difficult to account for with traditional fertilizer application methods. The common fertilizer application method is based on the reasonable assumption that, from a practical point of view, the soil nutrient supply to small fields of up to about 1 ha is more or less homogenous. Where on larger fields there are nutrient-related soil differences, the area can be divided into homogenous subunits of any suitable size and treated individually. With this modification, the common method of fertilizer distribution has been and still is successfully used in many parts of the world.

Much of the intrafield variability can be overcome by precision farming. This approach applies modern technologies to manage variability in space and time in order to improve crop performance and decrease nutrient losses. Precision farming is applicable to many aspects of crop production, such as soil fertility and plant protection management. The main objective is to produce uniform high yields over the whole field, economize on fertilizer and pesticide inputs, and create minimal undesirable effects on the environment. In order to be adopted widely, it needs to be efficient and profitable.

Precision fertilization presents a special method for distributing fertilizers according to the different needs of small plant populations caused by soil variability within a field. Such a concept is very promising for areas where fertilization practices have advanced over the years. It is based on:

➢ precise location control for both diagnosis and input application using systems such as the Global Positioning System (GPS);

➢ detailed assessment of soil fertility either by analysing distinct samples or recorded continuously by sensors of microlevel nutrient status and its variation in the field;

➢ comprehensive and rapid data processing;

➢ site-specific application of fertilizers to the small basic soil areas within the field.

The expected advantages of precision fertilization for the farmer are: (i) uniform nutrient supply to all parts of the field, which enables higher yields and product quality; (ii) savings in fertilizer rates; and (iii) lower nutrients losses. Although costly, modern precision fertilization is often profitable in commercial farming on a medium to large scale.

Prerequisites for precision fertilization

In order to practice precision fertilization, the technologies required are: a precise location control, a reliable assessment of microlevel variation, and equipment for site-specific applications, all coordinated by efficient computers using suitable software. For location control (knowing the exact position in the field), previous outmoded methods used for land survey have been replaced by the GPS, which permits the monitoring of even very small areas (100 m^2), which are called pedocells.

Assessment of the nutrient status of each pedocell is the backbone of precision fertilization. Without it, there can be no precise fertilizer application. Most weaknesses in the system are related to this central problem. Compared with cumbersome chemical soil testing, special sensors reacting to different light effects are much more efficient. However, sensoring of soil fertility aspects such as available nutrients is not yet possible. The equipment for precision nutrient application requires highly developed steering devices and devices for changing application rates quickly and distributing them accurately. For example, suitable centrifugal fertilizer distributors for quick changes in precise dosages are now available but they are expensive.

The absence of sensors for the actual diagnosis of soil nutrient status and the lack of inexpensive production of detailed nutrient maps will remain as obstacles in the adoption of precision fertilization. The preconditions for an efficient and cost-effective precision fertilization are: its capability to take into account large spatial differences in relatively small areas; simple provision of cheap diagnostic methods – preferably with sensors; and production of reliable soil fertility maps.

Precision soil fertility management

There are many possibilities and problems concerning the "precise distribution" of major nutrients. Precision farming offers great possibilities for improved nutrient supply to most plants by overcoming yield-limiting or fertilizer-wasting effects associated with natural or human-made variations within a field. Many aspects of precision fertilization have been discussed by Pierce and Nowak (1999).

The advantages of precision fertilization appear obvious and raise high expectations. However, the sceptical farmer who is advised to invest in modern precision technology would like to examine the system critically before adopting it, particularly the following aspects.

The relevant comparison of common fertilizer distribution with precision fertilization should not refer to uniform distribution of fertilizers on the whole field but to the customary method of differential fertilizer application. Where fertilization is based on soil testing, the principle of uniform fertilization applies only to uniform parts of a field from which separate test sample have been taken. This method takes into account the differences that are noticeable in the field, while ignoring small differences. Although this appears to be a rather crude method in comparison with precision fertilization, it is relatively effective.

Considering the inherent inaccuracies in soil sampling and soil testing, a detailed map would require an enormous number of samples. Where extrapolation procedures are based only on a few samples, this requires a sophisticated interpretation method. Although scientifically sound, both procedures have practical problems. For medium to high yields, a small surplus P and K application can be advantageous for adequate nutrient supply during nutritional stress. Because the available phosphate concentration is low and variable in time, there are no P-surplus problems for crops, and as there is hardly any leaching of phosphate, overfertilization of parts of the field is tolerable although not ideal or cost-effective.

Precision fertilization can be efficient and profitable where intrafield variability can be assessed reliably and economically. It will not be profitable where the diagnostic assessment remains expensive and unreliable and also where high level uniformity is neither required nor brings about significant yield increases. In most cases, it is not of much interest to smallholders with severe financial constraints in many developing countries. However, it is a valuable tool for large farms, organized plantations and for the large-scale production of high-value crops.

For an average farmer in many countries, the main question is not whether precision fertilization is useful or not but whether it is worthwhile. Many such farmers are in the very early stages of development in terms of scientific farming and optimizing plant nutrition. They are still some way away even from adopting blanket fertilizer recommendations made for their region or conventional soil-test-based fertilizer rates. It is for this reason that this guidebook does not include the nutrient details of precision fertilization. This in no way undermines the usefulness of precision farming.

GUIDELINES FOR THE APPLICATION OF ORGANIC MANURES

Application of solid manures

Bulky organic manures such as composts and FYM can be applied to all soils and almost all crops, as can oilcakes, recycled wastes and animal meals. In order to make best use of the slowly acting N, these should be applied a few weeks before sowing, spread uniformly over the field and immediately ploughed into the soil in order to avoid ammonia losses. Common application rates are about 20 tonnes/ha but range from 10 to 40 tonnes/ha. While large amounts are spread over the whole area, smaller amounts are preferably concentrated in plant rows or applied around the base of individual trees or bushes. Vermicompost is normally applied to the soil in the same manner as bulky organic manures. The commonly recommended rate for mature vermicompost is 5 tonnes/ha.

Many farmers use whatever quantities are available on the farm or in nearby areas. With 20 tonnes/ha of FYM, about 100 kg N/ha is added. In the first year,

20–30 percent of this N is utilized, but up to 40–50 percent can be utilized by the second year, including the residual effect.

Application of slurries

Slurry can be obtained from farm animals raised in organized dairy farms. Animal slurry is the major manure in many developed countries where cattle are raised on a large scale. Other forms of slurries are obtained from the treatment of sewage and from biogas plants.

Application of animal slurry

The common practice of spreading animal slurry on the soil surface results in substantial losses of ammonia where the slurry is not mixed immediately into the soil. N losses are reduced by modern drilling machines that place slurry a few centimetres into the, preferably, moist soil. In this respect, it is similar to suspension fertilizers.

The recommended application rates of animal slurry are related to the crops, e.g. 30–40 m^3/ha (75–200 kg/ha N) for winter cereals, applied partly in autumn and partly in spring; and 40 m^3/ha for silage maize in spring and the same on grassland for hay production. For accurate N application, the exact N concentration of the slurry should be known and special precautions must be observed where it is applied on growing plants both in order not to damage the plants and for health reasons. No slurry should be applied on vegetables intended for fresh consumption or on meadows at least one month before grazing starts. After that, it can be applied only if it is well fermented.

In some countries, legislation regulates the maximum rate of slurry application in order to prevent environmental damage caused by ammonia losses and the leaching of nitrate. It would be advantageous if slurry could be transformed into a solid product such as compost with more suitable application properties, but so far this has not been economically feasible.

Slurry obtained from biogas plants is also a kind of animal slurry as cattle dung is the most common feedstock used in biogas plants. It is a semi-solid product and is better than FYM as a manure because it is well digested and has a higher nutrient content. However, it is difficult to transport. In the case of small biogas plants (based typically on the dung of five head of cattle), the slurry is usually spread on the farmland near the biogas plant. An alternative method for using biogas plant slurry is to convert it into a compost. The use of biogas slurry in proper combination with mineral fertilizers is one of the major possibilities for INM.

Application of sewage (wastewater) and sewage sludge

In many countries, sewage sludge is rarely used directly as a nutrient source by applying it on bare soil. Because this procedure has health risks, wet sewage sludge is converted into a moist or dry solid product and possibly processed into sludge compost. Application rates of 2–3 tonnes/ha on a dry-matter basis are advisable, but they should not exceed 5 tonnes/ha within 3 years. As with any nutrient source, sewage sludge should not contain more than the critical concentrations of toxic elements and should only be applied to soils that contain such elements well below the critical toxic levels. This will prevent damage to soil health, crops, food quality and feed value. Farmers in developed countries have become less enthusiastic about using cheap city wastes as a nutrient source because of the ever-increasing regulations involved and the uncertainties about future regulatory aspects.

Wastewater reuse for crop irrigation and nutrient supply becomes particularly attractive where it is planned in conjunction with environmental safeguards. The wastewater must be treated and used in such a way that its content will not be hazardous to human beings or the environment. In order to protect public health, the effluent should either be treated properly before irrigation application, or its use should normally be restricted only to certain crops so that improperly treated wastewater does not come into contact with plants used for direct consumption as human food or animal feed. A suggested cropping list for irrigation with differentially treated wastewater for semi-arid tropical conditions in developing countries is as follows (Juwarkar *et al.*, 1992):

- primary treated:
  - cash crops: cotton, jute, sugar cane, tobacco,
  - essential oil crop: citronella, mentha, lemon grass,
  - cereals and pulses: wheat, rice, sorghum, pearl millet, green gram, black gram,
  - oilseeds: linseed, sesamum, castor, sunflower, soybean, groundnut,
  - vegetables: brinjals, beans, okra, etc. These should be cooked before eating;
- secondary treated:
  - all crops listed above,
  - all crops including vegetables that develop near or below the soil surface but are only to be consumed after cooking;
- secondary treated and disinfected:
  - all crops without restriction.

Optimal rates and intervals of wastewater application to agricultural soils should be determined primarily by crop needs and soil health considerations and not merely as an outlet for waste disposal. As with any other farm input, there is an optimal level that needs to be borne in mind for different soils and crops. Excessive loading with wastewater may lead to soil sickness, which can be corrected through adequate resting of the soil from crop production and use of soil amendments.

Application of green manure

Green manure can be either grown *in situ* and incorporated in the main field or grown elsewhere and brought in for incorporation in the field to be manured. Not all plants can be used as a green manure in practical farming. Some plants suitable for green manuring have been described in Chapter 5. Most plants used as green manures are legumes. As green manures add whatever they have absorbed from the soil, they also promote the recycling of soil nutrients from lower depths to the topsoil. The net gain is only in the case of biologically fixed N.

Green leaf manure consists of fresh green leaves of suitable plants grown on the bunds of the main field or elsewhere and brought in for incorporation in the soil. Green leaves of these plants are incorporated in the soil at or before planting the main crop.

In selecting a green manure crop, the most desirable characteristics are: (i) local adaptability of the plant; (ii) fast growth and production of a large amount of green matter (biomass)/unit area/unit time; (iii) tolerance to soil and environmental stresses, such as acidity, alkalinity, and drought; (iv) resistance to pests; and (v) easy decomposability – requiring least time between the incorporation and planting of the main crop. Where a green manure crop is raised before taking a wetland rice crop, it can be ploughed in even a few days before planting rice. Where the green manure is raised before maize, potato or sugar cane, it should be buried and incorporated in the soil 2–3 weeks before planting the main crop.

GUIDELINES FOR THE APPLICATION OF BIOFERTILIZERS

Biofertilizers can be applied to the seed, to the soil or to the roots of seedlings before these are transplanted in the main field. It is most important to know that not all biofertilizers are suitable for all soils and crops. Various biofertilizers have been described in Chapter 5. In general terms, the applicability and usefulness of biofertilizers for different crops can be stated as follows:

- cereals:
  - rice (wetland): BGA, *Azolla*,
  - others: *Azotobacter*, *Azospirillum*, PSB;
- pulses: *Rhizobium*, PSB;
- oilseeds:
  - legumes: *Rhizobium*, PSB,
  - non-legumes: *Azotobacter*, PSB;
- pastures, forages and fodders:
  - legumes: *Rhizobium*, PSB,
  - non-legumes: *Azospirillum*, PSB;
- forest trees:
  - legumes: *Rhizobium*,
  - casuarina: Frankia, PSB, mycorrhizae,
  - others: *Azotobacter*, mycorrhizae;
- others:
  - potato, cotton: *Azotobacter*, *Azospirillum*, PSB,
  - sugar cane: *Azotobacter*, *Azospirillum*, *Acetobacter*, PSB,
  - citrus: mycorrhizae, *Azotobacter*, PSM,
  - tobacco: *Azotobacter*,
  - plantation crops: *Azotobacter*, mycorrhizae,
  - vegetable crops, flowers/ornamental plants, spices: *Azotobacter*, PSB.

The most common method for the application of bacterial inoculants is by coating them on the seeds before sowing. Other methods include soil application by mixing the inoculum with organic manure and spreading the mixture on the nursery area, main field or in the furrows. Setts of sugar cane, cut tubers of seed potato and roots of seedlings can also be dipped in the biofertilizer slurry before planting in the main field. For example, cut tubers of seed potato can be soaked for 20–30 minutes in 50–60 litres of suspension containing 1 kg of biofertilizer.

Application of Rhizobium inoculant

Rhizobium inoculant is the most commonly used biofertilizer. It is specifically intended for application to legumes. It is very important to select the correct *Rhizobium* inoculant (Chapter 5). Generally, a significant beneficial effect from using *Rhizobium* biofertilizer can be expected where the native *Rhizobium* population is less than 100 cells/g of soil. It is important to check that the correct species of *Rhizobium* is being used for the crop to be treated and that the commercial inoculant is of acceptable quality and well within the stated date of expiry. The following biofertilizer application techniques have been adapted from Motsara, Bhattacharayya and Srivastava (1995). The procedure for inoculating the seeds of legumes consists of the following steps:

- First, a slurry of the biofertilizer is to be prepared. This can be done by adding 125 g of country sugar (unrefined cane sugar) to 1.25 litres of water and heating for 15 minutes. Where gum acacia has been added to the product as adhesive, farmers are advised to follow the instructions on the packet. As an alternative to country sugar, 500 g of gum arabic can be added, and the solution is cooled to room temperature.
- The inoculant (400–500 g) is mixed into the above sugar or gum-acacia suspension to form a slurry. To this, the seeds required to plant 1 ha are added and mixed thoroughly by hand. Finally, the seeds are dried in shade on a plastic sheet/paper and sown without delay.

Rhizobium bacteria are sensitive to low pH. Their tolerance to pH varies with species in the order: *B. japonicum* > *B. lupini* > *R. leguminosarum* > *R. trifolii* > *R. phaseoli* > *R. meliloti.* In acid soils, lime may have to be applied in order to create favourable conditions for their survival. Mo availability is also low in acid soils. As Mo is required for BNF, Mo sometimes has to be supplied as an external input. It can be added with the inoculum onto the seed. In areas where such cultures are not available, soil collected from another field under the same crop can be used.

The efficiency of BNF also depends on the adequate availability of nutrients that are required by the legume and the N-fixation system. Several plant nutrients in the soil can affect nodulation and N fixation:

- Ca and B have been shown to be involved in infection and nodule development.
- In moderately acid soils, the Ca requirement for nodule infection is higher than that of the host plant.
- B deficiency inhibits the formation of vascular strands from roots to nodules.

TABLE 37
Suitable quantities of *Rhizobium* inoculant and sticker for inoculating legume seeds

Legume	Seed weight	Inoculant	Gum arabic solution
	(g)		(ml)
Groundnut	100	10	4.0
Chickpea	100	7	3.5
Pigeon pea	100	8	3.5
Soybean	100	10	3.0
Lentil	50	5	2.0
Leucaena leucocephala	50	10	3.0
Green gram	100	9	3.5
Cowpea	100	8	3.5

Source: Motsara, Bhattacharayya and Srivastava (1995).

- The effect of P on N fixation is through its effect on overall plant growth.
- Mo, Fe and S are components of the nitrogenase enzyme, which is involved in the N-fixation process.
- Co is part of the cobamide coenzyme.
- Fe is a component of leghaemoglobin, which carries oxygen to the bacteria inside the cell.

Sowing during the hot period of the day should be avoided. The amounts of culture, water and sticker needed per hectare depend on the seed size and seed rate because the objective is to coat/cover all the seed with the biofertilizer slurry. There should be a minimum gap of 24 hours between seed treatment with a fungicide and biofertilizer in order to avoid any harmful effect of the agrochemical on the micro-organisms in the biofertilizer.

Table 37 provides a general idea of the suitable quantities of inoculant (biofertilizer) and sticker required for various legumes.

Preparation of methyl-cellulose solution for seed-coating

Seeds can also be coated with biofertilizer by using a 1-percent methyl cellulose solution for coating. To prepare the solution, methyl cellulose is weighed at the rate of 1 g/100 ml and sprinkled into about 50 ml of hot water (about 80 °C). This is stirred well and any lumps formed are broken. After it has dissolved, the remaining cold water (50 ml) is added while stirring to obtain the required volume. A fine gel is formed that can be coated on the seeds. First, a slurry is prepared by mixing and stirring the inoculant at the rate of 70 g in 300 ml of 1-percent methyl-cellulose solution. The thoroughly dispersed slurry is then poured over the correct weight of seeds (e.g. 300 ml/20 kg chickpea seeds) and mixed until all the seeds are coated. Mixing can be done in a vessel or on a plastic sheet. Any vessel contaminated with toxic materials or dust should be not be used for mixing. The seeds are dried in shade, kept away from direct sunlight, and sown as soon as possible.

Tree/legume seedlings can be readily inoculated in the nursery. A 50-g bag of inoculant is sufficient to inoculate 10 000 seedlings (regardless of species). This can be done by mixing the culture in cool water and using the suspension to irrigate the rooting medium of the seedlings.

Application through pelleted biofertilizer

Many bacteria are sensitive to acidic conditions and also to hot and dry weather. They can be protected from these adverse factors by application in pelleted form. If the inoculated seed is coated with powder lime, it gives good protection, especially where the soils are very acid, hot and dry. Pelleting can also help to protect the seeds from insects, especially seed-gathering ants. Calcium carbonate is the most common and beneficial of the many materials tested. Quicklime should not be used as it is highly toxic.

Seed pelleting with biofertilizer can be done as follows:

- The appropriate quantities of gum arabic and water to be used with the desired quantity of the particular seed to be pelleted are calculated. Gum arabic dissolves in cold water if left overnight and in hot water in about 30 minutes. The solution should not be boiled. The gum-arabic solution is cooled. The appropriate amount of inoculant is added to the solution and stirred to form a smooth slurry. This mixture must not stand for more than 30 minutes. Some gum arabic is acidic and will harm the bacteria unless the acid is neutralized by calcium carbonate as soon as possible.
- Small lots of seeds may be pelleted by hand, in a tub, bucket or on a smooth floor. For pelleting large quantities of seeds, a mechanical mixer can be used (seed drum, cement mixer, etc.). Vigorous agitators from the mixing equipment should be removed in order to prevent damage to the pellet coating.
- The seeds are poured into the mixer and then the gum inoculant slurry is added. The mixer is then rotated at high speed until all the seeds are coated. Without stopping the mixer, calcium carbonate is added all at once, and the mixer allowed to run until all the seeds have been pelleted.
- The mixer should not be cleaned between loads. After the whole job is done, the mixer is cleaned by running a load of water and gravel through it. Pellets are firmer if they are allowed to stand for 24 hours and these work better in a seed drill.
- The pelleted seeds are screened to remove any lumps in order to avoid clogging the seeding equipment. Where there is an excess of calcium carbonate powder, it is screened to prevent clogging of the seeding equipment.

Precautions

Rhizobium inoculation sometimes fails to give the expected results. This can be because of the following reasons:

- the soil already contains a sufficient population of effective and required strains of *Rhizobium*;
- poor quality of inoculum, which is unable to compete with the native bacteria;
- suboptimal (low dose) level of inoculum used;
- presence of toxic substances associated with seed-coat (e.g. phenolic compounds and condensed tannins);
- existence of biological antagonists, e.g. rhizophage, nematodes;
- inoculation applied with agrochemicals that are toxic to micro-organisms (e.g. thiram, bavistin and chlorpyriphos);

- poor soil conditions viz. acid soils (low pH), waterlogging, high soil temperature, etc.;
- low or excess soil moisture restricting the movement and proliferation of *Rhizobia*;
- nutritional stresses, e.g. deficiency of P, B and Mo.

Azotobacter

The application of *Azotobacter* inoculant involves making a slurry of the carrier-based biofertilizer using a minimum amount of water. The seeds are mixed with the slurry as in the case of *Rhizobium*, dried in shade and sown as soon as possible. For transplanted crops, the roots of seedlings can be dipped in the slurry for 20–30 minutes and then transplanted. In the case of sugar cane, *Azotobacter* application may be needed more than once during early growth. In this case, second and further treatments can be given by pouring the slurry near the rootzone. The slurry can also be mixed with FYM and applied near the rootzone.

Blue green algae (BGA)

BGA are a biofertilizer specific to wet paddy fields. The BGA can be inoculated in fresh form, dry form or as soil-based inoculum. Inoculation of fresh BGA is better than dry BGA or soil-based inoculum. This is because fresh BGA establish early in paddy fields and grow faster. Fresh BGA at the rate of 30–60 kg/ha and dry BGA or soil-based inoculum at 5–10 kg/ha is recommended for multiplication plots and transplanted paddy fields. Application of dried BGA flakes at the rate of 10 kg/ha is recommended for the main rice field. The flakes are to be applied ten days after transplanting rice. For best performance of BGA, the field should have an adequate level of available P. A thin film of water is maintained over the field. BGA multiply well in warm weather.

Azolla

As in case of BGA, *Azolla* is also used as a biofertilizer, primarily in wetland rice culture. It is in fact different from most other biofertilizers in that its biomass is incorporated in the soil just as in the case of a green manure. It can be used either as a conventional green manure before planting rice or grown as a dual crop along with rice and then incorporated in the soil while the rice is still growing.

Azolla as a green manure

The field is ploughed and levelled about 15 days before transplanting rice. It is subdivided into plots of 300–400 m^2 each. The subplots are flooded and puddled properly, after which 5–10 cm standing water is maintained. Fresh *Azolla* can be inoculated at the rate of 3–4 tonnes/ha (3–4 kg/10 m^2). After 2–3 weeks, the water is drained from the field and the green *Azolla* biomass is incorporated into the soil. Rice is transplanted within a week. For satisfactory N fixation, the soil should not suffer from nutrient deficiencies, particularly those of P, and the temperature as well as moisture should be optimal (Chapter 5).

Azolla as a dual crop

In this case, *Azolla* is inoculated in standing water at the rate of 3–4 tonnes/ha 1–2 weeks after transplanting rice. It grows fast, multiplies and fixes N while the rice crop is growing. Dry *Azolla* spores can be used as an inoculum at the rate of 5 kg/ha in transplanted rice fields. These are pre-soaked in water for 12 hours and inoculated in the rice field seven days after transplanting rice. After 3–4 weeks, the water is drained and the *Azolla* is buried in the soil where it is growing and incorporated with a weeder or other suitable implement. Repeated incorporation of *Azolla* is needed. As a dual crop, *Azolla* can be grown more than once for the same rice crop in order to obtain additional benefit. On decomposition, it releases the fixed N and other nutrients in its biomass for use of the rice crop.

Azolla can be grown as a dual crop even after it has been incorporated as a green manure before planting rice. Usually, the amount of inoculum recommended is 0.1–0.3 kg/m^2 (1–3 tonnes/ha) for multiplication plots and 0.5–1.0 kg/m^2 (5–10 tonnes/ha) for dual cropping.

Phosphate-solubilizing biofertilizers

For the application of phosphate solubilizers, the best method is seed treatment. Other methods such as seedlings and soil can also be used. For seed treatment, a slurry is prepared using 200 g of biofertilizer in 200–500 ml. of water. This is then poured slowly over 10–25 kg seeds. The seeds are mixed evenly to obtains a uniform coating of the seeds. The treated seeds are dried and sown immediately, as in case of N-fixing bacterial inoculants. For soil treatment, a mixture of 5–8 kg of biofertilizer with 100–150 kg soil or compost is prepared and applied by surface broadcast over 1 ha either at sowing or 24 hours earlier. For the treatment of seedlings, a suspension of 1–2 kg biofertilizer is prepared in 10–15 litres of water. The roots of seedlings from 10–15 kg of seed are then dipped into this suspension for 20–30 minutes and transplanted soon after.

Mycorrhiza

Mycorrhiza (VAM) is a mobilizer of soil nutrients and an enhancer of root reach for plant nutrients. Mycorrhizal fungal spores are used to produce the inoculum.

The inoculation of mycorrhiza for nursery plants involves sowing the seeds and raising seedlings or bare root cutting in plastic bags or pots. In all these methods, 4–5 g of appropriate VAM inoculum is placed 3–5 cm below the seed or the lower portion of bare root cuttings, followed by normal plant cultivation practices. In the case of application to seedlings grown on raised seed beds, the appropriate inoculum is applied by soil incorporation. About 6 kg of inoculum is mixed with soil sufficient for 25 m^2 and covered with a thin layer of soil. In most cases, the population of seedlings is sufficient for transplanting 1 ha. It is necessary to remove the inoculated seedlings from the raised seed beds carefully so that the mycorrhizae associated with roots are not affected and are transferred effectively along with the seedling to be transplanted. For optimal benefits, root treatment with a slurry of 250 g inocula in 1 litre of cow dung slurry can be given at the time of transplanting.

APPLICATION OF SOIL AMENDMENTS

Problem soils often require amendment before they can be cropped successfully and optimal use made of the plant nutrients applied. Liming of acid soils and reclamation of alkali soils are given here as examples.

Amendments for acid soils

Several liming materials have been described in Chapter 5. Generally, calcium carbonate is selected where it is readily available at reasonable cost. Where the soils also need Mg application, dolomite limestone is preferred. Basic slag and sugar-factory press mud from the carbonation plants also have a liming effect.

For some crops such as potatoes, liming ahead of planting is preferable because of their sensitivity to high soil pH after recent liming, which may cause scab owing to Mn deficiency. For "top-liming" of growing plants, only carbonate lime should be used in order to avoid leaf scorch. For grassland, lime is spread on top of short grass in spring, left to dissolve and allowed to be washed into the topsoil.

The liming material should be distributed evenly on the bare soil and then mixed well into the topsoil layer in order to achieve a uniform increase in soil reaction. Application after harvest but before tillage and sowing (sometimes termed "stubble liming") is the best procedure. Following this practice, the soil layer below the topsoil can also be ameliorated to a certain extent.

Liming of fields is generally required every 3–5 years or once in a crop rotation. It should be done on priority for crops such as sugar beet and oilseed rape, which do not grow well under acidic conditions and prefer higher soil pH. In general, except for crops such as tea, which must have an acidic environment, liming is recommended for bringing the pH towards neutrality and, in the process, improving the availability of several nutrients.

Amounts of lime required

The lime requirement cannot be calculated directly from the pH value because of the need to also neutralize reserve acidity, which is not reflected in the pH value. However, a knowledge of pH and soil texture can be used to approximate the amount of limestone needed. Generally, the target is to lime an acid soil to reach a pH of 6.5. Most soil-testing laboratories are able to provide information on soil reaction and soil texture. One method for determining the lime requirement is:

- Step 1: determination of H value from the pH measured in Ca acetate:
  - acetate pH of 6.5 corresponds to an H value of 3.5 meq/100 g;
  - acetate pH of 6.0 corresponds to an H value of 11 meq/100 g.
- Step 2: From the H value, the lime requirement to reach pH 7 (neutral) can be calculated:
  - 1 meq H/100 g = 0.84 tonnes/ha CaO for top 20 cm of soil weighing 3 000 tonnes/ha.

Table 38 provides a simple reference list for calculating the amount of lime required to treat acid soils of different textures. This amount is usually for treating the top 15 cm of soil. The amount of lime required will change proportionately as the depth of treatment changes.

TABLE 38
An example of the relation of soil pH and texture with lime requirement

Soil pH	Lime needed in soils of different texture		
	Sandy loam	Loam	Clay loam
		(kg/ha)	
5.0	5 550	6 000	6 450
5.2	4 650	5 100	5 500
5.4	3 750	4 200	4 650
5.6	2 850	3 300	3 750
5.8	1 950	2 400	2 850
6.0	1 050	1 500	1 950
6.2	650	850	1 050

Maximum amounts of lime

In Europe, the general advice is that the amounts of lime applied at one time must not exceed 2 tonnes/ha of carbonate lime on light soils and 3–5 tonnes/ha on medium and heavy-textured soils in order to ensure good mixing with the soil. Where a sufficient amount of limestone cannot be used, as in the case of wide-row crops, the furrows to be planted can be limed instead of spreading it on the entire field. This will economize on the lime required and still improve the pH in the rootzone.

Amendment of alkali (sodic) soils

At the global level, about 434 million ha of soils are affected by alkalinity. Such soils have a very large percentage of their cation exchange site occupied by the undesirable sodium ions (Na^+). In highly sodic soils, 70–80 percent of the exchange positions are occupied by Na^+ leaving few places for useful nutrient cations. Amendment of such soils is a prerequisite for efficient nutrient management and obtaining high yields. As an amendment process, steps are needed to remove excess Na^+ from the exchange complex and replace it with Ca^{2+} and make the soil normal. As the ESP increases, so does the pH. Therefore, soil pH is also used as an indicator to decide the quantity of amendment required.

Based on soil pH and texture, the amount of gypsum, a common amendment, is recommended (Figure 41). It is generally sufficient to incorporate gypsum in the top 10–15 cm of soil. Gypsum required to replace all the Na^+ ions is referred to as 100-percent gypsum requirement. The amounts

FIGURE 41
Relationship between pH and gypsum requirement in soils of different texture

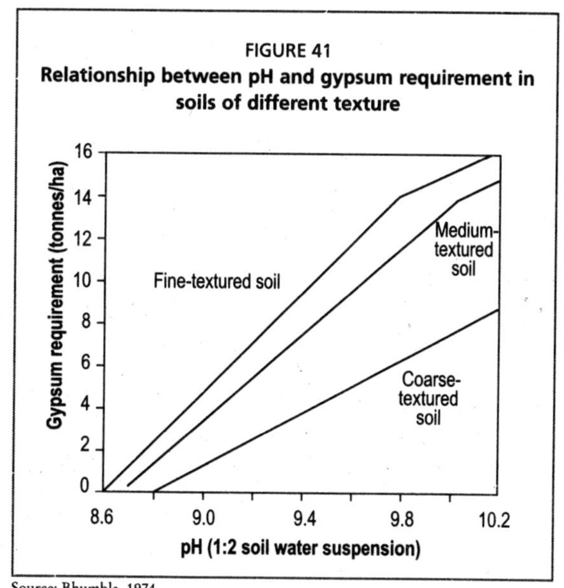

Source: Bhumbla, 1974.

required increase where soils are to be treated up to a greater depth (which is often unnecessary). Mineral gypsum ground to pass through a 2-mm sieve is efficient and cost-effective. Adequate availability of good-quality water is required for leaching during the reclamation process. Crops raised on amended soils benefit from green manuring, which is an important part of INM.

Iron pyrites and elemental S-containing compounds have also been used to amend alkali soils. The availability, efficacy and cost-effectiveness of the material (as also ease of application) determine the final choice of the material to be used. The reclamation and management of salt-affected soils has been discussed in detail by Gupta and Abrol (1990).

Chapter 8
Nutrient management guidelines for some major field crops

Practical recommendations and guidelines on nutrient management for specific crops are usually provided by the local research and extension services in each country. This is logical and also necessary because of the crop- and area-specific nature of such recommendations. The IFA (1992) has published examples of practical nutrient management guidelines for almost 100 crops in major countries where these are grown. There are also numerous publications on this aspect at regional and country level. Overall guidelines on the management of nutrients and their sources (mineral, organic and microbial) have already been provided in Chapter 7. The present chapter provides some crop-specific information on nutrient management including diverse nutrient sources as part of INM. Again, extension workers or farmers should seek the information relevant to their conditions from local sources and their applicability to local socio-economic conditions. The guidelines given below should be seen in the nature of illustrative information in order to appreciate the importance of balanced crop nutrition for sustaining medium to high yields of crops.

CEREALS AND MILLETS

Wheat (Triticum aestivum L.)

Wheat is the most widely grown cereal crop in the world. It is cultivated on almost 215 million ha out of 670 million ha under cereals. Wheat grain contains 70 percent starch and 12–18 percent protein. The highest grain yields are obtained with winter wheat. These range from 1 tonne/ha to more than 12 tonnes/ha, with a world average of about 3 tonnes/ha. High yields (up to 14 tonnes/ha) can be obtained from highly productive varieties with appropriate nutrient and crop protection management on fertile soils with adequate water supply. Globally, wheat yields have increased considerably as a result of breeding programmes that have incorporated the short-straw trait from Mexican varieties. Such varieties are more responsive to applied nutrients and are also more resistant to lodging as compared with the local wheat varieties.

Wheat can grow on almost any soil, but for good growth it needs a fertile soil with good structure and a porous subsoil for deep roots. The optimal soil reaction is slightly acid to neutral although it can be grown successfully in alkaline calcareous soils under irrigation. The water supply should not be restrictive and rains should be well distributed.

Nutrient requirements

The amounts of nutrients required can be derived from soil testing and the nutrient removal by grains and straw. A crop of winter wheat producing 6.7 tonnes grain/ha absorbs an average of 200 kg N, 55 kg P_2O_5 and 252 kg K_2O/ha. Under subtropical Indian conditions, a crop producing 4.6 tonnes grains + 6.9 tonnes straw absorbed 128 kg N, 46 kg P_2O_5, 219 kg K_2O, 27 kg Ca, 19 kg Mg, 22 kg S, 1.8 kg Fe, 0.5 kg Zn, 0.5 kg Mn and 0.15 kg Cu. The proportion of nutrients absorbed that ends up in the grains is 70 percent in the case of N and P and 20–25 percent in the case of K. For winter wheat, the nutrient requirement before winter is small. It is highest during the maximum vegetative growth in spring. More than 80 percent of the nutrients are taken up by ear emergence. Where organic manure is used, it should be applied before sowing or, if applied carefully, as slurry during early growth. Nutrient requirement varies considerably depending on the soil fertility, climate conditions, cultivar characteristics, and yields.

Macronutrients

In temperate regions, 25 kg N are required per tonne of grain containing 15 percent protein. Therefore, a yield of 10 tonnes will need 250 kg/ha N for the grains alone, and about 30–40 percent more for the total plant biomass, which results in a total amount of 350 kg N/ha. However, as fertile soils generally provide one-third of this amount, fertilizer amounts can be adjusted to N removal in grains. Ideally, N fertilizer applications to winter wheat (200–250 kg N/ha for high yields) should be split into several dressings as follows:

- in autumn: only 30 kg N/ha (or none where sufficient N is left from the previous crop);
- in early spring: about 120 kg N/ha (minus mineral N in soil, e.g. 30 kg/ha N);
- at beginning of tillering: about 30–50 kg N/ha;
- at ear emergence: 40–60 kg N/ha – this can be divided into two portions to enable a late foliar spray to improve protein content for better baking quality.

Wheat needs no special N fertilizer. However, for applications in spring with cold weather, quick-acting nitrate is superior to ammonium or urea. Placement of N fertilizers brings little or no advantage on most soils, except perhaps under low rainfall and in the absence of irrigation. One kilogram of fertilizer N produces about 15–25 kg of grain. Where yields are limited by climate or other constraints, the fertilization rate can be reduced in view of the lower requirements and the respective soil nutrient status.

Under subtropical conditions, the generally recommended amounts of N are 120–150 kg N/ha to irrigated HYVs, and about half of this to traditional varieties or where irrigation is not available. N application is generally recommended in 2–3 splits at planting, and one month and two months after planting. The basal dressing is generally given in the form of urea or through NP/NPK complexes. For top-dressing, any of the common N fertilizers are suitable but ammonium sulphate performs better than others on S-deficient soils. To unirrigated wheat depending solely on stored soil moisture and seasonal rainfall, N rates varying from 40 to 120 kg N/ha can be applied depending on stored soil moisture as described above (Figure 36).

Because an optimal supply of P and K is required for high yields, even during periods of water stress, these nutrients should be applied before sowing in spring or autumn unless there is danger of K leaching on sandy soils. As a rule, on fertile soils, nutrients applied to offset nutrient removal with grains and straw are sufficient. For a yield of 8 tonnes/ha of winter wheat, the recommended rates are: 90 kg/ha P_2O_5, 160 kg/ha K_2O and 25 kg/ha Mg. On deficient soils, the amounts added should be at least 30 percent higher, and on soils containing high amounts, about 50 percent lower than the values given above.

Deficiencies of nutrients other than NPK are likely to occur in poor soils, at high yields and with persistent use of NPK. S and Mg are the two most likely nutrients to be limiting. These can be applied prior to sowing or, in the case of S, through an S-containing N fertilizer in the standing crop. Where visible deficiency symptoms appear, water-soluble fertilizers or foliar sprays can be applied.

Micronutrients

For high yields, Mn and Zn may be in short supply in neutral to alkaline soils and Cu on sandy soils. Zn deficiency is generally a problem in coarse-textured soils under intensive cropping. Here, an application of zinc sulphate of 62.5 kg/ha once every 2–3 years is suggested. Zn deficiency can also be corrected by spraying 0.5-percent zinc sulphate (at a per-hectare rate of 2.5 kg zinc sulphate and 1.25 kg unslaked lime dissolved in 500 litres water). Generally, 2–3 sprays at 15–day intervals may be needed. In Mn-deficient soils, foliar spray with 0.5-percent manganese sulphate solution 2–4 days before the first irrigation and again 2–3 times at weekly intervals can be done on sunny days.

Rice (Oryza sativa L.)

Worldwide, rice occupies almost 150 million ha. A very high proportion of the world's rice is grown under the wetland system. This system consists primarily of submerged or waterlogged conditions for a major part of the growth period of the crops. Wetland rice soils vary greatly in their nutrient status. Regardless of their initial reaction, the pH of such soils moves towards neutrality after submergence. The general growth conditions and the fertilizer practices are influenced considerably by the anaerobic, reducing conditions in the flooded soil. These soils tend to have low organic matter and, therefore, they provide only a relatively small supply of N and P from mineralization unless green manured.

Nutrient requirements

Nutrient uptake and removal by rice is influenced strongly by the variety, season, nature and composition of the soil and the yield level. In order to produce 1 tonne of paddy (rough rice), the rice crop absorbs an average of 20 kg N, 11 kg P_2O_5,

30 kg K_2O, 3 kg S, 7 kg Ca, 3 kg Mg, 675 g Mn, 150 g Fe, 40 g Zn, 18 g Cu, 15 g B, 2 g Mo and 52 kg Si. Out of the total uptake, about 50 percent of N, 55 percent of K and 65 percent of P are absorbed by the early panicle-initiation stage. About 80 percent of N, 60 percent of K and 95 percent of P uptake is completed by the heading stage. The partitioning of uptake in the case of N and P is higher in grain than in straw (3:1), whereas greater proportions of K, Ca, Mg, Si, Fe, Mn and B remains in the straw. The S, Zn and Cu taken up is distributed about equally in straw and grain (Yoshida, 1981).

Macronutrients

There is a close association between the amount of N fertilizer applied to rice and the yield level. Yield responses of 20 kg or more of paddy or rough rice per kilogram of N are frequently obtained. The amount of N that can be applied to traditional, tall rice varieties is limited because of their susceptibility to lodging and low yield potential. However, the improved short HYVs that are resistant to lodging can benefit from a higher level of N supply (Figure 42). While traditional varieties could justify rates of up to 50 kg N/ha, 160 kg N/ha or more is recommended for HYVs under good management with assured water supply. The season of planting also influences the N requirement of rice. During the dry season, when abundant sunshine is available, the irrigated HYVs can justify 30–40 kg N/ha more than in the lower-yielding rainy season. Incorporation of a good green manure crop raised before planting rice can add 50–60 kg N/ha as well as a substantial amount of organic matter.

FIGURE 42
Average yields of 4 dwarf and 7 tall indica rice varieties as affected by N fertilization

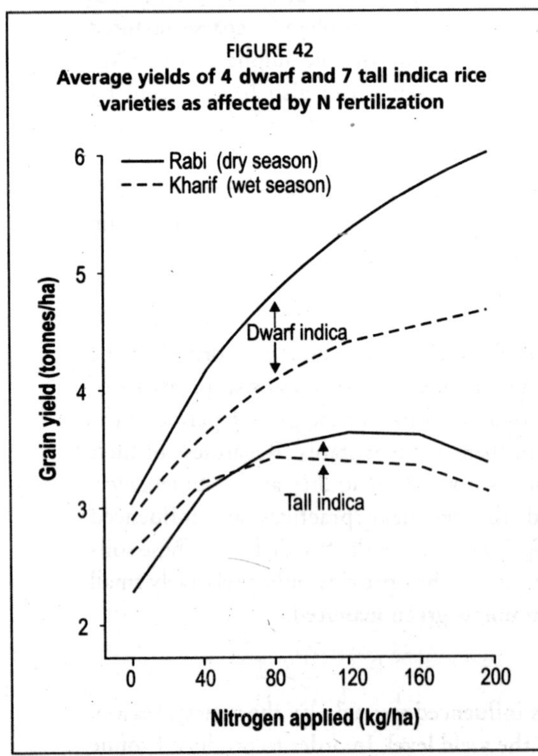

Source: Tanaka, 1975.

The timing of N applications is very important for improving the efficiency of N use by rice. The crop may require none or a modest basal application and up to three top-dressings in the standing crop in order to maintain the N supply throughout its growth. Split applications are especially important where total N requirement is high in order to avoid leaching losses (particularly on permeable soils).

The method of N application is also important for reducing N losses and improving the nitrogen-use efficiency the crop, which is often below 50 percent. The basal application should be worked into the flooded soil. The applications of ammonium or urea N should, where possible, be made into the reduced soil horizon. This is because broadcasting them into the floodwater is likely to result in high N losses. Placement of urea in the reduced zone can be facilitated by using urea supergranules. Nitrate-containing fertilizers such as AN or CAN are often less satisfactory for rice, particularly where given at or before planting. They can be used for top-dressing when crop uptake of nutrients is proceeding rapidly, and the topsoil is covered with a mat of roots, and thus, N losses are minimized.

Because upland rice relies mostly on rainfall and soil moisture reserves, rice yields are lower than in the case of wetland rice. As the soil under upland rice is not flooded, soil nutrient behaviour is similar to that in other upland cereal crops. Application of 50–100 kg N/ha can be justified, depending on yield potential. Total N should be split between a basal and a top-dressing. Owing to high leaching losses, upland rice can often suffer from N stress even where N is applied.

While the availability of soil P is improved by flooding, many old rice soils have a low P content because of crop removal over the years. This, together with the greater demand for P by improved varieties, makes adequate use of P fertilizer important. Optimal rates vary with local conditions, but 20–40 kg P_2O_5/ha is usually enough for traditional varieties and 40–80 P_2O_5/ha for improved varieties. In the intensive rice–wheat rotation, where wheat has been fertilized adequately, the rate of P application to rice can be reduced. This is because flooded rice can make better use of the residual P applied to wheat. Where two rice crops can be grown in succession within a year as in monoculture, the dry-season crop usually requires a higher rate of P application than does the wet-season crop. P should be applied as a basal dressing in order to promote root growth and tiller formation. Water-soluble P or a combination of water- and citrate-soluble P is normally most efficient for rice production. Many upland rice soils are low in available P, and moderate P applications are usually required.

The crop uptake of K is quite high but much of it remains in the straw. In traditional rice varieties, responses to K have usually been small. However, improved varieties usually respond to K, especially where given adequate N and P. Responses to K are generally greater on sandy soils. While 20–40 kg K_2O/ha may be sufficient for traditional varieties, improved varieties can justify the application of 60 kg K_2O/ha particularly on soils that are poor in K. On most soils, K fertilizer should be applied as a basal dressing. However, on free-draining sandy soils where leaching may occur, split application of K is being increasingly recommended. Potash fertilization should also keep in view the fact that, where K is cheaper than N and P, it can be equally profitable even at lower response rates.

S deficiency is becoming more widespread in rice. This is because of higher yields and, thus, greater S removals, the reduced use of organic manures, possible leaching of S and the widespread dominance of S-free fertilizers (urea, DAP and MOP) in the product pattern. Where either AS or SSP is a part of the fertilization schedule, the required S is often supplied through these sources.

Micronutrients

Owing to the intensification of rice production, micronutrient deficiencies are becoming more common. It is important to identify and correct them wherever they occur. Field-scale deficiency of Zn in rice was first discovered at Pantnagar in India. The deficiencies of Zn and Fe can occur fairly commonly in rice fields, especially on high pH soils, Fe more so in upland rice. Where Zn has not been applied to the nursery, 10–12 kg Zn/ha through zinc sulphate (21 percent Zn) can be applied before planting. It can be surface broadcast and incorporated before final puddling. Fe deficiency can be corrected by giving 2–3 foliar sprays of 1-percent ferrous sulphate at weekly intervals. Green manuring also reduces Fe deficiency.

Rice is unusual in responding to the application of Si (a non-essential beneficial element). Si in the form of soluble silicates and waste products containing Si is applied in some countries. It is thought that Si promotes growth by making soil P more readily available to the plants, by producing strong stems, by providing resistance against certain pests and by protecting the plant from Fe and Mn toxicity.

Organic and green manuring

The nutrient status of rice soils can be improved by applying organic manure a week or two before transplanting. Where adequate water is available, green manuring with a fast-growing leguminous plant is often recommended. A good green manure crop of *Sesbania* can add 50–60 kg N/ha where incorporated into the soil before planting rice. Details about green manuring have been provided in Chapters 5 and 7. Where a leguminous green manure such as *Sesbania* is planted before rice, it is sometimes recommended that the phosphate meant for application to rice be applied to the green manure instead. Adequate supply of phosphate also promotes greater N fixation.

Biofertilizers

There is a considerable scope for BNF in rice paddies by BGA and/or the *Azolla–Anabaena* association, which may supply up to 25–50 kg N/ha. Inoculation of the paddy-field with BGA can contribute 20–30 kg N/ha. Incorporation of *Azolla* biomass before or during the growth of rice can contribute similar amounts of N along with significant amounts of other nutrients that are present in its biomass. *Azolla* can also accumulate 30–40 kg K_2O/ha from the irrigation water. Information about the multiplication and inoculation with BGA and *Azolla* has been provided in Chapter 7.

Maize (Zea mays L.)

Nutrient requirements

A maize crop producing 9.5 tonnes of grain per hectare under North American conditions can remove the following amounts of nutrients through grain plus stover (IFA, 1992):

- macronutrients (kg/ha): N 191, P_2O_5 89, K_2O 235, MgO 73, CaO 57 and S 21;
- micronutrients (g/ha): Fe 2 130, Zn 380, Mn 340, B 240, Cu 110, Mo 9 and also 81 kg Cl.

Macronutrients

High yield in maize is closely associated with N application, but only where other inputs and management practices are optimal. N interacts positively with plant population, earliness of sowing, variety, weed control and moisture supply. Figure 43 shows an example of the mutual benefit from N fertilizer and enhanced plant population. However, neither higher plant population nor high levels of N alone will improve yields where a third factor is limiting. Where moisture supply is inadequate or uncertain, optimal levels of fertilizer as well as plant population will be below those required for top yields. Fertilizer can improve the utilization of soil water by increasing rooting depth. However, the best returns from N fertilizer are only obtained where the water supply, either natural or supplemented by irrigation, is adequate for full crop growth. Under good growing conditions, a yield response of 30 kg grain/kg N can be obtained.

Maize takes up N slowly in the early stages of growth. However, the rate of uptake increases rapidly to a maximum before and after tasseling, when it can exceed 4 kg N/ha/day. N fertilizer application is best scheduled in accordance with this pattern of uptake in order to avoid serious losses by volatilization or leaching and to ensure that N levels are high in the soil when the crop demand is also high. An application to the seedbed followed by a side-dressing when the crop is knee high, or for very high application rates two top-dressings (the second at tasseling) are usually recommended. The N application rates for rainfed maize are about half of those for the irrigated crop.

Fertilizer requirement in relation to yield level can be calculated directly from crop uptake of N only in specific regions because of the variations in soil N supply and the rather unpredictable efficiency of fertilizer N by the crop. However, N fertilizer requirement may be about 50 kg/ha with unimproved

FIGURE 43
Effect of N fertilizer on maize in relation to plant population

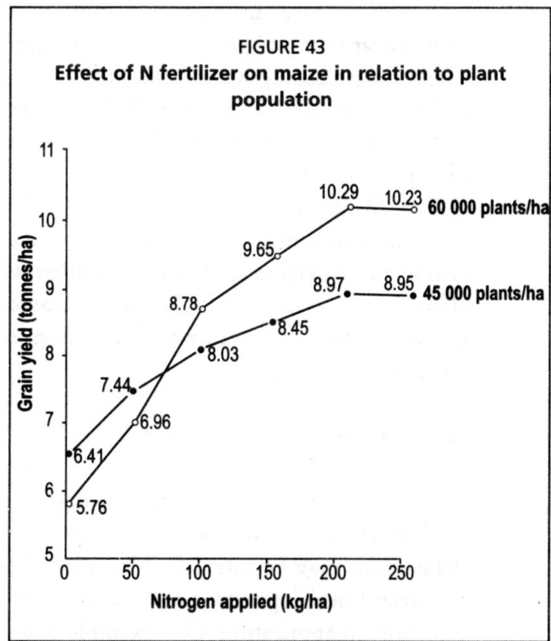

Source: Gros, 1967.

varieties under uncertain rainfall, rising with yield potential to 250–300 kg N/ha where yields of 12 tonnes/ha or more can be expected. Local recommendations on amounts of N should, as always, be based on local experimentation under the prevailing growing conditions. For irrigated HYVs of maize such as hybrids/composites in India, the general recommendation is to apply 60–80 kg N/ha to early-maturing varieties, 80–100 kg N/ha to medium-duration varieties, and 90–150 kg N/ha to late-maturing varieties.

Adequate P is very important for maize as the crop cannot readily take up soil P in the large amounts needed for optimal growth and high yield. Best results from N and other inputs will not be obtained without adequate P, which should be applied mainly in water-soluble form. Rates of P application should be varied according to soil test for available P and in relation to yield potential. These can be in the range of 30–100 kg P_2O_5/ha. Phosphate application in the highly successful maize production region of Illinois, the United State of America, is based on soil tests and crop removals with the twin objectives of building up the soil P level up to the optimum and replacing the P removed by maize at harvest (Table 33).

K is taken up in large quantities by maize but only a small proportion of total uptake is removed with the grain. While maize can obtain appreciable amounts of soil K, it is important to ensure that the overall supply is sufficient for high yields. Use of K fertilizer is especially important where high rates of N fertilizer are used and high yields expected. Recommended rates of K application are in the range of 30–100 kg K_2O/ha. Where the soils are supplied adequately with K, its application is advocated on the basis of soil analysis and yield potential.

In the intensive maize–wheat annual rotation, fertilizer recommendations in subtropical India suggest that the application of P and K to maize can be omitted where the preceding wheat crop has been regularly fertilized with these nutrients or 12–15 tonnes FYM/ha is applied to maize. Organic manures should be applied 3–4 weeks before planting maize. These can be surface broadcast followed by incorporation in the soil.

Both P and K are most effective where applied as a basal dressing before or at the time of planting through a seed-cum-fertilizer drill. Where suitable equipment is available, sideband application, together with a moderate rate of N will improve effectiveness on many soils. Where mechanical equipment for fertilizer application is not available, the fertilizer can be dropped in open furrows and covered with soil before planting.

Micronutrients

Maize can suffer from a number of micronutrient deficiencies. However, Zn deficiency is perhaps the most widespread problem. The problem is mostly on alkaline calcareous soils and soils with a low organic matter content. Zn deficiencies may be intensified by a high level of P supply from the soil and/or fertilizer. Local experience combined with soil and plant analysis can be used as a basis for Zn application. One example of a recommendation from India is to add 25 kg zinc sulphate (21 percent Zn) mixed with 25 kg soil along the row, followed by hoeing and irrigation. Where Zn deficiency symptoms are seen in the standing crop, foliar spray can be given at a per-hectare rate of 3 kg zinc sulphate + 1.5 kg of lime in 500 litres of water.

GRAIN LEGUMES

This section covers two important pulse crops. Major oil-bearing grain legumes, such as groundnut and soybean, are covered in the section on oil crops.

Chickpea (Cicer arietinum L.)

Chickpea is an important grain legume of the arid and semi-arid regions, where it is grown with or without irrigation. The grain contains about 20 percent protein and forms an essential part of human diet in many countries.

Nutrient requirements

A crop producing 1.5 tonnes of grain has been reported to remove the following amounts of major nutrients and micronutrients through total dry matter (Aulakh, 1985):

- macronutrients (kg/ha): N 91, P_2O_5 14, K_2O 60, MgO 18, CaO 39 and S 9;
- micronutrients (g/ha): Fe 1 302, Zn 57, Mn 105 and Cu 17.

A large part of the N is presumably derived from BNF.

Rhizobium inoculation

Being a legume, chickpea can benefit from BNF in association with *Rhizobium*. Therefore, inoculation with *Rhizobium* is often recommended to augment N supply by the soil. The benefit resulting from inoculation is broadly equivalent to the application of 20–25 kg N/ha. Details of the procedure for inoculation have been provided in Chapter 7.

Macronutrients

Even where the soil or the seed is treated with *Rhizobium* biofertilizer, an N application is necessary. This serves as a starter dose and meets the N needs of the crop until the N-fixation system becomes operational. For this purpose, 15–20 kg N/ha is generally recommended. In addition to N, application of 40–50 kg P_2O_5/ha is also recommended. The entire amount of N and P_2O_5 is normally given before planting. There is a strong positive interaction between the availability of moisture and nutrients. The benefits of supplying irrigation increase with increased nutrient application. In S-deficient soils, application of 20–30 kg S/ha through any of the conventional sulphate sources results in a significant increase in grain yields.

Micronutrients

In neutral to alkaline soils (where chickpea is usually grown), Zn and Fe deficiencies can be encountered. To correct Zn deficiency, soil application of zinc sulphate at a rate of 25 kg /ha is suggested under irrigated conditions. Fe deficiency can be corrected by providing foliar sprays with 2-percent ferrous sulphate solutions.

In B-deficient soils, application of borax can increase the yield by an average of 350 kg grain/ha.

Pigeon pea [Cajanus cajan (L). Millsp.]

Pigeon pea is an important grain legume crop. It is perennial in habit but often cultivated as an annual crop. The grain contains about 22 percent protein and forms an essential part of human diet in many areas.

Nutrient requirements

A crop producing 1.2 tonnes of grain has been reported as removing the following amounts of major nutrients and micronutrients through total dry matter (Aulakh, 1985):

- macronutrients (kg/ha): N 85, P_2O_5 18, K_2O 75, MgO 25, CaO 32 and S 9;
- micronutrients (g/ha): Fe 1 440, Zn 38, Mn 128 and Cu 31.

A significant part of this is presumably provided by BNF.

Rhizobium inoculation

Like other legumes, pigeon pea can benefit from BNF in association with *Rhizobium*. Inoculation with *Rhizobium* culture is generally recommended in order to augment soil N supply. The inoculation might result in benefits to the extent of 20–25 kg N/ha. Details of the procedure for inoculation have been provided in Chapter 7.

Macronutrients

Treatment of the soil or seed with *Rhizobium* biofertilizer, application of starter N dose of 15–20 kg N/ha, and 40–50 kg P_2O_5/ha are recommended. Often, for simplicity, the application of 100 kg DAP/ha is suggested, which delivers 18 kg N and 46 kg P_2O_5. The entire amount is normally given before planting. The need for K depends on the soil K status and yield potential of the cultivar. In S-deficient soils, application of 20–30 kg S/ha through any of the conventional sulphate sources results in a 10–15-percent grain yield increase.

Micronutrients

Deficiencies of B and Zn have been widely encountered in pigeon pea. These deficiencies can be corrected by the application of suitable carriers as per local recommendations. As an example, 5 kg Zn/ha can be applied to the soil through zinc sulphate.

OIL CROPS

Groundnut/peanut (Arachis hypogaea L.)

Groundnut, a legume, is major cash crop in India, China and the United States of America. It is also a traditional low-input crop grown in West Africa by smallholders. Its kernels contain an average of 25 percent protein and 48 percent oil. The kernels are used mostly as food in roasted or processed form by humans and also as a source of edible oil. It is well adapted to conditions ranging from semi-arid to semi-humid. The crop grows well on coarse-textured soils, which facilitate the development and growth of pods. After the oil has been extracted, the residue known as groundnut cake serves as an animal feed supplement and sometimes also as an organic manure.

Nutrient requirements

Nutrient removal by a crop producing 3 tonnes pods/ha in the United States of America was reported to be 192 kg N, 48 kg P_2O_5, 80 kg K_2O and 79 kg MgO (IFA, 1992). Nutrient removal per tonne of economic produce under north Indian conditions was of the following order (Aulakh, 1985):

- macronutrients (kg): N 58.1, P_2O_5 19.6, K_2O 30.1, Mg 13.3, Ca 20.5 and S 7.9;
- micronutrients (g): Fe 2 284, Zn 109, Mn 93 and Cu 36.

Rhizobium inoculation

Inoculation with *Rhizobium* culture is usually recommended, particularly where the crop has been introduced recently or has not been grown for several years, or where the native *Rhizobium* population is inadequate and/or ineffective. The groundnut–*Rhizobium* symbiosis can fix about 110–150 kg N/ha. Details of the procedure for inoculation have been provided in Chapter 7.

Macronutrients

Most of the N requirement of a groundnut crop is provided through BNF. Unless soil fertility is high, or organic manure has been applied, a starter dressing of 20–30 kg N/ha is needed to feed the crop until the nodule bacteria are fully established.

Groundnut needs P application for optimal yield and also for the optimal development of nodules in which BNF takes place. Phosphate requirements are normally in the range of 40–70 kg P_2O_5/ha. Generally, an S-containing fertilizer such as SSP is preferred as the source of P because it also provides 12 percent S and 19 percent Ca, both of which are very important for the development of pods and synthesis of oil. The K requirement of groundnut can generally be supplied by soil reserves, residues from previous crops and organic manure. However, potash application is needed on K deficient soils or for high yields under irrigated conditions. Recommendations range from 20 to 50 kg K_2O/ha. Fertilizers can often be sideband placed to advantage.

The nutrition of groundnut requires attention and action beyond supplying just N, P and K. The crop frequently requires supplementary applications of S and Ca. It can also suffer from Mg deficiency in acid leached soils. The S requirement depends on the S input through rainfall and whether or not previous crops have received S-containing fertilizers. S needs can be met by using AS, SSP, ASP, etc. Sources such as gypsum, pyrites and even SPM discharged by sugar factories based on sugar cane can also be used.

Groundnut is unusual in showing Ca deficiency. This can usually be overcome by liming the soil to pH 6.0. In some cases, it is necessary to apply additional Ca in the form of gypsum at the flowering stage. Foliar spray of a soluble Ca salt can also be effective. Ca deficiency can be accentuated by the use of excess K, so that an adequate Ca supply is particularly important where a large K application is made. In many groundnut-growing areas, application of 300–500 kg gypsum/ha is recommended for application at or before flowering. Sandy soils or acid soils may be deficient in Mg, which can be supplied by liming with dolomite. However, excess Mg has the same effect on Ca availability as excess K and, therefore, should be avoided.

Micronutrients

Depending on soil conditions, groundnuts are known to suffer from deficiencies of Mn, B, Fe and Mo. B deficiency, which causes internal damage to the kernels, may also occur on sandy soils, especially in dry conditions. It can be controlled by soil or foliar application of 5 kg borax/ha or two foliar sprays of 0.1-percent borax solution. Mn deficiency is usually attributable to overliming and is controllable by a manganese sulphate spray. Mo deficiency leads to reduced N fixation. As the Mo requirement is very small, it can be supplied as a seed treatment through sodium or ammonium molybdate at the rate of 0.5–1 kg/ha. Iron chlorosis is often observed where groundnut is grown in alkaline calcareous soils. This can be corrected by spraying a solution of 0.5–1-percent ferrous sulphate with 0.1-percent citric acid at 8–10 day intervals. Cultivars that are efficient users of Fe and tolerant of Fe deficiency should be preferred where such seeds are available.

Soybean [Glycine max (L.) Merr]

Soybean is a very energy-rich grain legume containing 40 percent protein and 19 percent oil in the seeds. The crop is adapted to a wide range of climate conditions. The highest soybean yields are produced in near neutral soils but good yields can be obtained also in limed acid soils. Under good growing conditions with adequate N fixation, grain yields of 3–4 tonnes/ha can be obtained.

Nutrient requirements

Total nutrient uptake by the plants per tonne of grain production can be taken as follows (IFA, 1992):

- macronutrients (kg): N 146, P_2O_5 25, K_2O 53, MgO 22, CaO 28 and S 5;
- micronutrients (g): Fe 476, Zn 104, Mn 123, Cu 41, B 55 and Mo 13.

Under conditions favourable for N fixation, a significant part of the N uptake can be derived from BNF.

Rhizobium inoculation

Inoculation with *Rhizobium japonicum* (now known as *Bradyrhizobium japonicum*) culture is often recommended particularly where the crop has been introduced recently or the native *Rhizobium* population is inadequate and ineffective. Under good conditions, the soybean crop will fix 100 kg N/ ha or more. Details of the procedure for inoculation have been provided in Chapter 7.

Macronutrients

N fixation can meet a large part of the N requirement of the crop, for which it is usually necessary to treat the seed with bacterial inoculant. The crop may respond up to the application of 100 kg N/ha in the absence of poor BNF. However, in most cases, a starter dose of 20–40 kg N/ha is recommended as it takes some weeks for the nodules to develop and N fixation to start. Large applications of N are needed where N fixation is very low.

Fertilizer P and K requirements of soybean should be based on soil test values. Typical application rates for soils of low nutrient status are 50–70 kg P_2O_5/ha and 60–100 kg K_2O/ha. In the soybean-growing areas of the United States of America, for an expected grain yield of 2.5–2.7 tonnes/ha, the recommended rates of P on low-fertility soils are 40–60 kg P_2O_5/ha, and 100–150 kg K_2O/ha on soils with a low to normal clay content. Application rates are higher at higher yield levels in soils with a high clay content. As an example, for each additional tonne of grain yield, an extra 10–15 kg P_2O_5/ha and 20–30 kg K_2O/ha is recommended.

Soybean responds to the application of Mg and S depending on soil fertility status and crop growth conditions. Significant responses of soybean to S application have been found in many field trials in India. In several cases, it may be advisable to apply phosphate through SSP so that the crop also receives an S application. Where DAP is used, gypsum can be applied to the soil before planting at the rate of 200–250 kg/ha.

Micronutrients

Depending on soil fertility status and crop growth conditions, responses have been obtained to the application of Zn and Mn. Application of 5 kg Zn /ha on coarse-textured soils and 10 kg Zn /ha on clay soils can remedy Zn deficiency. On Mn-deficient soils, the application of manganese sulphate at a rate of 15 kg/ha to the soils or 1.5 kg through foliar spray increases yield.

Oilseed rape (Brassica napus L.)

Among major oil crops, oilseed rape (canola) is of increasing importance. The oil extracted from the seeds containing 40 percent oil is used for salad oil, as a cooking medium and for fuel. The residues referred to as oilseed cake are protein-rich animal feed. In many parts of South Asia (including India) rapeseed mustard is an important winter-season crop that is grown either alone or as a secondary intercrop in wheat fields. The term rapeseed is a group name referring to various species of Brassica such as *B. juncea, B. campestris* and for rocket salad or *Eruca sativa* but not to *B. napus.* With new varieties of winter rape, including hybrids, high seed yields of 4–5 tonnes/ha are attainable compared with average yields of 3–3.5 tonnes/ha in Europe.

FIGURE 44
Nutrient uptake and growth of winter oilseed rape in different stages

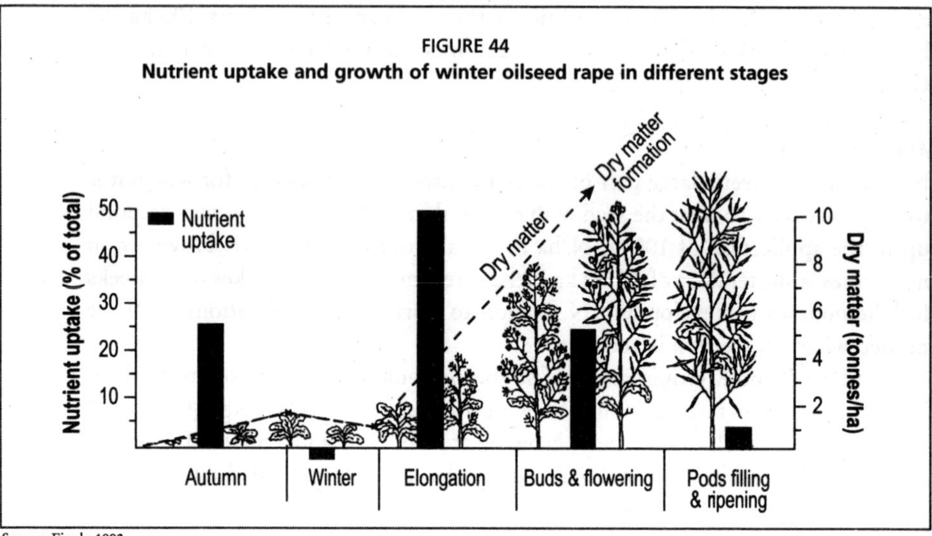

Source: Finck, 1992.

High yields are normally obtained on deep fertile topsoil without a compact layer to facilitate root growth, and a porous crumb structure of the uppermost soil layer for rapid germination of the small seed. This is assisted by a neutral soil reaction and sufficient organic matter for optimal biological activity.

Nutrient requirements

Oilseed rape needs an abundant and timely nutrient supply for good growth and high seed yield (Figure 44). The total nutrients absorbed by a crop producing 4.5 tonnes of seed per hectare are of the order (in kilograms): N 300–350, P_2O_5 120–140, K_2O 300–400, Mg 30–50 and S 80–100. The seeds contain the majority of nutrients except for K, which remains mainly in the straw. Out of the total nutrient uptake, about 20 percent takes place before winter and 50 percent in spring before flowering. In subtropical north India, the total nutrient removal per tonne of seed production by mustard was of the following order (Aulakh, 1985):

- macronutrients (kg): N 32.8, P_2O_5 16.4, K_2O 41.8, Mg 8.7, Ca 42.0 and S 17.3;
- micronutrients (g): Fe 1 123, Zn 100, Mn 95 and Cu 17.

Macronutrients

The N requirements are higher than the N removal figure of 30–35 kg N per tonne of seed. About 30–40 kg N/ha is sufficient for fertilization in autumn. Oilseed rape prefers nitrate N. However, ammonium nitrate is also a good source of N. N solutions and urea can be used except for the very early dose in spring. For the crop in the main growing season, about 250–280 kg/ha N are required from soil and fertilizer. The recommended N rates for seed yield of about 4 tonnes/ha are:

- a first application of 80–100 kg N/ha early in spring on soils that can supply about 40 kg N/ha;
- a second application of 60–80 kg N/ha at the start of elongation;
- a third application of 25 kg N/ha at the beginning of flowering for very high yields.

Correspondingly less N is required for lower yield levels. Oilseed rape tends to leave large amounts of N in the soil after harvest (both as nitrate and as crop residues). These may amount to more than 100 kg/ha N. With good N management, it is possible to keep the mineral N residue below 50 kg N/ha, which is tolerable from a pollution point of view, or to utilize the residual amount by the following crop.

In semi-tropical north India, the irrigated crop can respond to 240 kg N/ha on coarse-textured soils that are low in organic matter. Under dryland conditions, 30–50 kg N/ha is usually optimal. Application of N through AS or of P through SSP is advantageous in S-deficient soils. Response to P is determined by soil P status, moisture availability and yield level. As a general guideline, N and P_2O_5 are recommended in a ratio of 2:1.

The supply of major nutrients should be ample during the growing season, even during short periods of stress caused by dryness or cold. Application should be made at sowing, but a split application with part applied in spring is needed on light soils where losses may occur in winter. The amounts of P and K required depend on the nutrient removal and soil nutrient supply. For a high yield goal of 4.5 tonnes of seeds per hectare on a soil with an optimal nutrient range, the following application rates are suggested (in terms of kilograms per hectare): 80–100 P_2O_5, and 150–200 K_2O. On slightly deficient soils, the amount should be about 30 percent higher, and on soils in the sufficiency range about 50 percent less.

For a yield goal of 4.5 tonnes of seeds per hectare, on medium-fertility soils, the application of 30 kg Mg/ha is also suggested. Brassicas have the highest requirement of S among field crops. The optimal rate of S to be applied depends on the soil S status, yield potential and the level of N applied. In coarse-textured soils, 20–50 kg S/ha may be needed. Until about 1980, almost no fertilization with S was required in Europe because of the large amounts of S supplied through the atmosphere as a result of industrial pollution. In less industrialized parts of the world this was not so. Since atmospheric additions have fallen, S deficiencies have become widespread and rates of 20–80 kg S/ha are required in order to obtain 0.5 percent S in the young leaves. About 10 kg of S are required per tonne of seed yield. In the case of an acute deficiency, foliar spray with a soluble S fertilizer can be used as a quick remedy.

Micronutrients

Because oilseed rape has a B requirement that is at least five times higher than that of cereals, 0.5 kg B/ha should be applied in combination with other fertilizers on deficient soils. The Mn requirement is high and an application of about 1.5 kg

Mn/ha is recommended in many areas, and foliar spraying is effective. Because of the high soil reaction needed by oilseed rape, Mn availability is lowered and deficiencies frequently limit yields. Although only 10–15 g Mo/ha are required by oilseed rape, some soils do not supply this small amount. The need for Mo fertilizers must be based on diagnostic methods. Zn deficiency can be a problem that can be corrected by the soil application of 10 kg Zn/ha. Where the previous crop in the rotation has received Zn application or 10–15 tonnes of FYM/ha have been used, the application of Zn fertilizer can be omitted.

Sunflower (Helianthus annuus L.)

Sunflower is an important oilseed crop containing 40–50 percent oil in the seeds. Potential seed yields can reach 5 tonnes/ha but average yields are much lower. The roots of sunflower plants can reach down to a depth of 2 m.

Nutrient requirements

The total nutrient uptake by a sunflower crop producing 3.5 tonnes of seed per hectare can be of the following order (IFA, 1992):

- macronutrients (kg): N 131, P_2O_5 87, K_2O 385, MgO 70 and CaO 210;
- micronutrients (g): Fe 732, Zn 348, Mn 412, Cu 59 and B 396

Macronutrients

Sunflower hybrids may need an application of 75–80 kg N/ha under irrigated conditions, but 50–60 kg N/ha is adequate for the rainfed crop. Application of N in three splits is advantageous (50 percent at sowing, 25 percent at buttoning and 25 percent at flowering). Excess N increases the risk of disease and lodging, with a consequent reduction in oil content. Recommended rates of phosphate application are 60–80 kg P_2O_5/ha. In view of the very high removal of potash (particularly where the stalks are also removed), potash application is necessary. It should be based on soil tests and crop removal. The recommended rates of potash application range from 50 to 150 kg K_2O/ha. Application of FYM is commonly recommended.

In view of the high S requirement of the crop, S application is normally suggested, particularly on S-deficient soils. This can also be made by using S-containing sources of N or P. Alternatively, S-free fertilizers can be supplemented with gypsum.

Micronutrients

Sunflower is very sensitive to B deficiency on calcareous or sandy soils and under moisture stress. Therefore, special attention should be paid to B nutrition. B may be applied to the soil either at sowing time (1–2 kg B/ha) or at the ten-leaf stage as a foliar application of 500 g B/ha (0.1-percent B solution).

ROOT AND TUBER CROPS

Potatoes (Solanum tuberosum L.)

Nutrition of the potato crop is characterized by its shallow rooting habit and rapid growth rate. Therefore, high yields necessitate an adequate supply of nutrients throughout the growth period. Potato grows best on slightly to moderately acid soils although it can grow successfully in soils with a wide pH range.

Nutrient requirements

Nutrient removal data from a number of situations have been summarized by the IFA (1992). In one estimate from the United Kingdom, nutrient removal (in kilograms per hectare) by a crop producing 90 tonnes tubers/ha was: N 306, P_2O_5 93, K_2O 487, MgO 19 and CaO 10. Results from India show that nutrient removal by potato is higher in the hills than in the plains. In the hills of Simla, nutrient removal by a normal crop yielding 36 tonnes tubers/ha was (in kilograms per hectare): N 117, P_2O_5 32, K_2O 224, S 14, Ca 37 and Mg 63. In the plains, where the crop duration is shorter than in the hills, an adequately fertilized crop yielding 34 tonnes tubers/ha removed 135 kg N, 21 kg P_2O_5 and 157 kg K_2O (Grewal and Sharma, 1993).

Macronutrients

N application promotes early development of the foliage and, therefore, of the photosynthetic capacity during the growth period. However, excess N may delay tuber initiation and so reduce yield. The N requirement depends on many factors including soil type and previous cropping. A preceding legume or another crop with high residual effects, or an application of organic manure, can reduce fertilizer N requirements by 40–50 kg/ha. High-yielding, rainfed or irrigated potatoes in temperate regions, with a growing period of 150–170 days, respond to as much as 200–300 kg N/ha. Most recommendations for potatoes in tropical and subtropical areas are in the range of 80–150 kg N/ha. Recommendations for particular regions and conditions depend on the climate, growing season, soil type, cropping system and variety.

Potatoes utilize both ammonium and nitrate N, but show a preference for ammonium, especially in the early stages of growth. Usually, the entire N is applied to the seedbed. However, in high rainfall conditions, a split application may reduce leaching losses. N applications after the start of tuber development may delay crop maturity. In high rainfall areas, sources such as AS and CAN are superior to urea.

Potatoes need a good supply of readily available P because their root system is not extensive and does not readily utilize less available P forms. Water-soluble P is the most efficient source for potatoes. Moreover, many tropical potato-growing soils are acid and immobilize P fertilizer rapidly. Because of the low P-use efficiency of potatoes, P fertilizer applications need to be considerably higher than the 30–50 kg/ha of P_2O_5 taken up by the crop. Therefore, fertilizer recommendations range from 60 to 100 kg/ha P_2O_5 for most tropical areas. In some temperate regions, the P requirement can be in the range of 100 to 300 kg/ha P_2O_5 depending on soil P status. The applied P is used more efficiently by potatoes where P is sideband placed, especially at low or moderate P application rates.

K plays a major role in starch production by the potato crop. Potato plants well supplied with K are found to withstand frost better than low K plants. Fertilizer K requirement depends on soil type and organic manure application. Irrigation can improve the availability of soil K, and there can be varietal differences in susceptibility to K deficiency. Potash recommendations range from 60 to 300 kg K_2O/ha according to growth conditions and yield level. However, in most developing countries, they are between 60 and 150 kg K_2O/ha. Mg deficiency can occur on leached, sandy soils and may be intensified by large K fertilizer applications. It can be controlled by Mg applied in amendments such as dolomite or by Mg-containing fertilizer materials.

The source of K influences tuber quality as potatoes are sensitive to excess chloride, particularly where tubers are meant for further processing into crisps and other snacks. Hence, application of K through potassium sulphate is usually preferred to potassium chloride. Therefore, potassium sulphate can be recommended where the value of greater starch production exceeds the higher cost of SOP compared with MOP. Potato quality is also influenced by nutritional imbalances. Excess N can reduce tuber dry matter and cooking quality, while K deficiency or excess chloride can cause tuber blackening.

Micronutrients

Soil application or foliar sprays are the widely used methods for supplying micronutrients. The micronutrient needs of potato can also be met simply by soaking the seed tubers in nutrient solutions. The non-dormant seed tubers are soaked in 0.05-percent micronutrient salt solutions for three hours. Dipping seed tubers in 2-percent zinc oxide suspension is also effective for meeting the Zn needs of the crop (Grewal and Sharma, 1993). The high seed rate of potato makes it possible to supply the micronutrient needs of the crop through soaking. The deficiencies of Cu and Mn are controllable by soil or foliar application. The storage life of potatoes can be reduced where there is a B deficiency. Potato cultivars can differ markedly with regard to their sensitivity to micronutrient deficiencies.

Organic manures

Bulky organic manures and green manures have an important place in the nutrient management of potato. They add nutrients and also improve the physical environment for better plant and tuber growth. In spite of their low nutrient content, they help in fertilizer economy. The tuber yields obtained with the combined use of organic manures and fertilizers are higher than those with the use of fertilizers or organic manures alone. Thus, the combined use of organic and mineral sources of nutrients is essential for sustaining high levels of potato production.

Sweet potato (Ipomoea batatas Lam.)

Sweet potato, a perennial root crop, is used for food, animal feed and in industrial materials. China accounts for 80 percent of world production.

Nutrient requirements

Nutrient removal by a crop producing 14 tonnes of biomass per hectare (10 tonnes of tubers and 4 tonnes of leaves) has been estimated at (in kilograms per hectare): N 51.6, P_2O_5 17.2, K_2O 71.0, MgO 6.1, CaO 6.3 and Fe 0.8 (IFA, 1992).

Macronutrients

On most soils, N application increases tuber yield. However, excess N can stimulate foliage production at the expense of tubers and may also lead to tuber cracking. The full benefit from N application is only obtained where there is also sufficient K. It is usual to recommend about 50 kg N/ha, but less on soils well supplied with N. Because the crop removes more K than P, fertilizer K has a greater effect on yield than does P. Under average conditions, about 50 kg P_2O_5/ha should be applied, but this needs to be increased to 70–90 kg P_2O_5/ha on soils with a low P status. The crop needs a good supply of K and an N:K_2O ratio of from 1:1.5 to 1:2. A common recommendation is to apply 80–120 kg K_2O/ha. Potassium chloride can depress root dry-matter content. Where this is the case, the use of potassium sulphate or a mixture of the two sources is recommended. Sweet potatoes can suffer from Mg and S deficiencies, hence their inclusion in the fertilizer schedule may be necessary.

Micronutrients

Sweet potatoes can also suffer from B deficiency, hence corrective control measures may be necessary. Soil application rates range from 9 to 26 kg borax/ha. For foliar application, the suggested rate is 5–15 kg Solubor/ha at a maximum concentration of 2.5–5.0 percent (Shorrocks, 1984).

Cassava (Manihot esculenta Crantz)

Cassava is an important tuber crop of the tropics. It is normally grown at low levels of fertility. Seventy percent of the world's cassava production is used for food either directly or in processed form. Cassava plants have the ability to withstand drought conditions. This is because of their inbuilt mechanism to shed their leaves under adverse moisture conditions. Where raised on natural soil fertility, yields may be very low, but the crop responds well to fertilizer application and to a good moisture regime. While average tuber yields are often 10–15 tonnes/ha, modern varieties grown under good management can yield more than 50 tonnes/ha.

Nutrient requirements

Cassava removes large amounts of nutrients. A crop producing 37 tonnes of fresh tubers per hectare removes the following amounts of nutrients including those contained in tubers (IFA, 1992):

- macronutrients (kg/ha): N 198, P_2O_5 70, K_2O 220, MgO 47, CaO 143 and S 19;
- micronutrients (g/ha): Fe 900 (tubers only), Zn 660, Mn 1 090, B 200 and Cu 80.

Macronutrients

Cassava responds well to fertilizer N with an expected yield increase of 50 kg of tubers or more per kilogram of N applied. With insufficient N, individual tubers are thin and contain less starch. However, excess N may result in an excess of vegetative growth at the expense of tuber yield. A common recommendation is to use 40–80 kg N/ha depending on circumstances. On low-fertility soils, up to 120 kg N/ha can be applied where other growing conditions are favourable. The total N to be applied may be split between a basal application and a top-dressing.

Many soils on which cassava is grown are poorly supplied with P, and the crop has consistently shown considerable benefit from P fertilizer, even though cassava makes better use of soil P than do potatoes. Under most conditions, 40–80 kg P_2O_5/ha is suggested.

A good supply of K is essential for cassava, giving a benefit of up to 100 kg of tubers per kilogram of K_2O and helping to offset the very large removal of K in the tubers at high yield. K increases yield primarily by increasing tuber size. K-deficient plants can contain toxic levels of hydrocyanic acid (HCN) in the tubers. On soils of moderate K status, 100–130 kg K_2O/ha is recommended, with adjustments for different soil K levels. The optimal timing of K application depends on the K status of the soil, which also determines the amount of K to be applied. Generally, K application in two equal splits (50 percent as basal and 50 percent two months after planting) gives best results in terms of starch and dry-matter content. In general, an N:K_2O ratio of 1:1 is suggested.

Micronutrients

Deficiencies of Zn, Mo and B can occur in soils under cassava. With optimal NPK application, soil application of 12.5 kg of zinc sulphate increased tuber yield by 4.0 tonnes/ha; 1.0 kg of ammonium molybdate raised it by 2.8 tonnes/ha; and 10 kg of borax increased tuber yield by 3.1 tonnes/ha. Zn deficiency can be controlled by the application of zinc sulphate at a rate of 5–10 kg/ha at planting or by incorporating zinc oxide before planting. Under moderate deficiency, foliar application of 1–2-percent zinc sulphate may be effective, while under alkaline conditions, stake treatment by dipping in 2–5-percent solution of zinc sulphate for 15 minutes is recommended.

Organic manure

Cassava benefits from an integrated application of organic manures and mineral fertilizers, which produce an additive effect. Under tropical conditions in India, the impact of applying 12.5 tonnes FYM/ha on tuber yields was equivalent to that obtained with 100 kg fertilizer N/ha used alone. Neither FYM nor any of the nutrients (N, P or K) applied individually could increase tuber yields by more than 3 tonnes/ha, but the combined use of FYM + NPK through fertilizers produced a yield increase that was four times greater.

Liming of acid soils

Cassava is often grown in acid laterite soils of pH 4.0–4.5. In such soils, liming has a large beneficial effect on the yield and quality of cassava. In Kerala, the main cassava-growing state in India, liming increased the starch content of tubers and decreased their HCN content. The application of calcium carbonate or a combination of calcium carbonate and magnesium carbonate increased tuber yields substantially.

SUGAR CROPS

Sugar cane (Saccharum officinarum L.)

Sugar cane is a tropical grass that is grown primarily for the sugar content in its stems. Grown on a variety of soils, it grows best on well-drained loams and clay loams. It can grow well in soils of pH 5.0–8.0. Under very acid conditions, liming is necessary, especially to avoid Al toxicity. Because sugar cane has a long life cycle (10–24 months after planting), and in many cases successive harvests (ratoons) are taken, its nutrient management is more complex than that of annual crops. The crop benefits considerably from water and nutrient application.

Nutrient requirements

Under Brazilian conditions, the nutrient uptake per tonne of cane yield is as follows (IFA, 1992):

- macronutrients (kg): N 0.8, P_2O_5 0.30, K_2O 1.32, MgO 0.50, CaO 0.42 and S 0.25;
- micronutrients (g): Fe 31, Zn 4.5, Mn 11, Cu 2.0, B 2.0 and Mo 0.01.

Under Indian conditions, a crop yielding 100 tonnes of cane per hectare absorbed 130 kg N, 50 kg P_2O_5 and 175 kg K_2O. Even on a per-unit cane basis, nutrient uptake varies considerably depending on the climate, cultivar and available nutrient status even at comparable yields (Hunsigi, 1993). Sugar-cane trash is particularly rich in K (3 percent K_2O). It is invariably burned in the field to take a ratoon crop.

Macronutrients

N has a marked effect on cane yields, and an application of 250–350 kg/ha is common. In some situations and with some varieties, excess N depresses cane yield. Sugar content of the cane decreases with increasing N supply and the optimal rate is that which maximizes sugar yield (cane yield × sugar concentration). Excess N may also affect juice quality and sugar recovery. Suitable water management in the final stages of growth can minimize depressions in yield and quality at high N rates.

The requirement for N fertilizer varies with yield potential and, particularly in plant cane, with the soil N supply. Plant cane is able to draw on mineralized N in the soil, which can vary from 50–150 kg N/ha. As an approximate guideline, sugar cane requires 1 kg N/tonne of expected yield. For the ratoon crop, the soil N supply is lower and the rule of thumb is to apply 1.5 kg N/tonne of cane. Thus, for example, a plant cane yield of 100 tonnes/ha would require 100 kg N/ha and a ratoon cane yield of 140 tonnes/ha would require 210 kg N/ha. Much more N may be needed in soils that are very low in organic matter and for intensively grown crops. The recommended rates of N for sugar cane in various parts of India range from 100 to 300 kg N/ha for a 12-month crop.

N for plant cane is usually applied in split doses. The first application of 25–50 percent of the total is made in the planting furrow or broadcast a week or two after planting. The second application should be made during the period of rapid growth and nutrient uptake, one to three months after planting. Where labour is available, the total N can be given in three splits, but all within 100 days of planting. The splits can be given at tillering (45–60 days), formative stage (60–75 days) and grand growth stage (75–100 days). Later applications are often less efficient and may reduce sugar content. For the ratoon crops, N should be applied immediately, or within two months after cutting the previous crop.

More specific recommendations for N, P and K should be obtained from local sources and experience. Various systems of foliar diagnosis such as crop growing and DRIS (discussed in Chapter 4) have also been developed. These provide guidance on fertilizer requirements from the analysis of specified leaves or other organs at specified growth stages.

Phosphate stimulates root growth and early tillering and, therefore, should be applied at planting. Placing P in the planting furrow increases the efficiency of P uptake, especially on less fertile soils. However, many soils adsorb P rapidly so that availability of this initial application can be low for the ratoon crop. The ratoon benefits from an application of P immediately after cutting the previous crop. For soils of medium P status, an application of 100–120 kg P_2O_5/ha to the plant crop is frequently recommended, rising to 200 kg P_2O_5/ha on P-deficient soils. For the ratoon crop, 60 kg P_2O_5/ha will usually provide enough P to stimulate regrowth.

Sugar cane needs a good level of K for a number of reasons. The harvested crop removes very large amounts of K and high yields can remove as much as 400 kg K_2O/ha. K fertilizer increases cane and sugar yields in most cases. Adequate K counteracts the adverse effects of high rates of N on cane sugar concentration and juice quality. Typically, K applications are in the range of 80–200 kg K_2O/ha, but more K may be used on high-yielding, irrigated crops and lower rates on soils rich in available K. Potash nutrition can be monitored by soil and plant analysis, and supplementary applications made where plant K concentrations fall below a specified level.

Sugar cane is sensitive to S and Mg deficiencies. In recent years, owing to the dominance of S-free fertilizers and, hence, reduced S input, S deficiency has frequently been encountered in intensively cropped coarse-textured soils. This can be corrected by using S-containing fertilizers to supply N or P. Application of sugar-factory waste (press mud) from the sulphitation process or of adequate

FYM (15–20 tonnes/ha) can also supplement soil S supplies. Mg deficiencies can occur where soil Mg status is low. Conversely, on soils extremely high in Mg, excessive Mg uptake may suppress K uptake and induce a K deficiency.

Micronutrients

Deficiencies of Zn, Cu and Mn and lime-induced iron chlorosis can occur in sugar cane. These can be controlled by application of deficient elements as their sulphate salts or chelates. Iron chlorosis can be corrected by spraying 2.5 kg of ferrous sulphate in 150 litres of water twice at fortnightly intervals. Sugar cane, like rice, reacts favourably to soluble silicates on some soils, which probably also releases soil P. To correct Zn deficiency, soil application of zinc sulphate at a rate of 25 kg/ha can be made on coarse-textured soils.

Sugar beet (Beta vulgaris L.)

Sugar beet is an important source of sugar in many parts of the world as the roots contain 13–20 percent saccharose. It grows best on slightly acid to neutral soils of porous structure. Under very good conditions, beet yields of up to 80 tonnes/ha can be achieved as compared with an average yield of 35 tonnes/ha.

Nutrient requirements

Nutrient uptake by 10 tonnes of beet along with the associated foliage averages 40–50 kg N, 15–20 kg P_2O_5, 45–70 kg K_2O, 12–15 kg MgO and 5 kg S, out of which the beets contain about 50 percent. Where the leaves are incorporated into the soil after harvest, the nutrients thus recycled must be taken into account in estimating the fertilizer requirement of the next crop.

Macronutrients

The maximum nutrient demand by the crop occurs 3–4 months after sowing. Therefore, most of the recommended nutrients should be applied early, before sowing. In Germany, fertilizer recommendations have been developed for various levels of soil nutrient status. The rate of N is determined by the nitrate stored in the profile at the beginning of the season up to a depth of 91 cm. As an example, 200 kg N/ha is required for a yield of 50–60 tonnes of beets. Where the nitrate content of the soil is 70 kg N/ha, the N to be applied is 130 kg N/ha. For other nutrients on a soil of very low nutrient status and at an expected yield of 50 tonnes of beets per hectare, the recommended rates (in kilograms per hectare) are: 200 P_2O_5, 400 K_2O and 100 MgO. Where the soil nutrient status is high, per-hectare rates of 50 kg P_2O_5 and 100 kg K_2O are recommended.

Micronutrients

Deficiencies of B and Mn can occur because sugar beet has a high demand for these micronutrients, especially on soils with pH of more than seven. Where necessary, 1–2 kg/ha B and 6–12 kg/ha Mn should be applied before sowing, or these nutrients may be applied through foliar spray.

FIBRE CROPS

Cotton (Gossypium spp.)

Cotton, the major source of natural fibre, requires a warm growing season. It grows best on well-drained soils with good structure. Very acid soils need to be limed. An adequate moisture supply is essential, especially during flowering and boll development. Satisfactory rainfed crops are grown in many countries. Cotton is also well suited to irrigated conditions, where the highest yields are obtained. Good management, including timely sowing and effective weed and pest control, is necessary for high yields and for best response to fertilizers.

Nutrient requirements

Under Brazilian conditions, a cotton crop (*G. hirsutum*) producing 2.5 tonnes of seed cotton per hectare absorbed the following amounts of nutrients (IFA, 1992).

- macronutrients (kg): N 156, P_2O_5 36, K_2O 151, MgO 40, CaO 168 and S 10;
- micronutrients (g): Fe 2 960, Zn 116, Mn 250, Cu 120 and B 320.

Macronutrients

N application increases cotton yield by increasing the number and length of branches, and, therefore, the number of flowers, seed cotton yield and seed index. However, the amount of N to be applied depends very much on local conditions (including water supply). Excess N should be avoided as it may reduce yield and quality by overstimulating vegetative growth and delaying maturity. Recommended rates for rainfed cotton are usually 50–100 kg N/ha while most irrigated crops need 120 kg N/ha or more. In some intensively cropped, irrigated cotton-growing regions, N applications are as high as 300 kg N/ha. Soil and plant-tissue analysis for nitrate can be used to monitor the N status of the crop so that the N to be given as top-dressing can be determined. It is usual to split the N application, part being applied to the seed bed and part as a top-dressing at the start of flowering. Irrigated crops with high yield potential may receive two or three top-dressings.

P increases the yield of seed cotton, weight of seed cotton per boll, number of seeds per boll, oil content in seed, and tends to bring early maturity. P application should be related to soil P status. Recommended rates vary from 30 to 100 kg P_2O_5/ha. Highest rates of P are generally recommended for irrigated hybrids, and lowest rates or no P for rainfed traditional cultivars. At low to moderate yields, cotton can be grown without K application, but it should be applied for higher yields, particularly on low K soils. Recommended rates are similar to those for P at 30–100 kg K_2O/ha. In some parts of the world, P and K deficiencies occur where rapidly growing crops are furrow irrigated. This also leads to a loss of bolls in a syndrome known as "premature senescence".

Cotton is subject to a number of other nutritional problems. Mg deficiency can occur on acid sandy soils. This can be avoided by liming with dolomitic materials. Leaf reddening is sometimes attributed to Mg deficiency. This can be corrected by spraying a solution of 5-percent magnesium sulphate 50 and 80 days after sowing. S deficiency occurs fairly widely in North and South America and in Africa. As little as 10 kg S/ha is required to overcome it, for which any soluble S fertilizer or gypsum can be used.

Micronutrients

B deficiency on cotton has been reported in a number of countries. Fe and Zn deficiencies also occur. All are controllable by well-proven foliar or soil applications. Zn deficiency can be corrected by soil application of zinc sulphate at a rate of 25 kg/ha in coarse-textured soils or by giving three sprays of 0.5-percent zinc sulphate solution during 45 days growth. B deficiency can be corrected by spraying 0.1–0.15-percent B on the leaves at 60 and 90 days.

Jute (Corchorus olitorius L., Corchorus capsularis L.)

Jute is an important fibre crop in which the fibre is extracted from the stem. Of the two main types of jute, *Corchorus olitorius* L. is known as tossa jute while *Corchorus capsularis* L. is referred to as white jute. Jute prefers slightly acidic alluvial soils. Much of the world's jute production is in Bangladesh and India. Improved varieties are capable of yielding 3–4 tonnes of dry fibre per hectare, which is equivalent to 40–50 tonnes/ha of green matter.

Nutrient requirements

On the basis of nutrient uptake per unit of dry-fibre production, white jute has a 40-percent higher nutrient requirement than does tossa jute. Thus, tossa jute is a more efficient species for fibre production. This may be due in part to its deeper and more penetrating root system. Total nutrient uptake per tonne of dry-fibre production, by the two species of jute is as follows (Mandal and Pal, 1993):

- *C. olitorius* (macronutrients, kg): N 35.2, P_2O_5 20.3, K_2O 63.2, CaO 55.6 and MgO 13.3;
- *C. olitorius* (micronutrients, g): Fe 368, Mn 119, Zn 139 and Cu 18;
- *C. capsularis* (macronutrients, kg): N 42.0, P_2O_5 18.5, K_2O 88.5, CaO 60.0 and MgO 24.5;
- *C. capsularis* (micronutrients, g): Fe 784, Mn 251, Zn 214 and Cu 19.5.

An interesting feature of the jute plant from the nutrient management point of view is that a substantial amount of the nutrients absorbed are returned to the soil with leaf fall before harvest. In the case of tossa jute, the percentage of nutrients absorbed that are returned through leaf fall are: N 42, P 19, K 18, Ca 26 and Mg 21.

Macronutrients

The common per-hectare rates of fertilizer application to jute are: 30–45 kg N, 10–20 kg P_2O_5 and 10–20 kg K_2O/. In general, liming of acid soils and the application of 10 tonnes FYM/ha is recommended. Well-decomposed FYM is to be added 2–3 weeks before sowing. In K-deficient areas, K application increases yield and also reduces the incidence of root and stem rot. The Ca requirement in acid soils can be met from liming. In Mg-deficient areas, magnesium oxide can be applied at the level of 40 kg/ha either through dolomitic limestone or through magnesium sulphate. Where noticed, S deficiency can be corrected through the application of common S-containing N and P fertilizers.

Micronutrients
Positive results have been obtained in some cases from the application of B, Mn and Mo. However, micronutrient application should be based on soil nutrient status and local experience.

PASTURES

Permanent pasture and meadows

Areas used for grazing domestic animals cover large parts of the land surface, ranging from sparsely covered wastelands to very intensively managed pastures and meadows. Therefore, plant yields range from less than 1 tonne/ha to more than 15 tonnes/ha of dry matter. Grassland vegetation rarely consists of only one kind of grass, but is mostly composed of various grasses, a variety of herbs and often legumes, which supplies nutritious fodder for grazing animals. On some soils, animals may suffer from deficiencies even with abundant fodder. Extensively used grasslands, composed of native species, are limited in potential by low rainfall or adverse temperatures. The principles of grassland nutrition and some aspects of nutrient supply have been discussed in Chapter 7.

Nutrient requirements
Nutrient uptake under various systems of grassland and fodder production is substantial (IFA, 1992):

- ➢ temperate grasslands (permanent grass and sown grass or leys) for a dry-matter yield level of 10 tonnes/ha:
  - macronutrients (kg): N 300, P_2O_5 80, K_2O 300, MgO 34, CaO 84 and S 24,
  - micronutrients (g): Fe 1 000, Zn 400, Mn 1 600 and Cu 80;
- ➢ temperate grasslands (grass/legume swards) for a dry-matter yield of 8 tonnes/ha:
  - macronutrients (kg): N 320, P_2O_5 69, K_2O 240, MgO 33, CaO 189 and S 25,
  - micronutrients (g): Fe 1 500, Zn 260, Mn 880, Cu 80 and Mo 5;
- ➢ tropical grasses for a dry-matter yield of 8 tonnes/ha:
  - macronutrients (kg): N 170, P_2O_5 46, K_2O 240, MgO 34, CaO 28 and S 16,
  - micronutrients (g): Fe 640, Zn 240, Mn 560, Cu 56, B 160 and Mo 2.4.

Nutrient removal is minimal under grazing as considerable quantities of the nutrients absorbed by the plants are returned to the field in dung and urine. Where the fresh or dry biomass is removed for making hay or silage and off-site feeding, nutrient removal is much larger than under grazing and should be replaced.

Macronutrients

Annual N fertilizer application on grassland varies from 0 to about 1 000 kg N/ha but generally ranges from 50 to 350 kg N/ha. Legumes can supply up to 100 kg N/ha to a grass–legume mixture in temperate areas and 200 kg N/ha in tropical areas. The type of N fertilizer used is of minor importance. Applications of N should be made after grazing or cutting and possibly before the rains, especially with urea. Examples for N application in a temperate climate with good growing conditions are:

- pastures: 150–200 kg N/ha, split into portions of 60 + 50 + 40 + 30 kg N/ha;
- meadows: 250–300 kg N/ha, split as 100 + 80 + 60 + 40 kg N/ha (yield 8–10 tonnes/ha of dry matter).

Fertilization with other major nutrients such as P, K and Mg can be based on nutrient removals, which are small from pastures because of recycling but large from meadows where large amounts of nutrients are removed in hay or silage. On soils of high-fertility status, nutrients removed from the field should be replaced. On pastures, nutrient removals with 1 000 litres of milk are 2 kg each of P_2O_5 and K_2O, and 0.2 kg Mg. For intensive pastures, inputs of 20–30 kg/ha each of P_2O_5 and K_2O, and 3 kg/ha of Mg are suggested. On meadows for dry-matter yields of 10 tonnes/ha (12 tonnes of hay), about 100 kg/ha P_2O_5, 300 kg/ha K_2O and 35 kg/ha Mg are adequate. Any kind of P fertilizer can be used. Potash fertilizers should preferably contain some Na in order to meet the needs of animals. S deficiency is being recognized in many areas that do not receive S input through fertilizers or atmospheric pollution.

Micronutrients

Adequate Mo is essential for effective N fixation. Where Mo deficiency is recognized (often in acid soils), Mo should be applied, most conveniently in the form of fertilizers fortified with Mo, e.g. molybdenized SSP (0.02 percent Mo).

Organic fertilizers

Grasslands often receive abundant manure and slurry, but mainly as nutrient sources and less for the supply of organic matter. Single applications of slurry should not exceed 20 m^3/ha on sown pastures. Up to double these amounts are acceptable on meadows and pasture, but grazing should not take place for several weeks after slurry application.

Chapter 9
Economic and policy issues of plant nutrition

There are many complex economic and policy issues related to nutrient management. A detailed discussion of the subject is beyond the scope of this document and readers are referred to the publication on *Fertilizer strategies* (FAO/IFA, 1999). In view of the importance of the subject, some practical aspects are discussed here.

Before farmers can be convinced about applying a purchased input such as mineral or organic fertilizer, they need knowledge about such inputs and their effects on crop yield in both agronomic and economic terms. Once convinced of using fertilizers in principle, they have to make the complex decision on how much and which fertilizer to use. Their decision on whether to use fertilizer on a particular crop is generally based on some form of economic judgement that includes past experience from using such inputs, the cash or credit available, and probable produce prices.

While calculation of the economics of applying fertilizers is relatively straightforward, the economics of using nutrient sources such as animal manure, compost, crop residues, green manure crops and urban wastes is more complex. Critical elements in the calculation of the economics of using these products are their variable nutrient composition, their residual effect and the cost and availability of labour to access, process and apply them. These factors are often overlooked when advocating different nutrient management strategies.

For practical use, all agronomic data on crop responses to nutrients should always be subjected to economic analysis in order to account for differences in input and output prices and to address the basic issue of whether and to what extent fertilizer application will be profitable to the farmer. The discussion here uses mineral fertilizers as an example but the issues are also applicable to the other nutrient sources. Information on the factors that affect the returns from nutrient application is equally valuable in decision-making.

FACTORS AFFECTING DECISION–MAKING

The principal elements of production economics as applied to fertilizer use consist of:

- physical yield response to applied fertilizers, price of fertilizer and crop including transport, handling and marketing costs as also the cost of servicing a loan;
- the individual farmer's decision-making and risk-taking ability.

The following economic and institutional factors have been identified as important in influencing the economics of fertilizer use (FAO/FIAC, 1983):

- The price relationship between fertilizers and the crops to which they are applied together with the market outlook for these crops, which largely determines the profitability and incentive for using fertilizers.
- The farmers' financial resources along with the availability and cost of credit, which largely determine whether farmers can afford the needed investment in fertilizers.
- Conditions of land tenure, which determine the degree of incentive for farmers to use fertilizers.
- Adequate supplies and distribution facilities in order to ensure that the right types of fertilizers are available to farmers at the right place and right time.

Although the relative importance of these factors varies depending on local and seasonal conditions, they are interdependent to a considerable extent. Each of them can be influenced positively or negatively by the government policies, financing facilities and marketing systems in a country.

Farmers will apply plant nutrients only where their beneficial effects on crop yields are profitable. The decision to apply external plant nutrients on a particular crop will generally be based on economics (price and affordability), but conditioned by the availability of resources and by the production risks involved (Figure 45).

Ideally, farmers' pursuit of higher income through higher yield should be balanced against the need to maintain soil fertility and avoid soil degradation. Most farmers in developing countries have little choice except to face a certain amount of soil fertility depletion each year. Therefore, the profitability of adopting INM should be viewed over a longer term as improvements in soil conditions associated with superior NUE tend to become apparent only after several cropping seasons. Thus, apart from the physical response to the application of plant nutrients, certain economic and institutional factors are also important determinants for decision-making on fertilizer use.

FIGURE 45
An example of the decision-making process used by farmers

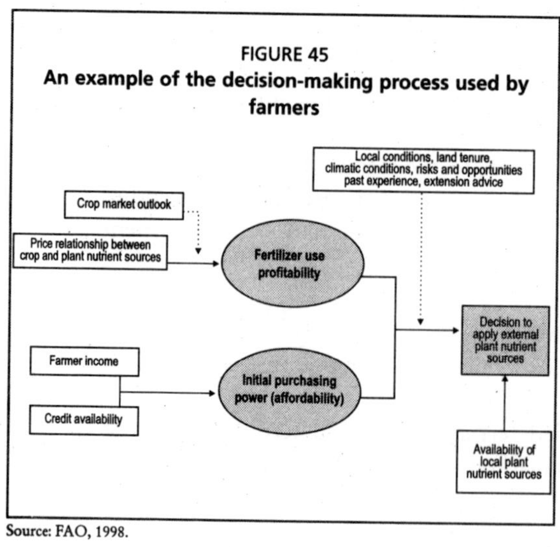

Source: FAO, 1998.

Small-scale farmers in harsh climates (drylands) and with scarce resources are compelled to look for short-term results when applying plant nutrients. Improved access to markets and low-risk production technology coupled with the removal of financial constraints and operational constraints (such as recycling of rainwater) will allow them to adopt plant nutrient management practices that are economically attractive and can support long-term sustainable crop production. It is not easy to have all these favourable conditions simultaneously.

Climate is one of the most difficult factors to take into account in deciding on nutrient additions to crops and pastures. In some developed agricultural areas, account is taken of soil moisture at planting and of the probability of rainfall using data provided by meteorologists. In irrigated areas, the availability of water can usually but not always be predicted.

In many developing agricultural areas, no such information is available and farmers must rely on their own experience and the experience of others. In this situation, the risk is much higher than in developed areas. As shown in Figure 35, the rainfall pattern has a major influence on crop response and, hence, economic returns to nutrient application. In the drought year, no fertilizer should have been applied, while in the year with above average rainfall, even the normal rate of application would have been insufficient for maximum yields. In those developing countries where irrigation facilities are well developed (e.g. India and Pakistan), the element of uncertain water supply is reduced. This allows farmers to invest in nutrients and target high yields. It is in such areas that the so-called green revolution took place and the productivity of irrigated cereals rose many times over in the period 1965–1990.

Yield maximization vs profit maximization

The basic requirement of profitable crop production is to produce an agronomic yield that can maximize net returns. Even the highest yield would not be of interest if its production were not cost-effective. Most farmers would like to maximize the net gains from whatever investment they can make in inputs. However, they should realize that top profits are possible only with optimal investment, correct decisions and favourable weather.

Whether a farmer aims for the maximum economic yield or maximum agronomic yield depends on circumstances. A farmer in a poor agricultural area with little or no purchasing power will generally try to produce sufficient food for family needs at the lowest risk. Such farmers are forced to operate at a subsistence level of farming. In these situations, maximum yields are not considered and even maximum economic yields are a distant goal. On the other hand, farmers in a developed area (even within a developing country) with access to cash and/or credit will generally try to maximize their return on invested capital and they are better equipped to take some risk.

The response function to fertilizer use is a basic tool that relates the amount of crop that can be produced in relation to the amount of fertilizer and other farm inputs applied. In other words, there will be a maximum obtainable amount of crop produce for any given amount of fertilizers and other farm inputs used. This is influenced considerably by the soil fertility status and this is why economically optimal rates of nutrient application should generally be based on soil tests and crop removals as discussed in Chapters 4, 6 and 7.

FIGURE 46
Response function of wheat grain yield to added N established in a field experiment

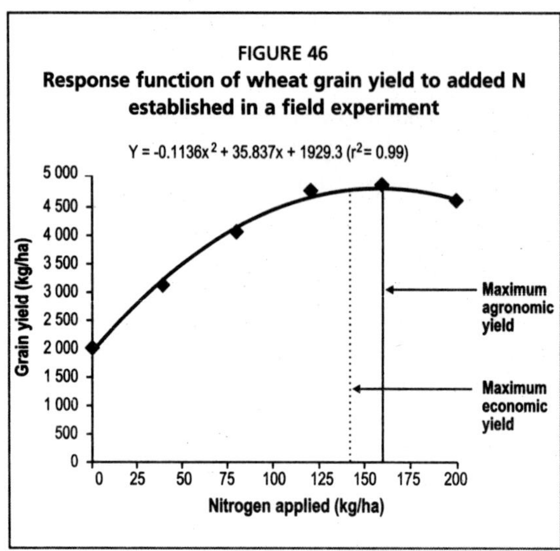

In theory, the determination of the response function should take into account all variables, such as the use of other inputs that influence crop yield. In a response function, crop yield is a function (f) of several factors: $Y = f(X_1, X_2.... X_n)$, where Y = crop yield, and $(X_1, X_2..... X_n)$ = inputs included within the response function as having the major influence on production.

However, normal practice in fertilizer response function studies is to restrict the variable inputs to the rate or level of fertilizer nutrient applied keeping all other factors constant. At the farm level, this can be a limitation as it does not take into account factors such as labour costs and weather fluctuations.

The important information supplied by the response function is the increment of crop yield (grain, tubers or fruits) obtainable from increasing levels of fertilizer application. This information is essential for determining the optimal fertilizer application rate (i.e. the most profitable level of fertilizer use). Such a level is not valid for all time even for a given crop on a given farm. It changes constantly depending on input costs, output price and the rate of crop response per unit of input.

The classical production function normally exhibits stages of increasing, diminishing and negative returns according to the law of diminishing returns whereby, beyond the initial linear range, successive increments of input result in a decreasing rate of response per unit of fertilizer applied. Farmers are interested only in the first and second stages of the response function. Their specific interest depends on whether their main consideration is maximization of profit or the rate of net return (BCR) from the money spent on fertilizer. This attitude is conditioned by the resources available and by their views on risk and uncertainty. Where the response function to a given input is known, as shown in Figure 46, it is possible to compute the economic and agronomic optimal application rates. Using N as an example, the response function is of the form:

$$Y = -0.1136X^2 + 35.837X + 1\,929.3$$

where Y = wheat grain yield valued at US$0.25/kg, and X = rate of N applied as fertilizer costing US$0.90/kg N.

In order to calculate the rate of N for maximum agronomic yield, the first derivative of the response function has to be set to zero: $dY/dX = 0 = -0.228X + 35.84$, $X = 157$ kg N/ha (for maximum yield).

The profit-maximizing optimal rate of N is calculated by setting the first derivative of the response function to the price ratio of the fertilizer to the grain price, i.e. (US$0.90/US$0.25): $0.228X + 35.84 = 0.90/0.25$; $X = 141$ kg N/ha (for maximum profit).

The yield-maximizing rate of nutrient application (any nutrient) will be somewhat higher than the profit-maximizing rate (157 vs 141). This is because the extra yield from maximum economic to maximum agronomic is uneconomic. Unless a farmer is aiming to win a highest yield competition, the profit-maximizing rate of nutrient application should not be exceeded.

While analysing the economics of fertilizer use, the principal considerations are the production increase attributed to fertilizer (or physical response), and the relationships between the cost of fertilizers and the price of produce. Where the objective of farmers is to obtain the economic optimal value from the use of fertilizer, their concern is to operate within the second stage of the response function where the yield obtained from a unit of fertilizer (the marginal yield) is increasing but at a decreasing rate.

Table 39 presents an example where the application rate of 150 kg N/ha is divided into six increments of 25 kg each. The example in Table 39 can be computed for any monetary unit. As illustrated in this table, each increment up to 125 kg N/ha produced sufficient crop to leave a net profit. As the number of units of N increased, the total crop yield also increased while the marginal yield increase per unit of fertilizer applied (column 5) declined. The marginal return from the fifth increment (from 100 to 125 kg N) was positive. However, the next increment (from 125 to 150 kg N) resulted in a net loss. This was because the 20 kg of grain it produced was not enough to pay for the 25 kg of N used to produce it. Hence, the marginal rate of return for the last increment was not favourable for going beyond 125 kg N/ha. The exact cut-off point would be the last kilogram of N that paid for itself. That would also be the profit-maximizing rate. It can be calculated for any situation.

In simple terms, the yield-maximizing dose (YMD) (close to 150 kg N in Table 39), is always somewhat higher than the profit-maximizing dose (PMD) (close to 125 kg N). The small portion between PMD and YMD consists of a positive but uneconomic response. For farmers in general, the PMD is of interest.

TABLE 39

The economics of incremental crop response to increasing rates of fertilizer application

N added	Yield	Each increment of N		Effect of each increment on yield		Net returns (value - cost)
			Cost	Crop	Value	
(kg/ha)	(kg/ha)	(kg/ha)	(Rs10/kg)	(kg)	(Rs)	(Rs)
0	1 500	0	–	–	–	–
25	2 200	25	250	+700	4 200	+3 950
50	2 750	25	250	+550	3 300	+3 050
75	3 150	25	250	+400	2 400	+2 150
100	3 400	25	250	+250	1 500	+1 250
125	3 550	25	250	+150	900	+650
150	3 570	25	250	+20	120	–130

Maximization of net returns or value–cost ratios

A question is sometimes raised as to whether a farmer should aim at maximum net returns from fertilizer use or at the maximum rate of gross returns as indicated by the value–cost ratio (VCR).

The decision by farmers to use fertilizer based on the VCR level depends on their own standard of profitability. However, the general rule is that a VCR of at least 2:1, i.e. a return above the cost of fertilizer treatment of at least 200 percent, is attractive to farmers. However, the absolute net return should also be considered because, at low application rates of fertilizers, the VCR may be very high owing to the small cost of the treatment and the associated high rate of response. However, at low application rates, the net return would also be small and unattractive to farmers. In addition, other factors should also be taken into account. These include the likelihood of the expected yield being obtained, produce storage facilities, an assured market for the crop, and the assured availability of the fertilizers to farmers. This aspect is discussed below.

FIGURE 47
Relationship between crop response and profit

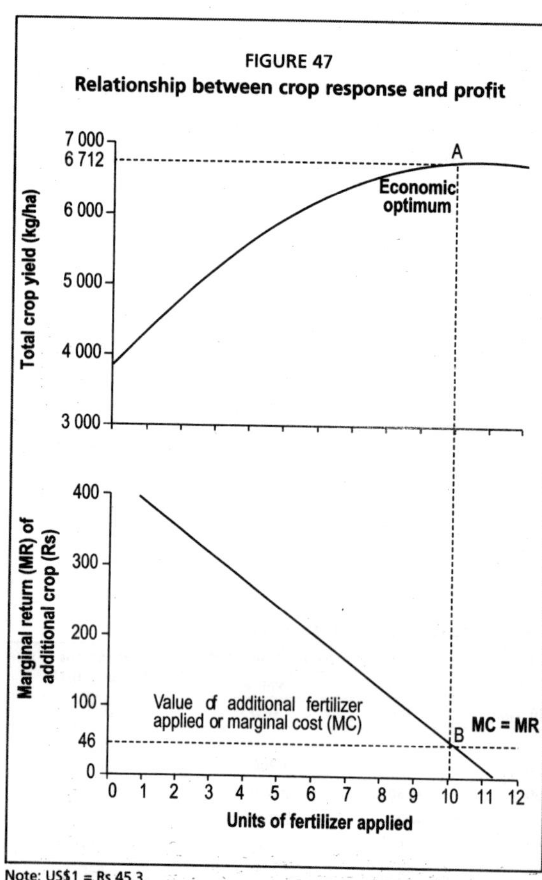

Note: US$1 = Rs 45.3.
Source: FAO/FIAC, 1983.

As the ratio of crop and fertilizer prices changes, the amount of fertilizer applied also needs to change in order to maintain optimal economic returns. The extent of the change depends on the shape of the response curve. This concept of economic optimum based on the rate of marginal return is further illustrated in Figure 47 using data from India. It is important that information on marginal yield and the prices of fertilizer and crop produce be available. Such computations can be made for any situation.

Most farmers, particularly in developing countries, often use less than the recommended fertilizer rates, and do so too in an imbalanced manner. This is because of a number of factors that include: their perception about the role or importance of each nutrient and its unit price; the anticipated yield increase; expected crop prices; cost and availability of fertilizers; level of financial resources and credit availability; land tenure considerations; the degree of risk and uncertainty, and the farmers' ability to bear them. Therefore, it is natural for farmers to be cautious and build in a fair safety margin when deciding the level of fertilizer to apply. Farmers can operate over a wide range of fertilizer application rates and benefit from them right up to the optimal level. In this respect, plant nutrient sources are very different and very flexible compared with other agrochemicals (pesticides and herbicides) that can only be effective where applied at a single critical rate.

Generally, farmers with sufficient resources can use fertilizer rates that are at or near the optimum in terms of economic returns. On the other hand, the rates of fertilizer application of interest to small-scale farmers with limited resources, who are concerned with the economic return on the money they spend on fertilizers, are those on the steeper part of the response curve where the BCRs (discussed below) are higher. However, such farmers will be sacrificing a considerable portion of the achievable yields and profits by operating below the optimal level.

ECONOMICS OF FERTILIZER APPLICATION

The required data sets

For a simple analysis, the minimum data required for economic analysis of fertilizer use consist of: (i) cost of fertilizer; (ii) value of the extra crop produced as a result of using the fertilizer; and (iii) the rate of increase in yield per unit of nutrient applied or the rate of response. For nutrients that leave a residual effect and benefit more than one crop, the cost of nutrient should be distributed among the crops benefited.

For a detailed economic analysis, the data set required is much larger and consists of:

- cost (expenditure):
  - cost of fertilizer (net),
  - interest on loan taken to buy fertilizer (until it is repaid),
  - transport charges of fertilizer to the village,
  - fertilizer application costs (labour, machinery and energy),
  - harvesting, threshing, winnowing and storage cost of extra crop produced by fertilizer use,
  - cost incurred in storage of produce,
  - cost of transporting the extra produce to the market,
  - direct and indirect marketing cost,
  - adjustment in fertilizer cost for residual benefit credited to next crop;
- income:
  - sale proceeds from main produce resulting from fertilizer use (grain, fruit, tubers, etc.),
  - sale proceeds from products resulting from fertilizer use (straw, stover, sticks, etc.);
- gross returns: sum of items under income;
- net returns: gross returns - cost;

- rate of gross returns: gross returns/cost (VCR);
- rate of net returns: net returns/cost (BCR).

Computation of economics

Apart from calculating the economically optimal nutrient application rates that are associated with maximum net returns, the rate of profitability of fertilizer use can be determined by using either the VCR or the BCR. The VCR is obtained by dividing the value of extra crop produced by the cost of fertilizer or any other nutrient source. The BCR is obtained by dividing the net value of extra crop produced (after deducting fertilizer cost) by the cost of fertilizer. Therefore, the VCR is an indicator of the gross rate of returns, while the BCR indicates the net rate of returns. In a simple way, BCR = VCR - 1.

Economic analysis can also be used to determine the units of crop produce required to pay for one unit of fertilizer nutrient or, alternatively, in a given price regime, the response rate required for a commonly accepted minimum VCR. Where three units of grain are needed to pay for one unit of nutrient, then a response rate of 6 kg of grain per kilogram of nutrient must be obtained for a VCR of 2:1. This also has implications for NUE as an improvement in efficiency will result in a higher VCR from the same investment.

Many fertilizer trials-cum-demonstrations do not permit the calculation of the response curve to the different nutrients owing to the design used. Nevertheless, where the range of treatments is wide enough, the net return and VCR can be determined. The example in Table 40 (based on FAO Fertilizer Programme data) illustrates this.

In the example in Table 40, the lowest N–P_2O_5–K_2O treatment (40–40–40) gave the highest response and highest net return with a high (but not the highest) VCR of 4.3. On the other hand, the highest N–P_2O_5–K_2O treatment (80–80–80) did not give the highest response or economic return. The highest VCR was obtained from the 40–0–0 treatment and its economic return was only slightly less than that from the 40–40–0 treatment. Assuming these results to be economically representative, the 40–40–40 treatment could be recommended for use by the better-off farmers and the 40–0–0 treatment by those with limited resources to purchase fertilizers. The real economically optimal rate is somewhere between the 40–40–40 and 80–80–80 treatments and this should be computed statistically. Depending on the

TABLE 40
Example of net returns and benefit-cost ratio as determined from the results of field trials

Treatment N–P_2O_5–K_2O (kg/ha)	Yield increase (kg/ha)	Increase (%)	Gross return	Cost of fertilizers (US$/ha)	Net return	VCR
Control	3 000	-	-	-	-	-
40–0–0	890	29	122.82	16.80	106	7.3
40–40–0	1 090	36	150.42	30.30	120	5.0
40–40–40	1 455	49	200.79	47.10	154	4.3
80–80–80	910	30	125.58	94.20	31	1.3

soil fertility level, it is possible to indicate that the profit-maximizing rate is either 80–60–60, 80–80–40 or 80–40–60.

Calculating the economics of residual value of nutrients

The application of a number of nutrients, particularly P, S, Zn and Cu, benefits more than one crop in succession. P is the best-known example among the major nutrients that leave a residual effect.

Where repeated applications of P are made, the P not used from the first application remains effective in the soil and can contribute to P supply to the following crop. In most cases, the economics of P fertilization in many developing countries continue to be worked out on a single-crop basis. Where a fertilizer trial with a nutrient such as P is conducted with repeated applications made in three successive years, the response curve appears to move to the left (Figure 48). The reason for this is that the residual P from the first-year application is contributing to the P supply in the later years. This implies that as the soil P status improves as a result of repeated applications, lower rates of P application are needed in subsequent years to obtain optimal yields. This allows for the exploitation of accumulated P on a limited scale. Such an increasing P status of soils should be reflected in a good soil test report so that the optimal P application rate can be adjusted. The same principle applies to all nutrients that leave behind a significant residual effect (Zn and Cu on a longer-term basis, and S on a relatively shorter-term basis).

Ideally, the contributions of residual P should be assigned a monetary value, and also an interest could be charged on the money locked in this P. This may not be acceptable in all cases, e.g. where the farmer argues that the freshly applied soluble P is more valuable (more effective) than the less soluble residual P. For practical purposes, it is necessary to know the number of crops that will benefit significantly and the quantum of benefit (response). Where four crops in succession benefit from an initial application and their successive share of the cumulative yield increase (crops 1 to 4) is taken as 100, then the cost of P fertilizer can be apportioned to each crop according to its contribution in the cumulative response.

FIGURE 48

Yield response to P application at a fixed site when P is re-applied in three successive years

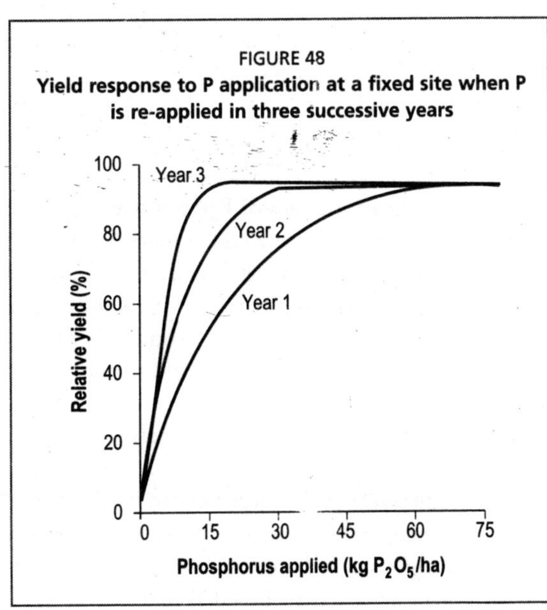

The challenge lies distributing the cost of a P application among various crops raised in a sequence that are the potential beneficiaries. Theoretically, if the effect of a P application last four years and the percentage share of each crop in the total yield increase obtained over four years is 50, 30, 15 and 10, then the cost of an initial P application can also be allocated in this proportion for economic analysis on a cropping-system basis. As an example, only 50 percent of the cost of P may be set against the crop receiving it because the remaining 50 percent of yield response is observed in the following three crops. This also helps in modifying the rates of application on builtup soils ultimately reaching P replacement (removal) values.

In one study (Goswami, 1976), the direct and residual response to P was evaluated in several systems of double cropping involving two crops in succession per year in India. Averaged over several field experiments with cereals, out of the total rotational response to P, the direct component was 60 percent and the residual component was 40 percent. Where P was added to the rainy season cereal and the winter crop was raised on residual P, the total rotational response consisted of 57 percent direct and 43 percent residual. Where the same amount of P was applied to the winter crop, the rotational response was made up of 63 percent direct and 37 percent residual. This shows that even in a single-year rotation, dividing the cost of P among the two crops is justified. Such partitions between direct and residual effects should be based on local research.

The effect of taking the residual effect into account in the economics of P fertilization is illustrated in this example using an application of 60 kg P_2O_5/ha to wheat in a wheat–rice cropping sequence:

- response of wheat to P (direct): 500 kg/ha;
- response of rice (residual): 300 kg/ha;
- value of wheat produced: US$66;
- value of rice produced: US$33;
- cost of 60 kg P_2O_5 through DAP: US$20;
- net returns from P use (basis: direct effect only): US$46 (66 - 20);
- net returns from P use (basis: direct and residual effects): US$79 (66 + 33 - 20);
- VCR with direct effect only: 3.3 (66/20);
- VCR with direct and residual effects: 5.0 ((66 + 33)/20).

The economics of P application also improve where the higher response is also for a crop that has a higher market value (e.g. wheat as opposed to millet, or oilseed as opposed to cereal). Thus, a beginning could be made towards economic analysis on a cropping-system basis by allocating only 60 percent cost of P fertilizer to the first (directly fertilized) crop. Otherwise, the returns from P application to the directly fertilized crop would suffer a penalty while the crop feeding on residual P would receive a bonus in terms of P residues. The detailed analysis should include more than one crop that benefits from the residual effect, as discussed below.

Calculating the indirect costs of applying fertilizers

Where fertilizers are applied to soils, many of them affect soil pH and other soil properties. Where acidifying fertilizers are applied that lower soil pH, the acidity produced has to be corrected by the application of liming materials. When the same fertilizers are applied to alkaline soils, the acidifying effect may result in additional benefits, such as increasing the availability of some nutrients (e.g. P and Zn). It is possible to ascribe a value to this effect although not directly but in terms of yield equivalent involved. In principle, this means that fertilizers should be costed not only for the nutrient they supply but also for their positive and negative effects on soil health.

Where AS is used as a source of N, the cost of lime needed to neutralize the acidity produced by the AS should be added to the cost of the AS. Similarly, where SSP is used as a fertilizer on S-deficient soils, its cost should be split between P and S. In the case of crops such as groundnut, where the Ca component of SSP also plays a role in pod formation, the cost of SSP should be split between P, S and Ca, particularly on acid soils. These are the issues that warrant examination where one moves from one-sided to multifaceted economics of nutrient application.

ECONOMICS OF ORGANIC MANURES AND BIOFERTILIZERS

The calculation of the economics of organic manures and biofertilizers is more complex than that of nutrients applied through mineral fertilizers, especially N (which leaves no or only a small residual effect).

Organic manures

Bulky organic manures have a more profound effect on improving soil physical properties than on nutrient supplies. The monetary value of improvement in soil conditions is not easy to estimate. However, the physical and chemical advantages of using organic manures are expected to be reflected in the crop yield increase. Therefore, it is simple to compute the economics of organic manures by treating them in the same manner as fertilizers that give both a direct and a residual benefit. It is easier to cost organic manures on the basis of cost of material plus application cost without splitting the total amount into individual nutrients. A further complication arises in trying to divide the cost of an organic manure among nutrients and organic matter, which primarily affects soil physical properties. The yield increase is expected to reflect the improvement in soil physical conditions as a result of manuring as well.

Green manures

Green manures bring in the organic matter produced as a result of photosynthesis but otherwise recycle the soil nutrients absorbed by them. Leguminous green manures do bring in a net N input. This can be costed in terms of equivalence of fertilizer N (if similar use efficiencies are assumed) or the cost of raising the green manure and the value of extra crop produced can be used for working out the economics. Here again, residual effects should be taken into account.

Biofertilizers

The economics of biofertilizers or microbial inoculants can be calculated either by costing the biologically fixed N in terms of the cost of fertilizer N that produces a similar yield increase, or by deducting the cost of inoculant plus its application cost from the value of extra yield produced. Residual benefit from the N fixed as a result of inoculation is not easy to compute except in terms of the value of extra crop produced.

It is necessary not to lose sight of the many ways in which a farmer can end up with low returns or even run into loss by using fertilizers. Prominent among these are: (i) continuous imbalanced nutrient application; (ii) growing low-yielding crop varieties; (iii) inefficient fertilizer use; and (iv) application of fertilizers without addressing other soil health constraints such as strong acidity or alkalinity. In order to maximize profits from fertilizer use, it is necessary to devote equal attention to factors and inputs other than fertilizers.

POLICIES FOR EFFECTIVE PLANT NUTRITION

Long-term planning and monitoring of the use of plant nutrients needs to aim at reconciling four objectives: (i) agronomic and economic efficiency to maximize agricultural output from available nutrient supplies; (ii) maintenance and enhancement of the production capacity of the natural resource base; (iii) consistency with a country's overall economic goals; and (iv) safeguarding the social security and livelihood-earning capability of the rural populations. Timely consideration of these issues is essential to planning and implementing a consistent and comprehensive policy both in the short term and the long term.

Fertilizer policies need to develop into INM policies so that diverse sources of plant nutrients find their rightful place in meeting the total nutrient needs of a country. Such a policy, besides serving as a tool for minimizing the depletion of soil fertility, would provide for a judicious use of the locally available manurial resources, maintain the soils in good health, ensure good yields on a sustained basis, and minimize the adverse impact of mineral nutrient resources on environment.

Planning

The efficient management of plant nutrients requires adequate involvement and planning in a wide range of areas. These tasks should ideally involve government, cooperatives and the private sector. A focal point for advice and planning on various sources of plant nutrients is essential for the establishment of a well-integrated plant nutrition policy including fertilizer policy. This should be well coordinated with the country's agricultural and food-security policies. An advisory unit with these functions could also provide required inputs for the formulation of a pricing and marketing policy. Such a unit could be made responsible for demand forecasting and identifying linkages with industry, research, extension services and farmers associations.

Assessment of nutrient requirements

An accurate assessment of plant nutrient requirements is the basis for planning the use of local sources of nutrients and for deciding upon domestic production and/or import of fertilizer products and raw materials, including the eventual use of foreign exchange to finance imports. Fertilizer demand projection is an assessment of the plant nutrient volumes that will be required to meet agricultural production targets. As against this potential demand, actual demand refers to the quantities that growers are likely to order. For example, in the case of N, in areas with sizeable acreage under legumes and wetland rice, the contribution of N from appropriate biofertilizers should be taken into account for finalizing total N needs. These should also take into account the nutrients expected to be available from organic resources on a realistic basis. Policies on the effective use of phosphates should include the use of a wide range of materials varying from fully water-soluble fertilizers to effective PRs depending on soil pH, crop duration, availability of local resources and distribution logistics. Similarly, policies concerning potash requirements should take into account the scope for recycling K-rich crop residues, organics and finished fertilizers.

Quality control

The setting up of fertilizer quality standards is an important part of fertilizer policies. Many countries have a fertilizer legislation in place and the machinery to enforce it. Fertilizer legislation deals with product specifications in terms of nutrient contents, inert material, physical properties, weight, packaging and labelling requirements, and the measures to enforce the legislation. Although the scope of fertilizer legislation varies from country to country, it usually has the following features:

- It defines the term "fertilizer" and provides a list of materials that can be labelled and sold as fertilizers. This means that no unlisted material can be labelled or sold as a fertilizer even though technically it may be an excellent fertilizer.
- It lays down the quality standards for the listed fertilizer products and specifies their physical and chemical properties in quantitative terms for maintaining quality. Apart from the nutrient content, specifications concerning the moisture content, particle size and the permissible limits of undesirable constituents are indicated.
- It lays down the packaging and labelling requirements and specifies the information to be provided on the fertilizer bag or any other type of packing.
- It lays down the procedures and regulations for the registration and licensing of the manufacturers, importers, and the distributors, along with the details relating to the mandatory information to be furnished by them to the regulatory authority at prescribed intervals or as and when required, besides identifying the personnel entrusted with the task of enforcing the legislation, their duties and their powers.
- It lays down the procedures for collection of samples, search procedures, disposal of substandard stocks, seizures of stocks, issue of notices in case of legislation violation, and initiation of legal proceedings.
- It specifies detailed standard analytical methods for fertilizer samples for quality checking.

At present, most such legislations are confined to mineral fertilizers. In several countries where quality standards have been or are being developed for organic and biofertilizers, these are not always a part of the legislation. In such situations, quality standards cannot be enforced by law – a situation that INM policies need to remedy.

Labelling

Product labelling is generally specified in the fertilizer legislation of many countries. It is essential to provide correct information about the product to dealers, extension workers and farmers. Labelling also permits the enforcement of fertilizer legislation. In several countries, detailed directives are given to manufactures as to what should and what should not appear on the label or the bag. Typically, the information to be printed on the fertilizer bags consists of: (i) name of manufacturer/importer; (ii) brand name and trade mark; (iii) name of fertilizer; (iv) nutrient content in percentage terms on a dry-weight basis (N-P_2O_5-K_2O), and (v) gross and net weight. In the case of phosphate, the total and water-soluble P_2O_5 contents are usually specified.

Most bags also mention the words "use no hooks" for the information of farm labour and others to ensure that the bag and the product inside is prevented from possible damage during handling. Labelling specifications can change to reflect changing needs. For example, until a few years ago, manufacturers of S-containing fertilizers (e.g. AS and SSP) in India were not allowed to print the S content of the fertilizer on the bag. This has now changed and printing the specified S content on the bag is compulsory. In the case of biofertilizers, the expiry date of the product is normally stated on the package. Where this is not done, it should be made compulsory.

Packaging

Packaging specifications are usually a part of appropriate legislation and quality control. Proper packaging should ensure ease of handling and transport, reduction of losses and ability to withstand unfavourable weather conditions, while keeping product prices affordable within the conditions and constraints of the distribution system. At the same time, it should convey the right information to the users. Fertilizer distribution systems and requirements for storage and transport (including humidity) determine the quality of fertilizer packaging. It also has to take into account the chemical and physical properties of the products and conditions of storage, especially at the end of the distribution chain.

Pricing and subsidies

Pricing is an important factor that affects the farmer's acceptance of a product in terms of the investment needed and returns expected. Input pricing always has to be viewed in relation to the likely prices of the output in order to see that their use is remunerative. The choice between produce price incentives and input subsidies to stimulate production has long been a controversial issue. The majority of developing countries in Asia provide subsidies to inputs such as fertilizers, amendments and power in agriculture, while developed countries support agriculture using other mechanisms, often with indirect or invisible effects, and not always designed to stimulate production.

Other incentives to fertilizer use take the form of guaranteed support prices for agricultural produce, duty-free imports of fertilizer and tax exemptions for credit and investment in fertilizers and crop production. Such measures affect the profitability of external nutrient application and provide the required economic motivation for increasing crop production.

Subsidies given directly or indirectly to farmers for fertilizers and other farm inputs have been the most important pricing policy factor in many developing countries. Their effect in increasing plant nutrient demand is readily and clearly identifiable. In some cases, a particular nutrient (most commonly N) is subsidized while other nutrients are not. This leads to a distortion in balanced nutrient application as farmers tend to apply the subsidized nutrient in preference to the costlier unsubsidized nutrients regardless of the nutrients needed by their soils and crops. This not only results in imbalanced and inefficient nutrient use, but it also promotes the mining of soil nutrient reserves of the nutrients not being subsidized, and consequently the depletion of soil fertility. In the long run, such subsidies become counterproductive as the depletion of other nutrients starts to limit crop response. Therefore, it is important that fertilizer policies do not treat each nutrient in isolation but rather take a holistic view.

Financing

Building up favourable conditions for adequate financial support to the fertilizer trade and distribution (besides credit facilities to farmers) should be one of the major objectives of an effective plant nutrition policy. Fertilizer demand is often highly seasonal. The pattern of seasonal demand is different for nutrients such as P and K, which are given before planting, compared with N, which can be given in several splits during crop growth. Therefore, the cash-flow requirement of fertilizer traders is high, involving considerable amounts of money for which adequate commercial credit should be available in order for the supplies to reach rural markets well ahead of the application season. In several cases, manufacturers or other suppliers provide input to distributors on credit for varying durations. The distribution credit (credit given to dealers) and the production credit (credit given to farmers) both have a very important role in the marketing and distribution of farm inputs. The interest rates and other terms and conditions laid down by financial institutions have a strong bearing on credit offtake. A major recent initiative in the area of farm finance is the provision of special credit cards to farmers in India.

Transport and storage

Adequate transport and storage are part of the essential infrastructure needed to ensure an efficient use of fertilizers. Product planning and movement into an area, keeping in view the soil nutrient deficiencies and cropping pattern, have a major effect on achieving balanced and efficient fertilizer use. This requires effective coordination between research, extension and trade. In order to promote INM, suitable transport and storage facilities are needed, especially in the case of biofertilizers. This is particularly important for the viability of microbial inoculants in tropical and subtropical areas. The costs involved are a relevant factor for establishing the priorities to be assigned to the use of alternative means of transport. These often depend on the distance between the production site and area of consumption. In several cases, a fertilizer bag may require manual handling six or more times between the factory and the farm. In such cases, handling costs can even exceed the transport costs. Sound logistics and efficient handling and transport of materials can lower the storage costs by reducing the storage period. Effective policies need to focus on developing an effective and efficient transport and storage network to serve the needs of the region.

Marketing

The establishment and strengthening of a viable agricultural-input marketing system should be one of the major objectives of a plant nutrition policy. Fertilizer marketing normally involves three or four stages starting from the factory or port before the material reaches the farmer. The actual system used varies from one country to another and even from one company to another. In most cases, the marketing chain consists of: producer – wholesaler – retailer – farmer. The number of links used in the marketing chain is generally fewer in the case of private companies than in the case of government or institutional agencies. Fertilizer marketing systems should basically satisfy the farmer's requirements while being profitable for the marketer. These systems require the careful design and implementation of policies, in which the right balance of government and private participation in the production, import and distribution of fertilizers has to be found. This is a critical issue that is highly dependent on national economic and political conditions in many developing countries.

Effective marketing systems should promote efficient fertilizer use through balanced supplies backed by good extension advisory services. Farmers should be encouraged to heed soil-test-based recommendations and translate these into the right fertilizers with the help of extension services and industry agronomists. As the use efficiency of plant nutrients also depends on the status of other production inputs, a very positive development in several countries is the establishment of multi-input distribution enterprises and farm service centres. Such initiatives, in which a range of inputs (along with nutrient sources) and services are available to the farmers under one roof, need to be encouraged by policy-makers and financial institutions.

Extension and training

Extension and training systems consisting of demonstrations, training sessions, training materials and extension efforts on efficient crop production and nutrient management techniques are essential components of farm-support policies. Policy measures, especially for the developing parts of the world, need to have a strong orientation towards the augmentation of extension and training facilities. Extension requirements must be assessed and the services established properly on a country-specific basis in order to match the technological level and experience of the farming community. Adequate technological packages, including the balanced use of mineral fertilizers, as part of INM and basic knowledge of the economics of fertilizer use have to be introduced. Farmers should be brought to appreciate the contribution of sources other than mineral fertilizers and how these can be used for adjusting fertilizer recommendations. Research and extension efforts should provide motivation for farmers' increased participation in the development, testing and adoption of new technologies. They should also provide for receiving and taking into account feedback from the field on a regular basis.

It would be desirable to train farmers so that they can compute the nutrient balances of their farms. By doing so, they could adopt such INM practices that would minimize the depletion of their soils and also use locally available nutrient sources most productively in a pre-planned manner. Large-scale efforts would be needed to train extension field staff in the area of INM so that the essential expertise could be provided to the farmers. All such technologies to be transferred must meet the criteria of being technically sound, practically feasible, economically attractive, socially acceptable and environmentally safe.

Chapter 10

Plant nutrition, food quality and consumer health

GENERAL ASPECTS

Good quality is important in almost all harvested crop products be they food, fodder or industrial raw materials. Because high-value food or feed is an essential precondition for the health of humans and domestic animals, the influence of plant nutrient supply on the quality of foodstuffs is of considerable importance. Farmers want to produce good-quality products and sell them for a remunerative price while consumers want to buy nutritious and safe food as cheaply as possible.

The concept of quality is variable and any discussion on the subject should be based on a terminology that can distinguish between: (i) commercial quality, which determines the market price of the product, and (ii) nutritional quality or value, commonly called food value, which is relevant for health. Although the two concepts partly overlap, the respective priorities, namely, monetary vs health aspects, set them apart.

Commercial quality

The commercial quality of a product defines the price at the market and is based on easily recognizable properties that, to a certain extent, also indicate its food value. The price of food for direct consumption depends mainly on easily detectable characteristics. Food should appear attractive, clean, fresh and without blemishes. Usually, farm produce is classified according to the desired properties into commercial grades that determine the price paid to the farmer and, finally, by the consumer. Maintaining quality is also important with respect to the safe storage, ability to withstand transport and shelf-life of fresh foods and grains. This is to ensure that the product does not deteriorate because of any physical or biochemical defects.

In the case of products used for industrial processing, the specific concentrations of important ingredients, such as sugar, starch, protein, fat and oil, are important. Commercial quality requirements depend on the specifications of the output from the processing factory and they are assessed for special product properties based on easily measurable analytical data. The main features of commercial quality are: (i) external features, such as size, cleanliness and freshness; (ii) sensory features, such as taste, smell and colour; (iii) keeping quality and shelf-life during storage and transport; and (iv) concentration of special important ingredients, e.g. protein concentration for baking-quality wheat, and ingredients for industrial processing (starch, sugar and oil).

Food quality

The nutritional value, commonly called food quality, includes all substances that contribute to complete nutrition of humans and animals. Consumers desire attractive, wholesome, nutritious food that is free of harmful substances. The nutritional value of food is determined by adequate concentrations of about 50 essential ingredients required by humans and also several beneficial substances that must be taken up in balanced proportions and at regular intervals. According to medical expertise, about half of all human diseases are caused by inadequate or imbalanced nutrition. Therefore, special attention should be given to the concentrations of essential and beneficial substances in food.

Food quality should go beyond the supplies of energy derived from starch, sugar, oil and fat and the "pleasure" value derived from the taste and smell of food. Its main emphasis should also be on the essential and beneficial components required for the building and functioning of humans and animals. As sufficient uptake of these nutrients is a prerequisite for good health, their concentration in food is an important index of nutritive value. Food quality also includes safe food, which refers to the absence of health-harming substances. Good food should not contain: (i) excesses of plant nutrients that may be dangerous to health; (ii) toxic heavy metals from soils or from nutrient sources; (iii) toxic organic compounds, e.g. from organic waste materials; and (iv) radioactive contaminants.

The "health" value of foods is complex and remains hidden for consumers. Moreover, the damaging effects of poor food quality on health mostly appear over a long period of time and consumers tend to neglect this aspect. However, it should be of central importance for their present and future well-being.

Consumers rarely base their decisions on the actual nutritive value of the food but on easily perceivable food properties, such as taste and appearance. However, such perceptions can be misleading and harmful to the health in the long run. Taste is subjective and, hence, not suitable for objective food evaluation. In recent times, the aspect of "safe" food (not containing health-damaging or toxic substances) has been gaining more importance than the nutritive value in many developed countries. Consumers are becoming very sensitive to this aspect and some prefer certified safe food, produced in reliable production systems.

Importance of food quality

The quality of food products depends on many factors. It is influenced primarily by: (i) genetic factors that determine the basic quality, specific to the kind of crop; (ii) climate factors, such as light, temperature and water supply, that enable plants to approach their genetic potential; and (iii) an adequate and balanced supply of all plant nutrients, often achieved by external nutrient application through fertilizers and manures (discussed below).

In many developing countries, the importance of food quality is generally underestimated because the need for a sufficient quantity of food has often been considered more important than good quality. With increasing income levels and a better understanding of the role of nutritional factors, it is becoming increasingly clear that high food quality is as important as food quantity. Even with sufficient food, deficits of essential food components can cause malnutrition and other health problems. Typical examples are diseases resulting from deficiencies of protein, vitamins and mineral nutrients. Several deficiency diseases are widespread in developing countries and constitute a serious obstacle to the full development of their human potential.

Poor protein quality and a deficiency in total protein typically appears in small children after weaning, when their diet includes food that is rich in starch but poor in protein, such as that from cassava and other starchy foods. The resulting protein deficiency disease, called "kwashiorkor" (first described in Ghana), is a very serious illness and makes the person prone to infectious diseases. This health problem is more prominent in SSA than in Asian countries, where baby food is based mainly on protein rich cereals and pulses.

It is being increasingly recognized that a lack of mineral nutrients is responsible for special diseases with far-reaching consequences on the health of humans and animals. There are widespread and growing deficiencies of some micronutrients, such as Fe and Zn. In Southeast Asia and SSA, more than 75 percent of the population appear to be affected by Fe deficiency, half of them to the extent of having anaemia (Graham, Welch and Bouis, 2001). Although not always detected, vitamin deficiencies appear to be even more common. These lower human resistance to several infectious diseases. They are widespread in many developing countries.

Perceptions of food quality

In addition to the unsatisfactory comprehension and evaluation of food quality by many consumers, food quality is also an area of many prejudices as many people have their own personal experiences about the relationships between eating and health. Several common questions are regularly raised on these issues. Some such questions followed by their answers are given below.

Question: Does food quality increase or decrease with the adoption of modern crop-production technologies, especially with respect to mineral fertilizers?

Answer: Although critics claim that the increased use of mineral fertilizers reduces crop product quality, this is not the case. Most such critics oppose anything produced by using fertilizers because of their opposition to manufactured inputs in general. Most fertilizers are derived from natural products, which are concentrated and processed only to be more effective. Moreover, nutrients in all sources whether organic or mineral must be converted finally into inorganic ionic forms (Table 6) in order to be usable by plant roots. Phosphate and potash fertilizers are obtained from natural products such as PR and salt deposits. Although mineral fertilizers are produced in factories, they are basically derived from natural minerals. Even nitrogenous fertilizers, although largely synthetic chemicals, obtain their N from atmospheric air and finally deliver it in the same mineral form (nitrate) as do "natural" organic manures. Synthetic nitrate is completely identical to nitrate derived from humus. Thus, the argument of organic farming that synthetic N fertilizers should not be used in order to obtain a high-food quality is not justified (discussed below).

Question: Do so-called intensive production methods aimed at high yields inevitably lower food quality?

Answer: Regardless of the yield level or intensity of cultivation, not all the valuable components of a crop product can be increased simultaneously. Where the starch concentration of grain is increased, the protein concentration or another component may be lowered, or vice versa. Even an increase in the total amount of vitamins per plant may result in lower percentage concentrations owing to the dilution caused by relatively higher starch and protein concentrations or biomass. The dilution effect is principally important for quality considerations. However, its consequences should not be interpreted as a negative effect of yield-improving measures on quality, especially as this plays only a minor role in the medium yield range. With high yields, some components may be lowered to some extent by dilution, whereas others are increased. Higher yields contain greater total amount of nutrients even if their concentration is lower (total = concentration × weight). A well-known example of the dilution effect is the consumer experience that small fruits often taste better than large ones. This is because of a lower concentration of aromatic components that have not increased as much as the fruit weight.

Question: Can food quality decrease although the crop product quality increases?

Answer: Ideally, these two concepts should be identical. However, there can often be differences between them. Agriculture is responsible only for crop product quality, not for the changes in quality that occur during food processing in factories or during cooking in the kitchen. For example, whereas agriculture produces higher vitamin B_1 (thiamine) concentrations in wheat grain (higher crop-product quality), consumers obtain less vitamin B_1 (lower food quality) because of their preference for white bread. In Europe, the concentration of vitamin B_1 in bread made from wheat grain is now much lower than what it was decades ago. This decrease is not due to the increased use of mineral fertilizers. In wheat grain, the concentration of vitamin B_1 is connected closely to the protein concentration. With higher N fertilization, the concentrations of both have increased. The decline in vitamin B_1 in white bread is the result of the increasing refining of flour, where the starch-containing flour is separated from the bran, which is rich in valuable substances such as vitamin B_1 and minerals. The bran is used for animal feed. Therefore, consumers of brown or whole-wheat bread receive more vitamin B_1 than do consumers of white bread. If consumers prefer whiteness to nutrition, it is their choice, albeit not a nutritionally sound one. Awareness of such factors can influence the type of flour used for bread-making. However, agricultural practices can be modified to meet the requirements of food processing, e.g. using SOP instead of MOP for potatoes.

PLANT NUTRITION AND PRODUCT QUALITY

Because only properly nourished plants can provide products of overall high quality, any fertilization that improves the supply of plant nutrients from deficiency to the optimal range raises the amount of nutritional substances. However, it is impossible to increase the concentrations of all valuable substances simultaneously.

The nutrient supply required for high crop yields and for good food quality is nearly similar. In certain cases, e.g. baking quality of cereals or additional nutrient supply for highly productive animals, high-quality food and feed is produced by keeping supplies of some plant nutrients in the luxury supply range.

The relationship between nutrient supply and the resulting change in quality of crop products has largely been established. In assessing the effects of added nutrients on produce quality, it should be remembered that: (i) increasing the nutrient supply from deficiency to the optimal range usually results in better produce quality; (ii) increasing supplies from optimal to the luxury range may increase, maintain or decrease quality; and (iii) extreme increases in supplies into the toxicity range reduce quality and must be avoided. Nutrients differ in their roles in plant production and produce quality. Such effects are discussed in brief below.

Nitrogen supply and product quality

The addition of N generally has the greatest effect on plant growth and also considerable influence on product quality, especially through increases in protein concentration and its quality. It also increases the concentration of several other valuable substances. However, where the N supply is excessive, harmful substances may be formed that decrease quality. Various N compounds in plants are important for quality assessment. The manner in which these are affected by N supplies is summarized below:

- Nitrate: Form of N taken up from soil; basis for protein synthesis; nitrate concentrations of plants are generally low, but it may be accumulated.
- Crude protein: This is an approximate measure of protein and some other N compounds. Crude protein concentration = N concentration × 6.25. The concentration of crude protein in wheat grain may be raised from 10 percent to more than 15 percent, thus improving the "baking quality" of the flour.
- Concentration of pure protein increases up to the optimal N supply level despite some counteracting dilution effect. Pure protein can be divided into several fractions:

- Prolamine and gluteline (low-value protein). Gluten is important for baking quality. N supply increases the prolamine content in grains, thus increasing the gluten concentration of grain kernels, which improves baking quality.
- Albumin and globulin (high-value protein), containing many essential amino acids. The concentration of albumen, which has high nutritional quality, increases with the concentration of pure protein.

➢ Essential amino acids: Nine protein constituents that are vital for humans and must be contained in food. Their concentration determines the biological value of the protein, expressed by the Essential Amino Acid Index (EAAI). Vegetable proteins have values of 50–70 percent compared with 100 percent in case of egg protein. The concentration of essential amino acids often increases up to the optimal N supply level, but it sometimes decreases through dilution, especially where there is luxury N consumption.

➢ Amides: These are important storage forms of N (e.g. asparagine or glutamine) found in leaves and vegetative reserve organs. Amides have only small nutritional value for humans, but, if heated, may produce substances with an undesirable odour. They can be a source of protein for ruminants.

➢ Amines: Various N-containing compounds present in small concentrations in plants. Some, e.g. choline, have important functions, whereas others, e.g. nitrosamines and betaine, are unwanted.

➢ Cyclic N compounds such as chlorophyll; N-containing vitamins such as vitamin B_1; alkaloids, such as nicotine in tobacco; purine derivates, such as theobromine in cocoa.

Where N supplies are excessive, some unwanted N compounds may accumulate in vegetative plant parts. These are primarily the unutilized nitrate and amines. Nitrate can accumulate in leaves, especially where light intensity is reduced. Concentrations of nitrate-N (in dry matter) in vegetables should not exceed 0.2 percent in salad vegetables or 0.3 percent in spinach because of the risk of nitrite formation. Nitrite, which usually occurs in insignificant amounts, can be formed in leaves under reducing conditions, e.g. where spinach is stored without access to air. When food high in free nitrate is consumed, it may cause methemoglobinaemia. The best way to keep nitrate concentrations in vegetables low is to restrict N fertilization to a medium level and to apply total N in splits.

Nitrosamines are formed from nitrite and secondary amines and some are carcinogenic (e.g. diethylnitrosamine). Their concentration in plants is normally insignificant and not a health problem. Betaine is an important constituent of the so-called "detrimental nitrogen", which interferes with the crystallization of sugar from the juice of sugar beets and, thus, reduces sugar yield.

An increase in N supplies also causes several types of changes in other substances, e.g.: (i) the concentrations of carotene and chlorophyll increase up to the optimal N supply; (ii) the concentration of vitamin B_1 in cereal grains increases until luxury N level; (iii) the concentration of vitamin C (ascorbic acid) decreases owing to the dilution effect; (iv) the concentration of oxalic acid, a harmful compound, increases in vegetables leaves (for human consumption) and in sugar-beet leaves (used as fodder for cattle), especially after fertilization with nitrate-N; and (v) the concentration of HCN in grass increases slightly – while its normal concentrations appear to promote animal health, higher doses are toxic.

Thus, the concentrations of all N fractions increase with higher N supply, but in different ways. The highest biotic value is obtained in the optimal supply range. Luxury N supply improves only certain quality components and this is often accompanied by quality reductions of other kinds. Thus, intensive fertilization of cereals with N may improve baking quality, but it lowers the average protein value.

Because plants normally absorb nitrate independently of the source from which it is applied, a direct influence of the form of N applied cannot be expected. However, where ammonium is applied and managed so that this is the form taken up by the plant, the nitrate concentration in leaves can be kept low. This can be achieved also by using slow-release fertilizers and nitrification inhibitors wherever their use is feasible and economic. Other influences observed on the qualitative composition resulting from the application of different N forms are mainly caused by side-effects, such as changes in soil pH.

Phosphorus supply and product quality

Owing to its many important roles in plant metabolism, the supply of P plays a central role in crop quality. Important quality indicators with respect to P are: (i) the P concentration and the composition of the plant P fraction; (ii) the concentration of other valuable substances that increase with better P supply; and (iii) the concentration of toxic substances that are often lower with increased P supply.

The major P-containing compounds that are important for crop quality are:

- Phosphate esters: These are the products of phosphorylization, i.e. bonding of phosphate anions as phosphoryl group ($-H_2PO_3$) to organic molecules like sugars ($R-O-H_2PO_3$).
- Phytin: This is the main organic form of phosphate storage (Ca-Mg-salt of phytic acid, i.e. inositol hexaphosphoric acid). Phytin is the main P reserve of seeds and can constitute up to 70 percent of total P. The proportion of phytin in vegetables such as potatoes is about 25 percent, and phytin, like inorganic P, is utilized by all animals, but best by ruminants. However, for humans, phytic acid may reduce the bioavailability of Fe and Zn.
- Phosphatides or phospholipids: These are important constituents of cell membranes that contain phosphoryl groups (e.g. lecithin, a glycerophosphatide). These form only a small portion of total plant P.

The P concentration of food and fodder is an important quality criterion because insufficient P intake causes "bone weakness" and deformations, which were common in cattle before the use of mineral P fertilization. In contrast to N, the P supply to crops remains in the "normal" range and rarely reaches the luxury range on most soils. In other words, there is practically no danger of overfertilization with P, which may cause problems owing to excess phosphate in food or feed.

When the P supply increases from deficiency to the optimal level, the total P concentration increases in the vegetative and reproductive parts, thus improving crop quality. The concentration of nucleic P increases only slightly, while the concentration of phosphatide-P remains approximately constant, and both occur in low concentrations. There is also a higher concentration of other value-determining substances, such as: (i) crude protein in green plant parts and essential amino acids in the grains; and (ii) carbohydrates (sugar and starch) and some vitamins, e.g. B_1. Seed quality improves with P nutrition, which results in greater seedling vigour. On the other hand, the concentration of some other substances such as nicotine in tobacco, oxalic acid in leaves or coumarin in grass can be reduced.

Potassium supply and product quality

Among plant nutrients, K is very closely associated with crop quality. It is required for good growth as well as for good crop quality, plant health, tolerance to various stresses and seed quality. By greatly affecting enzyme activity and through osmotic regulation, K affects the entire metabolism of the plant, especially photosynthesis and carbohydrate production. It improves the quality of several products including tubers, fruits and vegetables.

Increasing K supplies to plants up to the optimal level brings about the following changes:

- The concentration of carbohydrates increases owing to intensified photosynthesis, which results in larger concentrations of sugar, starch, fibres (cellulose), and also of vitamin C.
- The concentration of crude protein is reduced although the total amount is increased. This results from the dilution effect owing to the relatively greater increase in carbohydrate content. However, the more valuable fraction of pure protein may sometimes increase.
- The concentration of vitamin A and its precursor, carotene, increase.
- Losses of starch-containing tubers, such as potatoes, during storage are reduced through the prevention of decomposition of starch by enzymes.
- Unwanted "darkening" of potatoes is reduced. This phenomenon is caused by the formation of melanines and is particularly pronounced where K is deficient. Proper K supplies also prevent "black spotting" of potatoes upon cooking.

Unlike P, the K concentration is not a quality-determining component. Food usually contains more K than is required by humans or animals. Luxury supply of K in leaves may occur as a result of high K uptake. This is not detrimental but excess absorption of K by plants tends to reduce the uptake/concentration of Ca, Mg and Na, resulting in an imbalanced supply of these regulators of cell activity. K-induced Mg deficiency can decrease crop quality. On grassland, this can result in Mg deficiency in grazing animals.

Some effects of K fertilizers on crop quality are not caused by K itself but by the accompanying anion such as chloride or sulphate. Application of potassium sulphate results in a higher starch concentration in potatoes than where potassium chloride is applied. This is because chloride disturbs the transport of starch from the leaves to the storage organ (tubers). Similarly, in the case of cigarette tobacco, potassium sulphate is the preferred source of K over potassium chloride because excess chloride can reduce the burning quality of the leaf.

Calcium supply and product quality

A good Ca supply is essential for osmotic regulation and pectin formation. The Ca concentration of food and fodder is important for a proper balance of the major cations. Adequate supplies of Ca prevent a number of crop quality problems, such as inner decay of cabbage, brown spot and bitter pit in apples, and empty shells in groundnuts. Although Ca supply may not increase the oil content in groundnut, the total oil yield increases as a result of the favourable effect of Ca on kernel yield. Many of the benefits of liming on crop quality stem less from Ca itself but more from indirect effects caused by changes in soil pH that increase the supplies of other elements.

Magnesium supply and product quality

A good supply of Mg increases the concentration of carbohydrates and also chlorophyll, carotene and related quality components that are important for grazing animals. The Mg concentration is an important quality criterion because the major cations (K, Ca and Mg) should be balanced in order to ensure the best nutritional quality in cereals. Adequate Mg increases grain size and boldness. It is also reported to increase the oil content in oilseeds. For example, excess K in grass can result in Mg deficiency leading to hypomagnesaemia or grass tetany in grazing animals.

Sulphur supply and product quality

As S is an important constituent of some essential amino acids (cystein, cystine and methionine), S deficiency lowers protein quality. About 90 percent of plant S is present in these amino acids. Some plants (crucifers) contain S in secondary plant substances, e.g. oil, whose synthesis is inhibited where S is deficient. Mustard and onions rely for pungency and flavour on S-containing substances and these are also useful for increasing resistance against infections in the plant. An adequate supply of S improves: oil percentage in seeds; seed protein content; flour quality for milling and baking; marketability of copra; quality of tobacco; nutritive value of forages; grain size of pulses and oilseeds; starch content of tubers; head size in cauliflower; and sugar content and sugar recovery in sugar cane.

Micronutrient supply and product quality

Because micronutrients are involved in many metabolic processes, their adequate supply is a precondition for good food quality, especially with respect to the concentrations of proteins and vitamins. A survey of micronutrients in staple foods has been provided by Graham, Welch and Bouis (2001). The total concentration of the individual micronutrients is an important index of food and feed quality. However, some compounds containing micronutrients are utilized only partly by humans and animals.

Because the concentrations of micronutrients are not determined routinely, their average concentrations are often considered for nutritional purposes although these may give only an approximate idea of actual concentrations. For example, in leafy vegetables, a wide variation may occur. The following concentrations (in milligrams per kilogram of dry matter) range from marginal deficiency to luxury supply but are not toxic: Fe 20–800, Mn 15–400, Zn 10–200 and Cu 3–15. The consequences for health are clear. If a person is to be supplied with vegetables rich in Fe for better blood formation, then products with higher Fe concentrations are certainly preferable. Micronutrient concentrations should not be increased up to the toxicity level. Toxic concentrations are not only detrimental as such, but also negatively affect the composition of organic food constituents. The following comments on individual micronutrients relate to food quality:

- B is required in good supply for fruit and vegetable quality. B deficiency causes spots and fissures that substantially reduce produce quality and market value.
- Cu is required in optimal amounts for high concentrations and quality of protein and also to avoid spottiness in some fruits. A shortage of Cu partly combined with Co deficiency in grass retards the growth of grazing animals, and metabolic disorders manifest in the so-called "lick disease".
- Fe in green-leaf vegetables such as spinach is an important source of Fe for humans. Soils with high pH tend to produce products low in Fe.
- Mn raises the concentrations of some vitamins, such as vitamin A (carotene) and C, in food and fodder crops. For good fertility, grazing animals require Mn concentrations that are about double those required for optimal grass growth.
- Mo deficiency decreases protein content and quality because of the important functions of Mo in BNF and N metabolism. Mo is also involved in the formation of healthy teeth.
- Zn is connected with plant growth hormones. Therefore, a good supply is required in order to obtain full-sized products, as in the case of citrus fruits. Compared with Cu, the optimal range of Zn is large but its toxicity can become a problem on soils with excessive Zn.

Excess micronutrients reduce food quality properties. However, this rarely is the case on most soils. An excess of chloride can aggravate salinity problems, adversely affect salt-sensitive crops and lower the quality of crops such as potato, tobacco and grapes.

Effect of toxic substances on crop quality

Good-quality food implies not only high concentrations of valuable substances but the absence or the presence of only insignificant concentrations (far below the critical toxicity limit) of harmful inorganic and organic substances. People want safe food that has no harmful components and does not cause health problems. There are increasing cases of pollution-related effects and risks associated with toxic substances that are taken up from the soil and endanger crop product quality.

In fact, there have always been problems with natural toxic substances in soils in certain areas. For example, high concentrations of Al are found in plants on very acid soils. These cause damage to plants and possibly also health problems to animals. However, with proper soil fertility management practices such as liming, the Al concentrations in plants can be kept at a low and insignificant level. Other substances in toxic amounts occur locally in small areas with high natural soil concentrations, e.g. Se, As and Ni, and these can cause health problems. A large-scale As-related toxicity problem has been reported in Bangladesh, where tubewell waters high in As are used for irrigation.

The danger to health from pollution can be either from the polluted atmosphere or from products used as soil amendments and nutrient sources that may contain harmful substances. A source of major concern is the disposal of toxic wastes and effluents on agricultural lands disregarding optimal application rates without adequate and proper treatment. An element of major concern is Cd. Its concentration is 0.1–2 mg/kg in normal soils and about 0.05–1 mg/kg in plants. On heavily polluted soils, plant concentrations of more than 5 mg/kg may be reached, which is a toxic level in food products. However, Cd concentrations in plants do not depend entirely on total Cd concentrations in soil, but on available concentrations, which are largely determined by soil reaction. Therefore, on acid soils, the Cd concentrations can be reduced to a certain extent by raising the soil pH by liming. Other toxic heavy metals are Pb, Cr, Ni and Hg; none of these should be allowed to reach toxic levels in food.

Prevention of the accumulation of dangerous substances in crops in order to ensure safe food is of great importance. The potential problems related to organic toxic components are also a cause of great concern. Some potentially dangerous compounds are decomposed and, thus, eliminated in biologically active soils. However, some persistent ones are liable to be taken up by plant roots and may endanger food safety. Serious problems arise from the recycling of urban or industrial waste materials, which may be polluted by heavy metals and possibly by some toxic organic substances. While it is desirable to recycle these materials in order to preserve plant nutrients, strict limits must be set on such substances in order to ensure food safety because of the long-term effects caused by their accumulation in soils.

The responsibility for preventing these effects from occurring is not primarily that of agriculture but of municipal authorities and the industries that generate such wastes laden with undesirable elements and wish to dispose of them. Agriculture should use only safe urban and industrial waste materials that are practically free of toxic substances in order to promote sustainable crop production and to produce the secure food demanded by urban consumers. The need for "safe" waste materials will increase in future with growing urban populations. In fact, whether wastes are treated or not, these should be certified as "fit for agricultural use" before being applied to the soils.

The potential danger from radioactive materials should also be taken into consideration. Consumers need to realize that a certain level of radioactivity in food is unavoidable. Some natural substances like radioactive potassium (potassium-40 or ^{40}K) are ubiquitous in soils, plants, food, animals and humans, and are not harmful. However, excessive radioactivity in soils via heavy atmospheric pollution with strontium (Sr, e.g. strontium-90 or ^{90}Sr) or uranium (U) isotopes from deliberate or accidental nuclear reactions should be avoided. Radioactive fallout is absorbed by both roots and leaves. A useful countermeasure against their uptake from soils is the stronger fixation of the radioactive substances in soils and, thus, a decrease in their uptake. Higher phosphate and sulphate levels in soils are advantageous for this purpose because strontium phosphate and strontium sulphate are less available to plants than are Sr^{2+} ions.

CONSUMER HEALTH ISSUES AND FOOD QUALITY

High-quality nutrition is an important precondition for the health of humans and animals. It appears that about half of all diseases are caused by nutritional disorders. However, the consequences of many disorders remain hidden because of the complexity of the relationship between food quality and health and because of the time lag between cause and effect. Agriculture that produces healthy food contributes to the prevention of diseases and this aspect is often underestimated.

The effects of food quality on health can be assessed by determining the value of the ingredients in food products or by medical indices of health status where nutritional disorders are not directly observed. The problem of the latter is that of latent (slight or hidden) deficiencies, which occur much more frequently than do acute (visible) deficiencies.

Humans health based on essential nutrients in food

Similarly to essential nutrients in plants, essential nutrients in food also play an important role in the growth and development of humans. The progressive decrease in the incidence of tuberculosis in the United Kingdom between 1880 and 1940 is a good historical example of the effects of better plant nutrition on human health, and this decrease is attributed partly to the improvement in food quality resulting from the introduction of fertilizers.

The ingredients that determine the nutritive value of food are:

- Essential substances: In addition to carriers of energy like starch, sugar and fat, about 50 other components must be present in food for good nutrition and health:
  - Amino acids: These are the building blocks of proteins. Out of 21 amino acids, there are nine that cannot be produced by the body and must be obtained from food. These essential amino acids are: leucine, valine,

TABLE 41
Essential mineral nutrient elements besides N and S, daily requirements and the effects of deficiencies

Mineral nutrient	Daily adult requirements	Major deficiency symptoms in humans and domestic animals
Na + Cl	5 g	Dehydration (salt-loss syndrome), disturbance of kidney function (excess Na can aggravate hypertension)
P	1.5 g	Weakness of bones, skeleton deformities, rickets
K	2 g	Disturbances of growth and fertility, weakness of muscles, but K deficiency is rare
Ca	1 g	Bone stability reduced, neuromuscular disturbances
Mg	0.3 g	Cardiac insufficiency, grass tetany in cattle
Fe	10 mg	Anaemia (widespread, especially in women)
Zn	15 mg	Disturbances of body growth, healing of wounds, hair growth
Mn	3 mg	Disturbances of growth and fertility, skeletal deformities
Cu	2 mg	Anaemia, reduced fertility, damage to coronary blood vessels
I	0.2 mg	Disturbances of thyroid function (goitre problem)
F	2 mg	Caries (tooth decay).
Mo	0.1 mg	Dental caries
Se	0.05 mg	Necrosis of liver, eye damage
Co	-	Deficiency of vitamin B_{12}

lysine, iso-leucine, threonine, phenylalanine, tryptophane, methionine and histidine (only for children).

- Essential fatty acids: Linoleic acid, linolenic acid and arachidonic acid are lipid constituents and a person's daily requirement is about 7 g. Supplies of essential fatty acids do not appear to be a major problem in most cases.
- Vitamins: There are about 15 vitamins. The four fat-soluble vitamins are A, D, E and K. There are more than 10 water-soluble vitamins such as B_1, B_2 (riboflavin), B complex (a group of vitamins), B_6, B_{12}, C and H (biotin).
- Several mineral nutrients (listed in Table 41).

➢ Beneficial substances: Among the several plant constituents contributing to health and well-being are:

- Aromatic substances for good taste.
- Substances for better mechanical functioning of intestines, e.g. cellulose.
- Special ingredients, e.g. resistance-improving substances (antibiotics).

Major nutrients

The primary constituents of major nutrients for humans are C, H, oxygen, N, P and S (the same as plant nutrients). These form bulk of the carbohydrates, proteins, fats, oils and vitamins.

Daily protein requirements for humans are about 1 g/kg of body weight. Supplies of protein, especially of essential amino acids, that must be obtained from food appear to be about adequate in developed countries except in cases of unusual eating habits. In contrast, protein deficiency, especially among infants, is common in many developing countries with poor food supply. It results from both quantitative undernourishment with protein and inadequate protein quality (often a deficiency of lysine) and leads to kwashiorkor disease. Better N nutrition of crops resulting in more and improved protein supply to the population could be an effective measure for controlling the deficiency.

Vitamins

Vitamin A, derived from the photosynthetic pigment carotene, occurs mainly in green leaves, carrots, milk, and egg yolk. Lack of vitamin A is the most important cause of blindness in childhood and is still prevalent in some parts of South Asia. A good supply of vitamin A can be obtained from eggs and milk, but a sizeable portion of the population relies mainly on vegetable sources.

Vitamin B_1 (thiamine) occurs primarily in the germ of grain kernels. Its concentration increases with increasing N supplies. Lack of vitamin B_1 is associated with the disorder beriberi. The disorder can cause severe damage to the heart and muscles. The main problem of supplying thiamine to the population is not the production of foodstuffs (e.g. rice) rich in thiamine, but the trend towards refining, which often results in consumption of only the inner part of the rice grain leaving the germ of kernel out (e.g. polished rice). The technique of parboiling rice is helpful in retaining vitamin B_1.

Lack of vitamin B_2 (riboflavin) appears to be the most widespread deficiency and is often associated with insufficient protein intake. The acute symptoms do not appear very serious, but people become more prone to sickness in general.

Niacin (nicotinic acid) is a vitamin in the B complex group. It is found in meat, milk, eggs and wheat germ. People in areas where maize is the main food source are at risk of developing pellagra (a skin problem) and encephalopathy (mental illness) because maize is low in niacin. Furthermore, the niacin in maize cannot be absorbed in the intestine unless the maize is treated with alkali.

Vitamin C (ascorbic acid) occurs especially in fresh fruits and leaves. Some fruits such as citrus, guava and aonla (Indian gooseberry, *Emblica officinalis* Gaertn.) are exceptionally rich in vitamin C. Additional vitamin C beyond the daily requirement seems to improve resistance to several diseases, including the common cold.

Supplies of vitamins in food appear to be largely adequate in developed countries. Acute deficiencies (avitaminoses) have become a rarity, but hidden deficiencies (hypovitaminoses), mainly of vitamins A, B_1 and C, are common in certain population groups. In most cases, this lack of supply is not caused by their shortage in food but by consumers' eating habits, e.g. a preference for refined food, from which vitamins are partly removed. In developing countries, acute and hidden vitamin deficiencies are widespread and these may increase in the future. The consequences of such deficiencies are considerable, mainly in terms of reduced resistance to many diseases.

Minerals

While major mineral nutrients, such as Na, P and Ca, etc. have bee well studied, some micronutrients (often called trace elements in medical publications) have only recently attracted the attention of nutritionists and biochemists. Mineral nutrients are present in food either as salts or as organic compounds. Not only are their concentrations important, but so too is their bioavailability (the portion that is absorbable and utilizable). Moreover, substances that inhibit (e.g. phytic acid) or promote (e.g. some vitamins) nutrient bioavailability should be taken into account, as should certain antagonistic effects between minerals. The requirements for minerals and some pathological effects of mineral deficiencies in humans are listed in Table 41.

Two examples can illustrate the consequences of mineral nutrient deficiencies and their amelioration on health:

- Phosphate and bone stability: In the nineteenth century, P deficiency was widespread in Central Europe and so was "bone weakness" in cattle. In an area of Austria with a severe phosphate deficiency in the soil, people (especially women) had deformed bones (rickets). However, after several years of phosphate fertilization, these symptoms of deformities disappeared, resulting in considerable health improvement.
- Molybdenum and teeth stability: In about 1950, it was noticed that the teeth of children in Napier, New Zealand, were healthier than those of the children in the nearby town of Hastings, where there was a high incidence of tooth decay (caries). Investigation into causal factors showed that this was not caused by a lack of fluoride in the drinking-water. The difference was caused by a differential Mo supply to vegetables grown in gardens. The Mo supply was adequate in Napier but deficient in Hastings. Insufficient Mo in the vegetables resulted in weak teeth because Mo is required for the formation of stable dental enamel, which is a fluorapatite.

Resistance-improving substances

The resistance capacity of the human body to pathogens is one of the major determinants of health. It is improved by good nutrition, which in turn is enhanced by the intake of quality food, primarily obtained through a proper plant-nutrient-management-based crop production system. Well-nourished people, especially children, suffer much less from infectious diseases and have a much lower mortality rate than do malnourished persons.

Resistance-improving substances can be mentioned as beneficial food ingredients. They are produced by certain fungi in fertile soils, composts, etc., where their concentrations are about: 5 mg/kg streptomycin, 0.1 mg/kg terramycin, and 0.02 mg/kg aureomycin. These antibiotics are taken up by plants and occur in low concentrations in the leaves, where they apparently act as protective agents against certain infections. Humans and animals may probably derive a certain natural resistance by eating these foods.

Animal health and feed quality

The relationship between food quality and health is best demonstrated by grazing animals. In contrast to humans, who generally have a variety of food, grazing animals are restricted to the fodder present in the pasture. The key to animal health

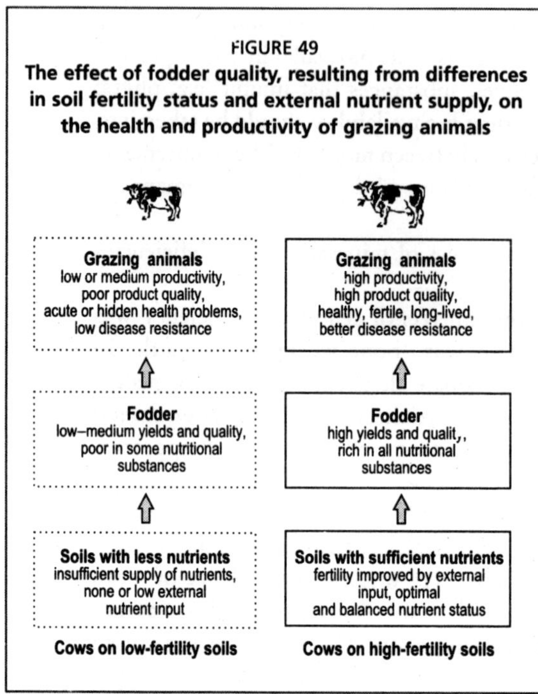

FIGURE 49
The effect of fodder quality, resulting from differences in soil fertility status and external nutrient supply, on the health and productivity of grazing animals

Source: Finck, 2001.

is an adequate and balanced supply of mineral nutrients obtained from the grassland fodder. Much information on human mineral-nutrient needs has been obtained from animal nutrition. Nutrient management of fodders, grasslands and pastures has been discussed in Chapters 7 and 8.

The nutrient supply level of soils, whether good or poor, is reflected in feed quality and has direct effects on animal growth, health and fertility and also on the quality of products such as milk, meat and wool. With grazing animals, the causal chain of soil – plant – animal is demonstrated clearly. This is further established through the well-known examples of phosphate deficiency in many countries and those of Cu and Co deficiencies in Northern Europe and Australia.

Major issues relating plant nutrient management and feed quality with animal health and productivity are:

- Low soil fertility of many grazing lands: Because grazing lands are often on soils with marginal fertility, their fodder productivity is often low or medium, and so is the quality. Figure 49 summarizes the relationship between soil fertility, fodder quality and the response of animals in terms of production and health. Salient examples of this kind are found in countries with large areas under severe nutrient deficiency. Co deficiency is an example. Solutions to such problems can be achieved through appropriate nutrient management.
- Higher nutrient requirements of very productive animals: For cattle with a high milk production, higher amounts of mineral nutrients are required. This should be taken into account either by feed improvement via fertilization or by supplementary feeding to promote animal health and fertility.
- Additional nutrient needs of animals: Animals require more essential nutrients than do plants. These nutrient elements should also be considered in evaluating feed quality. For example, Se deficiency does not affect plant growth but it causes serious health problems in grazing animals, particularly sheep. In such situations, the addition of such missing nutrients may not improve plant production but it will improve animal health and productivity.

When considering food quality, not only rational but also emotional aspects are encountered. Discussion on food quality should be on a rational scientific basis and not based on general beliefs or prejudices that originate from philosophical, religious or other ideas.

Most agricultural production systems with efficient plant nutrient management produce high-quality food. However, many consumers do not derive the full benefit from their food because of eating habits that neglect quality. This is unfortunate because high-quality food is produced for the consumers' benefit.

Consumers can rarely evaluate the nutritional quality of the food they purchase even where they are aware of the principles and facts of food quality. Many consumers would be happy if they could obtain certified, good-quality food produced in production systems designed for this purpose. They are even prepared to pay higher prices for such reliable food as this is considered a kind of "insurance" for good health.

The question arises as to whether food production systems can be adjusted to such demands of the consumers. Food of good quality can be produced on fertile soils using good crop management. This occurs on the majority of farms worldwide where adequate and balanced fertilization through integrating various sources is a part of crop production.

In order to produce acceptable and certifiable quality foods from a plant nutrition point of view: (i) the supply of nutrients from internal and external sources (INM) should be based on good soil nutrient supply, which is evaluated by diagnostic methods and on the nutrient demands of the crops; (ii) nutrient deficiencies should be overcome by appropriate fertilization with the goal of obtaining food with a high concentration of valuable components while avoiding quality problems caused by unwanted excess of nutrients; and (iii) there should be no harmful substances in the food.

Chapter 11

Plant nutrition and environmental issues

The influences of nutrient management on the environment relate to two broad issues. The first issue concerns the interaction of plant nutrient status with various soil and climate stresses, as discussed in Chapter 6. The present chapter examines the second issue relating to the effect of nutrients or other constituents of fertilizers and manures on environment quality, pollution, human health, etc.

TABLE 42
Environmental issues in land and water development for 13 Asian countries

Environmental issue	Frequency of occurrence
Low fertility and imbalanced nutrition	13 (all countries)
Population increase, water and wind erosion	12
Land-use policies, sedimentation and siltation	11
Deforestation, waterlogging, shifting cultivation, land conversions	10
Salinization	9
Drought, acidity	8
Pollution, acid sulphate soils, organic matter depletion	7
Desertification, overgrazing, landslides	6
Poor crop management	5
Peat soils	4

Source: FAO/RAPA, 1992.

Depletion or improvement in soil fertility is also a part of environmental degradation or improvement. Nutrient depletion from soils is a major form of soil degradation (FAO, 2003d). On a global scale, soil fertility depletion is far more widespread than is soil fertility improvement. Nutrient depletion destroys the productive capital of the valuable soil resource. Depletion of soil nutrients is caused primarily by negative nutrient balances, faulty nutrient management strategies and a lack of resources for investment in soil-fertility-enhancing inputs.

In a survey of 13 Asian countries (Bangladesh, China, Democratic People's Republic of Korea, India, Malaysia, Myanmar, Nepal, Pakistan, Philippines, Sri Lanka, Thailand and Viet Nam), soil nutrient depletion coupled with imbalance in soil fertility was the most frequently mentioned issue identified with land and water development in all the countries (Table 42).

BASIC EFFECTS OF NUTRIENT MANAGEMENT ON THE ENVIRONMENT

Nutrients added through fertilizers, manures and composts can have negative as well as positive effects on the environment depending on how poorly or properly these inputs are managed. The added nutrients may be absorbed by crops, immobilized by the soil or lost from the soil system. Depending on the nutrient and various conditions, these can be lost to the atmosphere by volatilization, lost through soil and water erosion, lost from the soil profile by leaching. Leached N can also be lost to the atmosphere through denitrification.

Positive and negative effects of nutrients

Positive effects

The positive effects of nutrients on the environment are:

- Efficient use of plant nutrients ensures that yields are higher than those obtained on the basis of inherent soil fertility by correcting either an overall deficiency or an imbalance of nutrients.
- Nutrients removed from the soil through harvesting and export of produce can be largely replenished through various types of recycling in order to maintain and enhance the production potential of the soil.
- By increasing yields per unit area from suitable arable land, application of plant nutrients allows land of low quality, e.g. land susceptible to erosion, to be withdrawn from cultivation. This reduces the overall pressure on land, including deforestation and overgrazing on non-cropped areas.
- Efficient use of plant nutrients eases the problem of erosion control on the cropped area because of the protection provided by a dense crop cover.
- Balanced plant nutrition also results in an increased addition of organic matter through greater leaf residues, and root and stubble biomass.
- Where balanced fertilization is practised, there is greater N uptake by crops and less nitrate is leached down the profile for the pollution of groundwaters or further loss through denitrification.
- INM promotes the correct management of all plant nutrient sources on the farm and helps reduce the losses of plant nutrients to the environment.

Negative effects

The negative effects of plant nutrients on the environment need to be considered both at high and low input levels.

At high levels of input use, the nutrients applied to the soil are not taken up completely by the growing crop even under the best conditions. Out of the remaining fractions, the soil constituents are able to bind and immobilize most of them so that they do not move freely with soil water and create possible negative impacts on the environment (water and air). Nitrate and, to a lesser extent, sulphate and B are not held strongly by the soil and can leach down with percolating waters and contribute to the undesirable enrichment of water. Phosphate generally moves very little way away from the site of application. Where it does, it is mainly through soil erosion or surface runoff. Over a period of years, phosphate applied through fertilizers or organic manures can move to deeper layers of coarse-textured soils in high rainfall areas. If it exits the soil profile and moves into waterbodies, its concentration increases and it can lead to excessive growth of algae, etc. and result in eutrophication to the detriment of other organisms. The relative importance of these phenomena depends on the physico-chemical and biological reactions in which the nutrients take part. Chapter 4 has presented details of the dynamics of individual nutrients in soils.

Table 43 summarizes the environmental problems associated with fertilizer use and general strategies to minimize them. Most of the problems, except those associated with Cd, are largely caused by the incorrect use of nutrients and their poor integration with other production inputs. This implies that most of the problems observed can be controlled if appropriate measures are taken.

The negative effect of levels of input use can be summarized as follows:

- The constant removal of crop produce without sufficient replenishment of plant nutrients exported by the crop causes a steady decline in soil fertility. This mining of plant nutrients, leading to severe depletion of soil fertility, is also a kind of soil degradation and a major environmental hazard in a number of developing countries (Table 42). The use of low levels of input places additional stress on soil nutrient supplies, resulting in excessive mining of soil nutrients and in depletion of soil fertility, leading to land degradation.
- To the extent that land and labour resources are available, low crop yields resulting from nutrient depletion force farmers to cultivate land under forests or marginal soils that are subject to erosion or desertification and, therefore, not normally fit for cropping. Bringing unsuitable land into cultivation promotes land degradation.
- Large areas of soils in the tropics are inherently poor in soil nutrients and suffer from problems of acidity, salinity, alkalinity and Al toxicity. Such soils can be made productive with appropriate amendments and a basic input of plant nutrients. Low or zero use of plant nutrients on such soils prevents the development of agriculture on a sustained basis. Organic recycling can only

TABLE 43
Environmental problems associated with fertilizer use and possible solutions

Problem	Cause mechanism	Possible solutions
Groundwater contamination	Leaching of weakly held nutrient forms such as nitrate (most important), chloride, sulphate and boric acid.	Balanced use of fertilizers; optimal loading rates of animal slurry, organic manure and wastewaters; improved practices for increasing N efficiency; including use of nitrification inhibitors, coated fertilizers and deep placement of N fertilizer supergranules where economic; integrated N and water management.
Eutrophication	Nutrients carried away from soils with erosion, surface runoff or groundwater discharge.	Reduce runoff, grow cover crops, adopt water harvesting and controlled irrigation, control soil erosion.
Methaemoglobinaemia	Consumption of high nitrate through drinking-water and food.	Reduce leaching losses of N, improve water quality.
Acid rain and ammonia re-deposition	Nitric acid formed by the reaction of N oxides with moisture in the air, ammonia volatilization and sulphur dioxide emissions.	Reduce denitrification, adopt proper N application methods to reduce NH_3 volatilization, correct high soil pH, increase CEC by organic additions.
Stratospheric ozone depletion and global warming	Nitrous oxide emission from soil as a result of denitrification.	Use of nitrification inhibitors, urease inhibitors, increase nitrogen-use efficiency, prevent denitrification.
Itai-itai (ouch-ouch) disease	Eating rice and drinking water contaminated with Cd.	Soil management such as liming or water control in rice fields, monitoring Cd content of PR and finished fertilizers.
Fluorosis in animals	Ingestion of soil or fertilizer treated with high fluoride PR.	Monitor the F content of PR applied directly to acid soils.

Source: Modified from Pathak *et al.*, 2004.

partially solve the problem as the biomass produced on poor soils is itself extremely poor in essential plant nutrients.

Effective management practices can prevent or remedy the negative effects of the applications of plant nutrients, both at low and high levels of input. Optimal fertilization can overcome the problem of nutrient depletion and of mining soil fertility. Judicious management of plant nutrients can prevent pollution, mainly through practices that reduce losses of nutrients into the aquifers or the atmosphere. This can be achieved through balanced, timely, targeted fertilization such as SSNM combined with other practices (e.g. improved varieties, water management, and plant protection) that stimulate maximum uptake of plant nutrients by the crop. At the same time, due attention should be given to controlling losses through soil erosion, runoff and land management.

The excessive use of inputs is not advised under any circumstances by scientific farming. High-input application is only justified where the nutrients are balanced and used efficiently. These are also justified only where the crop varieties grown can use the "high input" to achieve high production. Towards this end, farmer education is of utmost importance because these measures have to be taken by individual farmers, often on very small landholdings. INM is an excellent approach for such improvement at all productivity levels if farmers are advised properly.

ENVIRONMENTAL ASPECTS OF PLANT NUTRIENTS

Nitrogen

Nitrogen losses

Of all the inputs, N additions have had the single largest effect on crop yields and also have contributed most to environmental concerns, discussions and problems. Added N that is not absorbed by the crop or immobilized by the soil can be lost from the soil by various means. These include: leaching of nitrate to groundwater; and volatilization of ammonia into the atmosphere and as nitrous oxide (NO) to the atmosphere resulting from denitrification of nitrate by soil organisms. In addition to these, soil and applied N can also be lost through soil erosion and surface runoff.

The magnitude of these losses varies greatly between systems and environments. It is necessary to be aware of the validity of various estimates and the errors associated with them, as highlighted by the relative errors associated with the computation of N and P balances on farms in the Netherlands. For example, the error associated with fertilizer input was 1–3 percent, that with manure input was 10–20 percent, but errors of 50–200 percent were associated with losses through leaching, runoff or volatilization (Oenema and Heinen, 1999).

Mineral fertilizer supplies about 50 percent of the total N required for global food production. Global fertilizer N consumption was 84.7 million tonnes N in 2002 (FAO, 2005). The contribution of N through other crop production inputs is estimated as: BNF, about 33 million tonnes; recycling of N from crop residues, about 16 million tonnes; animal manures, about 18 million tonnes; and atmospheric deposition and irrigation water, about 24 million tonnes (Smil,

1999).Of the about 170 million tonnes N added, about half is removed from the fields as harvested crops and their residues. The remainder is incorporated into SOM or is lost to other parts of the environment, for which global estimates of individual loss vectors are highly uncertain (Mosier, Syers and Freney, 2004). About 47 percent of the applied mineral N (39.8 million tonnes) is lost to the environment every year (Roy, Misra and Montanez, 2002).

Fertilizers, organic manures, crop residues and crop management (as also the water input) have a major influence on N losses. In flooded-rice cultivation, it is common that 20–30 percent of the applied N is unaccounted for (lost) after crop harvest. Often, a sizeable portion (30–50 percent) of the applied N remains in the soil and only a small proportion of this is recovered in the following crop. Except for the natural leaching of soil nitrate as a result of rain and snow, most other reasons can be attributed to inadequate fertilization practices and poor water management.

Nitrate leaching

Nitrate is not bound by soil particles and remains in the soil solution where it moves freely with the soil water. Even where the N is applied in the ammonium or amide form, soil bacteria readily transform it under aerobic conditions to nitrate. Given that most N fertilizers are readily soluble, there is generally an excess supply of N immediately after application. The amount that is not taken up by the plant or immobilized by the soil is susceptible to loss. Considerable quantities of nitrate can also be lost from the mineralization of SOM, organic manures, animal slurry and crop residues. This generally occurs soon after harvest. Losses from animal manures are important contributors to nitrate losses in some areas. Leached nitrate can originate from any potential source.

Nitrate lost by leaching or transported in surface runoff can result in increased nitrate concentrations in drinking-water, eutrophication of surface waters and increased production of NO. It has been estimated that the groundwater under some 22 percent of the cultivated land in the European Union (EU) has NO_3^- concentrations exceeding the EU upper limit of 20 mg/litre. Similar high concentrations are found in many parts of the United States of America and other countries. Factors contributing to nitrate leaching to groundwater are:

- coarse-textured or extensively cracked soils;
- high concentration of nitrates in the soil profile as a result of excessive applications of N through fertilizers and manures;
- heavy rainfall that moves nitrates downward;
- restricted plant rootzone (due to plant species, time of year) to intercept nitrates for crop use;
- high water table;
- uncontrolled flood irrigation.

Not all of the above conditions have to be met for nitrate leaching to occur. However, nitrate leaching is at its maximum where all these factors exist and minimum where the reverse is the case. A deep and extensive root system enables crops to utilize N more efficiently, thus minimizing the risk of leaching. Leaching losses of N can be very high where N is applied to crops that have a shallow root system or that contain a small amount of N in the produce.

Nitrate leaching has another associated negative effect. When leached, all anions (nitrate, sulphate and chloride) take along with them equivalent amounts of cations. Therefore, nitrate leaching can deplete the soil of exchangeable cations such as Ca^{2+}, Mg^{2+} and K^{+}. The total N loss through leaching consists not only of N loss but also basic cations, which can increase soil acidity.

Emissions of ammonia

Ammonia volatilization from soil and vegetation contributes about 21 million tonnes/year of N (Smil, 1999). The global ammonia loss from mineral fertilizers is estimated at 11 million tonnes N (14 percent of mineral N-fertilizer use) (FAO/IFA, 2001). The loss from animal manure is about 8 million tonnes N/year (23 percent of animal manure N use). The global NH_3 loss from the use of mineral N fertilizer in wetland rice cultivation amounts to 2.4 million tonnes (20 percent of the 11.8 million tonnes of N applied to wetland rice). In grasslands, the annual global use of mineral N fertilizer is 4.3 million tonnes, with estimated loss rates of 13 percent for developing countries and 6 percent for developed countries (FAO/IFA, 2001).

The highest emissions of ammonia are in regions with intensive animal production activity (Europe), widespread use of urea (India,) and application of ammonium carbonate fertilizer (China). The dominant source of ammonia emission is animal manure as about 30 percent N in urine and dung is lost through this route.

Ammonia volatilization losses from surface-applied urea can amount to 25 percent on pastures and up to 50 percent in flooded rice. In a study on perennial dairy pastures in southeast Australia, losses of up to 45 percent of applied N have been recorded, and the magnitude of loss was affected by the N source used (Eckard *et al.*, 2003). Ammonia volatilization losses could be substantially reduced in summer by applying ammonium nitrate rather than urea. However, the approximately 45-percent cheaper unit price of N in urea compared with ammonium nitrate favours urea application on an agro-economic basis.

Factors favouring ammonia volatilization are:

- high soil pH (> 7.0);
- soils high in calcium carbonate (lime);
- soils with low retention ability for ammonium, e.g. low clay content, low organic matter, low CEC;
- high soil or atmospheric temperature;
- liquid fertilizer applied onto dry soil;
- high wind velocity and/or highly aerated soils;
- high rate of fertilizer or manure application;
- shallow (< 2 cm) depth of incorporation/penetration.

In arable soils, ammonia volatilization can be severe from surface applied urea that is not incorporated on neutral to alkaline soils during hot and dry periods. Such losses can be reduced substantially by incorporating urea in a moist but not very wet soil. Ammonia that is volatilized into the atmosphere returns back to earth with rain and snow as a part of the N cycle.

Volatilization of ammonia from liquid animal manure represents a significant cause of N loss. The magnitude of this loss depends on a number of factors including the method of application. In Canada, Manitoba Agriculture, Food and Rural Initiatives (http://www.gov.mb.ca) estimated the losses as shown below:

- broadcast, no incorporation for 2–3 days: N loss, 25–35 percent;
- broadcast, followed by incorporation within 2 days: N loss, 15–25 percent;
- broadcast, no incorporation on cover crops: N loss, 35 percent;
- injection: N loss, < 2 percent);
- irrigation within 3 days: N loss, 25–35 percent.

Where time to incorporation exceeds three days, N losses can be 40–60 percent with broadcasting and 60–80 percent with irrigation. For solid manure, volatilization losses from broadcasting may be less than those reported for liquid manure.

Emissions of nitrogen gases

Emissions of N gas in elemental form or as various oxides such as nitrogen dioxide (N_2O) and NO_2 occur on a large scale. Large amounts of the inert N_2 gas are emitted as the end product of denitrification. However, apart from reducing the nitrogen-use efficiency of crops, it does not have any negative environmental impact.

Both NO and N_2O are produced by soil microbes breaking up nitrate under conditions of low oxygen supply (waterlogged soils). The process is known as denitrification. Factors conducive to denitrification are: (i) soils with high organic matter (5 percent or greater); (ii) limited oxygen, due to high water content, rapid respiration or compaction; (iii) neutral or alkaline pH (7.0 or greater); and (iv) temperatures above 20 °C. N gases released by denitrification react with volatile organic compounds in sunlight to form ozone (O_3). This is the principal gas that shields the earth surface from ultraviolet radiation from outer space but which can be damaging to crops at low concentrations.

Denitrification losses as gaseous dinitrogen (N_2) amount to about 14 million tonnes/year, and N_2O and NO from nitrification/denitrification contribute about another 8 million tonnes N to the total loss (Smil, 1999). One study (FAO/IFA, 2001) estimates the global annual N_2O and NO emissions from agriculture as 3.5 and 2.0 million tonnes, respectively. The mineral fertilizer induced emissions for N_2O and NO amount to about 1.25 million tonnes/year, while the figure for animal manure induced emissions is about 0.32 million tonnes/year.

It is estimated that N_2O contributes 5–6 percent to the present greenhouse gas effect. Chemodenitrification (denitrification without microbial activity) requires low pH, but may be significant in freezing soils with high salt concentrations and high nitrite content. Denitrification cannot take place without nitrate. It can be prevented by avoiding high applications of N to arable areas with high water tables, by avoiding intermittent ponding, by the use of nitrification inhibitors and by deep placement of fertilizer/supergranules where feasible.

Phosphorus

Phosphate occurs in soil in both organic and inorganic forms that differ greatly in terms of their solubility and mobility. P applied through mineral fertilizers is in inorganic forms of varying solubility. Even at optimal rates, the use of mineral fertilizers and organic manures can lead to a buildup of soil P over time. The P thus retained is beneficial rather than harmful as it improves soil fertility and crop productivity.

The $N:P_2O_5$ ratio in most animal manures is about 1:1 whereas plants remove about 2.4–4.5 times more N than P_2O_5. Such residual organic forms of P are free to move with soil water in much the same way as nitrate and they can be leached. In this respect, these are different from fertilizer P or the more stable forms of organic P that are a part of SOM. On the other hand, inorganic forms of P are bound strongly to clays and oxide surfaces in acid soils, and precipitated as relatively insoluble calcium phosphates in alkaline soils. These bonding and precipitation mechanisms keep the P concentration in the soil solution at a low level; hence, leaching and surface runoff of phosphate in solution does not generally contribute to eutrophication. However, P bound to soil particles can be lost through soil erosion.

The P that can contribute to the enrichment of waterbodies, and hence lead to eutrophication, is a combination of the P that is attached to soil particles less than 0.45 µm in size that are transported during soil movement. Figure 50 shows the movement of P in surface water flow. The risk of P losses to the environment through surface runoff is greatest on sloping lands, and where the fertilizer is surface applied and then followed by rainfall or irrigation.

Most governments have set limits on the concentration of P in waters. In the United States of America, the Environmental Protection Agency has recommended a limit of 0.05 mg/litre total P for controlling eutrophication in streams that enter lakes and 0.1 mg/litre for total P in flowing streams. It has not been possible to prescribe safe P concentrations in runoff leaving a field because of the considerable P transfers that occur between the field and the waterway. Grassed riparian strips are recommended for trapping particulate P.

FIGURE 50
The movement of P in surface water flow

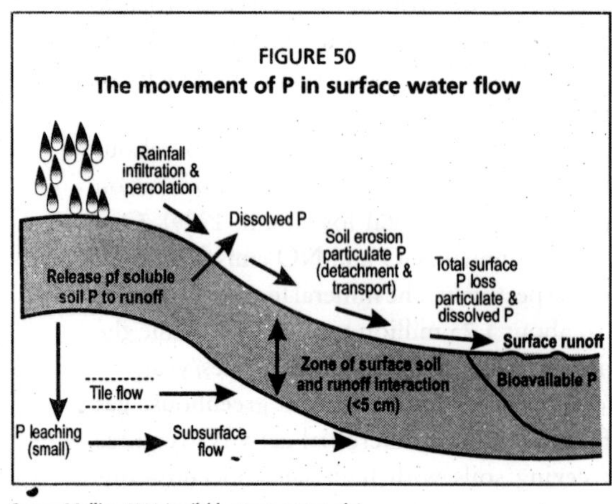

Source: Mullins, 2001 (available at www.ext.vt.edu).

Phosphate leaching is only a problem on soils that are well supplied or oversupplied with P, especially where they have inadequate capacity to immobilize P. Maintenance of good soil cover is the best protection against such losses. Subsurface leaching of P can take place where: (i) P is in soluble organic form, as in manure; (ii) the capacity of the soil for binding inorganic P has been exceeded; and (iii) a preferential flow of water through channels and cracks in the soil prevents contact with the adsorption sites in the soil (Laegreid, Bockman and Kaarstad, 1999). With good nutrient management, the phosphate losses to the environment can be kept low and with in a tolerable range.

Other nutrients

Losses of K, Ca, Mg and S to the environment are not considered very important. Deficiencies of some or all of these nutrients result in poor plant growth and the increased risk of soil erosion. Losses of basic cations can occur along with the leaching of anions such as nitrate and chloride. In general, leaching losses are greater where soluble nutrients are not fully utilized by the crop and the soil particles do not have sufficient capacity or reactive surfaces to adsorb them. K can be lost through leaching from coarse-textured soils under heavy rainfall or flood irrigation. The loss of K through leaching and erosion is a waste of resources but it is not known to constitute any environmental or health hazard.

Sulphate is relatively more mobile than nitrate or chloride but much less so than phosphate. S that has leached from the topsoil and accumulated in the subsoil can be utilized by deep-rooted crops in a later season. However, S can be lost through leaching in shallow soils or soils without sufficient retention capacity, but it is not associated with environmental or health problems. Unlike nitrate, the World Health Organization (WHO) has made no recommendations regarding the limits of sulphate concentration in drinking-water. In highly reduced soils, S can be lost to the atmosphere as hydrogen sulphide (H_2S).

B dissolved in soil water occurs as the water-soluble boric acid (H_3BO_3), which can be lost by leaching. The pumping of B-rich groundwaters for irrigation is not advised as it can add excess of B to the soil, leading to possible B toxicity. The upper limit of B in irrigation water in heavy-textured soils is 2 ppm B for semi-tolerant crops and 3 ppm B for tolerant crops. For coarse-textured soils, these limits are 3 ppm B and 4 ppm B, respectively (Yadav and Khera, 1993).

All nutrients can be lost by surface runoff and water and wind erosion where the nutrients are soluble and the soil particles containing them are detached and transported. Although these are a loss to the site from where they are removed, a significant part of such losses can be intersite transfers to the extent these are deposited at another site along the way. Many alluvial soils owe their fertility to the soil brought in with surface runoff, e.g. during floods.

Soil contamination from nutrient sources

In addition to the essential nutrients applied through minerals, finished fertilizers and manures, incidental additions of undesirable substances can also take place.

TABLE 44
Chemical analysis of potentially hazardous elements in sedimentary phosphate rocks

Country	Deposit	Reactivity	P_2O_5 (%)	As	Cd	Cr (mg/kg)	Pb	Se	Hg (µg/kg)	U (mg/kg)	V
Algeria	Djebel Onk	High	29.3	6	13	174	3	3	61	25	41
Burkina Faso	Kodjari	Low	25.4	6	< 2	29	< 2	2	90	84	63
China	Kaiyang	Low	35.9	9	< 2	18	6	2	209	31	8
India	Mussoorie	Low	25.0	79	8	56	25	5	1 672	26	117
Jordan	El Hassa	Medium	31.7	5	4	127	2	3	48	54	81
Mali	Tilemsi	Medium	28.8	11	8	23	20	5	20	123	52
Morocco	Khouribga	Medium	33.4	13	3	188	2	4	566	82	106
Niger	Parc W	Low	33.5	4	< 2	49	8	< 2	99	65	6
Peru	Sechura	High	29.3	30	11	128	8	5	118	47	54
Senegal	Taiba	Low	36.9	4	87	140	2	5	270	64	237
Syrian Arab Republic	Khneifiss	Medium	31.9	4	3	105	3	5	28	75	140
Togo	Hahotoe	Low	36.5	14	48	101	8	5	129	77	60
Tunisia	Gafsa	High	29.2	5	34	144	4	9	144	12	27
United Republic of Tanzania	Minjingu	High	28.6	8	1	16	2	3	40	390	42
United States of America	Central Florida	Medium	31.0	6	6	37	9	3	371	59	63
United States of America	North Carolina	High	29.9	13	33	129	3	5	146	41	19
Venezuela	Riecito	Low	27.9	4	4	33	< 2	2	60	51	32

Source: Van Kauwenbergh, 1997.

PR is the basic raw material used in the production of phosphate fertilizers. In the mineral form, it contains a wide range of both useful and potentially harmful elements that may persist through the manufacturing process. Generally, PR of sedimentary origin, which constitutes about 85 percent of world reserves, contain higher concentrations of these elements.

All PRs contain hazardous elements including undesired heavy metals, e.g. Cd, Cr, Hg, Pb, and radioactive elements, e.g. U, that are considered to be toxic to human and animal health (FAO, 2004b). The amounts of these hazardous elements vary widely among PR sources and even in the same deposit. Table 44 shows the results of a chemical analysis of potentially hazardous elements in some sedimentary PR samples. Ranges in the concentration of potentially useful and harmful elements in PRs have also been summarized in Table 45.

TABLE 45
Range in concentration of potentially useful and harmful elements in phosphate rock

Potentially useful elements	Range of concentration (mg/kg P)	Potentially harmful elements	Range of concentration (mg/kg P)
Cobalt	5–42	Arsenic	30–150
Copper	104–756	Cadmium	0.9–600
Manganese	50–2 500	Chromium	6–4 600
Molybdenum	20–70	Lead	7–180
Nickel	11–590	Mercury	0.2–12
Selenium	15–213	Thorium	28–1 528
Zinc	35–6 040	Uranium	49–1 100
		Vanadium	25–5 660

Source: Laegreid, Bockman and Kaarstad, 1999.

Undesirable heavy metals can also originate from finished fertilizers and organic manures (Table 46).

Many studies have been conducted on the potentially harmful effects of these incidental additions of elements in the diets of humans and animals and have concluded that they pose no danger, perhaps with the exceptions of Cd and the radioactive elements thorium (Th) and U.

TABLE 46
Total content of undesirable heavy metals in some fertilizers and manures

Fertilizer/manure	Cd	Cr	Pb
		(mg/kg)	
Urea	< 0.1	< 3	< 3
Triple superphosphate	9	92	3
Potassium chloride	< 0.1	< 3	3
Cow manure	1	56	16
Sewage sludge	5	350	90

Source: Webber and Singh, 1995.

Cadmium

Among the hazardous heavy metals in PRs and finished P fertilizers, Cd is probably the most researched and of greatest concern. This is because of its potentially high toxicity to human health from consuming foods derived from crops fertilized with P fertilizers containing a significant amount of Cd. In addition to Cd being added mostly through phosphatic fertilizers, significant additions to agriculture can be made through animal manures, sewage sludge and industrial effluents (Table 46). The Cd added to soil is bound strongly to soil particles and its availability to plants increases with decreasing pH. Similarly, Cd availability increases with decreasing SOM. Both high soil moisture and salinity increase Cd availability to plants, whereas high Zn concentrations decrease Cd uptake. Leafy vegetables accumulate more Cd than other food crops.

Cd ingested by animals and humans accumulates in the kidneys, where it may result in the organ dysfunction. It is recommended that the daily intake of Cd by humans should not exceed 40 µg, of which less than 5 percent is absorbed by the body. Various countries have either voluntary or mandated concentrations of Cd in fertilizers, and these are constantly under review. The reactivity of the PR influences the availability of Cd to the plant. Thus, a PR with a higher reactivity and Cd content can release more Cd than one with a lower reactivity and/or low Cd content for plant uptake. In addition to PR reactivity and Cd content, plant uptake of Cd also depends on soil pH and crop species.

Fluorine

Most PRs also have high concentrations of fluorine (F), which is a part of the apatite minerals. Fluorine content often exceeds 3 percent by weight (250 g F/kg P). Excessive F absorption has been implicated in causing injury to grazing stock through fluorosis. However, the concentrations of F in herbage were generally found to be less than 10 mg F/kg and it was concluded that plant uptake of F is unlikely to lead to problems for grazing animals in most soils. However, caution is needed in case of ingestion of soil by animals or ingestion of fertilizer material. Thus, there is a need to monitor the F additions through PRs to acid soils on a long-term basis (FAO, 2004b).

Radioactive elements

Th and U have higher concentrations in many PRs than in soil. Some PR sources may also contain a significant amount of radioactive elements compared with others, e.g. 390 mg U/kg in Minjingu PR (the United Republic of Tanzania) versus 12 mg U/kg in Gafsa PR (Tunisia). As Minjingu PR is highly reactive and agro-economically suitable for direct application to acid soils for crop production, there can be concern over the safety of using it.

K contains 0.012-percent radioactive isotope potassium-40 (^{40}K), which is constantly decaying. The addition of ^{40}K through fertilizers replaces this decaying material. The ^{40}K contained in K fertilizers may be considered undesirable and it needs to be monitored. Theoretically, application of 20 kg K/ha mixed into the top 10 cm of soil adds about 0.16 percent K annually. However, analyses of soil samples from long-term experiments where K fertilizers have been applied have detected only slight or no accumulation of these radioactive elements. In none of the experiments were there detectable increases in the concentration of these elements in the plant material.

MINIMIZING THE NEGATIVE ENVIRONMENTAL EFFECT OF NUTRIENT USE

Improving fertilizer-use efficiency

The negative effects of plant nutrients on the environment are mainly the result of undesirable losses of N through various means and losses of P through surface runoff and soil erosion. The nutrients thus lost enter the atmosphere (in the case of N) and waterbodies (in the cases of N and P). Most of such losses can be reduced by management practices that minimize the negative effects on the environment. These negative effects are not caused by any fundamental properties of these elements but as a result of their interaction with soils and plants under human intervention. Where such losses are small, the negative effects on the environment are also minimal.

N losses can be reduced significantly by adopting practices that improve N utilization by crops and N conservation in the soil. Towards this goal, the integrated management of N with water and balanced nutrient application are of utmost importance for increasing nitrogen-use efficiency. This requires that N application rates not be excessively above the optimum whether delivered through mineral fertilizers or organic manures. In the case of P, appropriate soil and water conservation measures, application rates based on soil P levels and best methods of application are very important.

The practices that can lead to improved nitrogen-use efficiency are listed below. These are also practices that will reduce N losses as efficiency and losses are inversely related:

- Matching N application rates with the nature and yield potential of the crop.
- Ensuring a good crop stand and optimal plant population.
- Correcting all nutrient deficiencies in order to provide balanced nutrition.
- Distributing of total N to be applied in splits of 25–40 kg N/ha during crop growth.
- Increasing the number of splits in coarse-textured soils and high rates of N.
- Increasing the number of splits in the case of long-duration varieties.

- Synchronizing N application with moisture availability either through rainfall or irrigation.
- Using nitrification inhibitors where economical and feasible with N fertilizers.
- Avoiding overirrigation.
- Withholding N application during attacks by pests and diseases.
- Applying pre-plant N below the soil surface for dryland crops raised on stored soil moisture.
- Minimizing surface application of urea and ammonia fertilizers to alkaline soils.
- Deep placement of supergranules in flooded-rice fields.
- Minimizing nitrate fertilizers to flooded-rice soils.
- Following INM practices, e.g. combined application of mineral fertilizers with organic/green manures.
- Preferring S-containing N sources in soils that are also deficient in S.
- Adopting conservation tillage and residue recycling to control surface runoff and promote infiltration.
- Using organic manures to improve infiltration and enhance WHC.

Advances in agricultural technologies (e.g. improved soil sampling and analysis, better plant diagnostic methods, less soil-degrading tillage methods, use of starter fertilizers, and better timing and placement of nutrients) now enable farmers to apply nutrients with greater accuracy, minimizing or avoiding altogether any damage to soil, water, and air. For example, maize farmers in the United States of America increased yields by 40 percent and nitrogen-use efficiency by 35 percent between 1980 and 2000. One of the factors that made this possible was balanced nutrient application and correction of nutrient deficiencies.

It is known that nitrogen-use efficiency declines markedly where P, K or any other nutrient needed is omitted from the fertilization programme. This is demonstrated in Figure 51 and Table 27.

Balanced fertilization can have dramatic effects on soil NO_3-N concentrations, as shown by a study in Kansas, the United States of America (Figure 52). Where N was applied without P, there was a dramatic and dangerous accumulation of NO_3^- in the soil

FIGURE 51
Impact of balanced and integrated nutrient management on the nitrogen-use efficiency by maize and wheat in an alluvial soil in Punjab, India

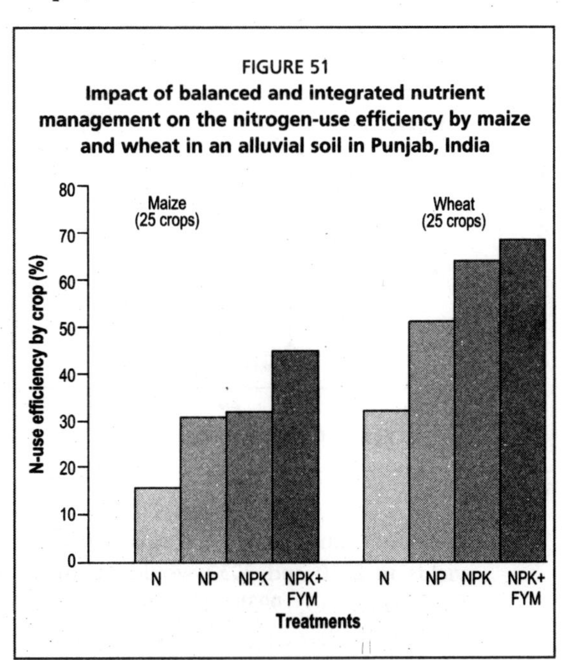

Source: Brar and Pasricha, 1998.

FIGURE 52
Impact of balanced fertilization on soil NO_3-N concentrations in Kansas, the United States of America

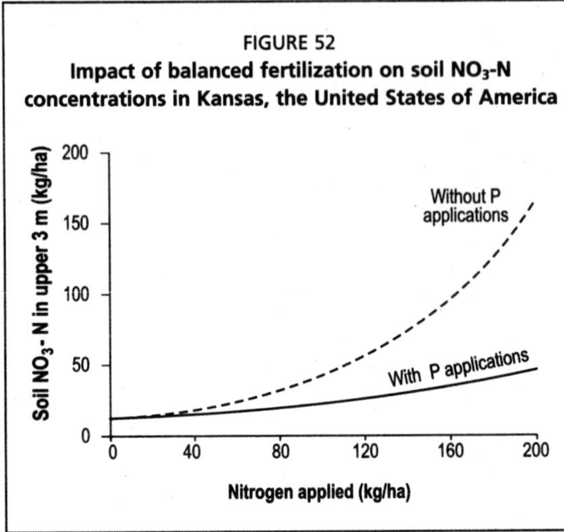

Source: Stewart, 2002.

FIGURE 53
Importance of efficient fertilizer use in achieving required cereal yield levels in Asia

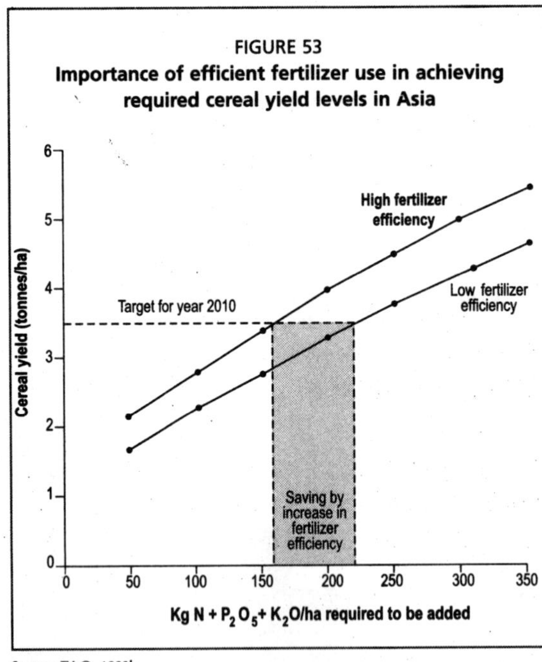

Source: FAO, 1993b.

profile. Where N was applied with P, the accumulation was low and within the range useful to plants. In the intensively fertilized region of Punjab, India, balanced nutrient application significantly reduced the amount of N in the rootzone after harvest. When only N was applied to wheat or maize, a substantial portion of it was found as nitrate N in the soil up to 2 m depth. However, when 60 kg P_2O_5 and 30 kg K_2O/ha were also applied along with 120 kg N/ha, there was little nitrate N that could potentially leach below the rootzone. Moreover, as the irrigation rate was increased but its frequency decreased, more nitrate N leached to deeper soil layers (Singh, 1996).

In addition to minimizing negative environmental effects, the efficient use of applied nutrients has another very important payoff in terms of reducing the amounts of nutrients required to achieve a given production goal. An analysis of nutrient requirement in Asia shows that with low fertilizer efficiency and associated innovations, developing Asia will be able to meet the minimum cereal yield requirement in 2010 (3.5 tonnes/ha) with 230 kg/ha of nutrients from fertilizers and in 2030 (5.5 tonnes/ha) with 475 kg/ha of nutrients from fertilizers. With high fertilizer efficiency and associated innovations, the corresponding nutrient requirements for the stated yields in 2010 and 2030 would be 160 kg/ha and 380 kg/ha of nutrients from fertilizers (FAO, 1993b). Therefore, the efforts of agricultural research and extension services, emphasizing fertilizer efficiency at farm level, can probably lead to a saving of 70 kg/ha (N + P_2O_5 + K_2O) by 2010 and 95 kg/ha by 2030 (Figure 53).

Managing nutrients to minimize losses

Efficient use of fertilizers and manures ensure that minimum amounts are left to be lost permanently from a site. Developments of nutrient budgets are the most practical way of preventing losses of nutrients to the environment. This, together with an understanding of the loss processes, can help to reduce losses to an environmentally acceptable level or even eliminate them. Table 47 summarizes the conditions favouring N losses and general strategies for minimizing them. Some guidelines for minimizing N losses are also provided in Table 43.

Losses of P to the environment can be reduced by: (i) avoiding excessive application rates of animal manures and slurries; (ii) soil and water conservation measures to reduce surface runoff and soil erosion; and (iii) balanced nutrient application to enhance crop utilization of available P.

TABLE 47
Conditions favouring N losses and general strategies for minimizing such losses

Channel of N loss	Conditions that favour loss of N	Strategies for minimizing N loss
Volatilization	Sandy soils	Mix fertilizers with soil
(loss as ammonia)	Ammonium or urea fertilizer left on soil surface	Drill basal dose for upland crops, follow N broadcast by hoeing, light irrigation. etc.
	Alkaline soils/over liming	Use gypsum, pyrite and organic manure
	Shallow N application in flooded-rice soils	Practice split application of N
	Hot dry period	Use USGs in medium–fine textured soils (deep placement in rice)
Leaching	Sandy soils	Add organic matter
(loss of N from rootzone with drainage water)	High rainfall areas	Split application of N (more splits at higher rates of N)
	Heavily irrigated fields (more water/irrigation)	Controlled/light irrigations (less water per irrigation)
	Heavy N applications or all N as basal	More splits of N for long duration crops/ varieties and in high rainfall areas
	Unbalanced fertilizer application leading to poor utilization of N	Balanced fertilization to ensure better utilization of applied N fertilizer
		Use soil-cured urea or neem coated urea
Denitrification	Conditions favouring movement of nitrate into lower depths, compact pockets	Improve drainage and soil aeration, avoid soil compaction
(Gaseous loss owing to biological or chemical decomposition of nitrate)	Waterlogged soils, poor soil aeration	Adopt practices to conserve N in ammonium form in reduced soils (flooded rice)
	Addition of nitrate N to waterlogged soils	Use non-nitrate sources for basal application
	Surface application of N to flooded rice soils	Place USG or NH_4-N 10–15 cm deep in flooded-rice soils
	High temperature	
	Acidic pH (for chemical denitrification), non-acidic condition (for biological denitrification)	Lime acid soils
Erosion/runoff	Sloping lands	Contour cultivation
(loss of N through surface flow due to heavy rains, over irrigation or soil erosion)		Land levelling
	Lack of soil cover	Minimum/zero tillage
	Poorly levelled fields	Suitable moisture conservation practices (ploughing before rain, bunding, mulching etc.)
	High level of tillage	Incorporate fertilizer in soil
	Inadequate moisture conservation	Controlled and light irrigations

Guidelines for the best agricultural practices to optimize fertilizer use in Asia and the Pacific were developed several years ago (FADINAP, 1993). Similar practices for Europe are also available (IFA/EFMA, 1998). The objectives of these guidelines are:

- to integrate the principles of economic crop production with environmental protection;
- to create public confidence that farmers use fertilizers responsibly;
- to provide planners and policy-makers with a sound understanding of the role of fertilizer in sustainable systems of crop production.

The need for widespread dissemination and adoption of best agricultural practices cannot be overemphasized. When this happens, nutrient management will be based on scientific findings, it will be efficient, profitable and associated with minimum adverse effect on the environment, a concern common to all sources of nutrients be they mineral fertilizers or organic manures.

Glossary

Acid-forming fertilizer

A fertilizer that leaves behind an acidic effect in the soil (reduces soil pH). Such fertilizers, which lack a metallic cation, are generally acid forming. Their continuous use makes a soil acid (lowers pH) and reduces soil quality and, hence, productivity. The excess acidity can be neutralized by lime application. This is generally of practical importance in the case of nitrogenous fertilizers. Examples: ammonium sulphate, ammonium chloride, anhydrous ammonia and urea.

Agricultural liming material

Material containing oxides, hydroxides and/or carbonates of Ca and/or Mg, used for neutralizing the acidity of the soil. Its use is referred to as liming.

Alkaline (or basic) fertilizer

A fertilizer that leaves behind an alkaline reaction in the soil (raises soil pH). Examples: calcium nitrate, sodium nitrate. Opposite of acid-forming fertilizer.

Ammoniated superphosphate

A product obtained from superphosphate treated with ammonia or solutions containing free ammonia. The end product provides extra N but, in the process, its total P content and also the water solubility of this P are reduced.

Ammonium chloride (sal ammonia or muriate of ammonia)

Ammonium salt of hydrochloric acid containing 25 percent N in ammoniacal form. Formula: NH4Cl. An acid-forming fertilizer.

Ammonium citrate

A compound, the solution of which is used to determine the available phosphate content of fertilizers usually consisting of water-soluble and citrate-soluble phosphate.

Ammonium molybdate

An important molybdenum fertilizer containing 52–54 percent Mo. Formula: $(NH_4)_6Mo_7O_{24}\ .4H_2O$. It can be applied either to soils and seeds, or through foliar spray. Standard specifications of ammonium molybdate based on Indian experience are:

- molybdenum (as Mo), percent by weight, minimum: 52.0;
- matter insoluble in water, percent by weight, maximum: 1.0;
- lead (as Pb), percent by weight, maximum: 0.003.

Ammonium nitrate

A product obtained by neutralizing nitric acid with ammonia. Formula: NH_4NO_3. It is usually in a granular or prilled form, and coated with a suitable material to prevent absorption of moisture and caking in storage. Fertilizer-grade ammonium nitrate has a total N content of 33–34.5 percent, of which 50 percent is present as ammoniacal-N and 50 percent as nitrate-N. It leaves behind an acidic effect in the soil.

Ammonium phosphate

Group of solid fertilizers, manufactured by reacting ammonia with phosphoric acid. Type of compound formed depends on the amount of ammonia that is reacted with phosphoric acid. Two important ammonium phosphates are: (i) mono-ammonium phosphate or MAP ($NH_4H_2PO_4$), containing about 11 percent N and 52 percent P_2O_5; and (ii) di-ammonium phosphate or DAP [$(NH_4)_2HPO_4$], typically containing 18 percent N and 46 percent P_2O_5.

Ammonium phosphate sulphate

An important complex fertilizer containing N, P and S. Typical grades are 16-20–0–15 percent and 20–20–0–15 percent in terms of N + P_2O_5 + K_2O + S. It is essentially a factory-made complex consisting of 60 percent ammonium sulphate and 40 percent ammonium phosphate. Useful for basal dressing to provide N, P and S, all of which are present in water-soluble, plant available form.

Ammonium sulphate (AS)

Traditionally, the best-known N and S fertilizer. Formula: $NH_4(SO_4)_2$. It contains about 21 percent N (all as ammonium) and 23–24 percent S (all as sulphate). Its specific gravity is 1.769, its bulk density is 720–1 040 kg/m^3 and its angle of repose is 32–33 °. It is an acid-forming fertilizer because it lacks a metal cation. Highly soluble in water, it can be produced through various processes and used directly or as an ingredient of fertilizer mixtures. It is used as part of the basal dressing or as top-dressing to provide both N and S. Ammonium sulphate should not be mixed with PR or urea.

$$\underset{\text{Ammonia Sulphuric acid}}{2NH_3 + H_2SO_4} \longrightarrow \underset{\text{Ammonium sulphate}}{2(NH_4)_2SO_4}$$

Standard specifications of ammonium sulphate based on Indian experience are:

- moisture, percent by weight, maximum: 1.0;
- ammoniacal-N, percent by weight, minimum: 20.6;
- sulphur (as S), percent by weight, minimum: 23.0;
- free acidity (as H_2SO_4), percent by weight, maximum (0.04 for material obtained from by-product ammonia and by-product gypsum): 0.025;
- arsenic (as As_2O_3), percent by weight, maximum: 0.01.

Ammonium sulphate nitrate (ASN)

A fertilizer containing 26 percent N and 15 percent S, both in soluble and plant available form. It is a double salt of ammonium sulphate and ammonium nitrate in which 75 percent of total N is present as ammoniacal-N and 25 percent as nitrate-N. Agronomically, it is comparable to ammonium sulphate, except for the more mobile nitrate-N component in ASN.

Ammonium thiosulphate

A liquid fertilizer containing 12 percent N and 26 percent S (thio refers to S). Fifty percent of the S is in the sulphate form and the rest is in elemental form. It can be used directly or mixed with neutral to slightly acid P-containing solutions or aqueous ammonia or N solutions to prepare a variety of NPK + S and NPKS + micronutrient formulations. It can also be applied through irrigation, particularly through drip and sprinkler irrigation.

Aqueous ammonia

A solution containing water and ammonia in any proportion, usually qualified by a reference to ammonia vapour pressure. For example, aqua ammonia has a pressure of less than 0.7 kg/cm^2. Commercial grades commonly contain 20–25 percent N. It is used either for direct application to the soil or for preparation of ammoniated superphosphate.

Apatite

Common name of the major P-bearing compound in PR (used as raw material in the manufacture of phosphate fertilizers). General formula: $Ca_{10}(PO_4, CO_3)_6 (F, OH, Cl)_2$. Depending on the dominance of F, Cl or OH in the apatite crystal structure, it is known as fluorapatite, chlorapatite or hydroxyapatite.

Ash

The mineral residue remaining after the destruction of organic material by burning. Ash of plant residues or wood is usually a rich source of K.

Azolla

A floating freshwater fern. It fixes N in symbiotic association with the cyanobacterium (BGA) *Anabaena azollae.* Cultivation of *Azolla* in Viet Nam and China began during the Ming dynasty (1368–1644). *Azolla* is distributed in both temperate and tropical rice-growing regions. One crop of *Azolla* can provide 20–40 kg N/ha to the rice crop in about 20–25 days.

Benefit–cost ratio (BCR)

The ratio of the value of extra crop produced (minus cost of fertilizer or any other production input) to the cost of fertilizer. It indicates the rate of net returns from the use of an input and, hence, is an important indicator of the degree of profitability from input use. If a fertilizer costing US$50 produces extra crops worth US$150, then the BCR = (150 - 50)/50 = 2. A useful decision-making tool before investing in an input. BCR = VCR - 1.

Biofertilizer

A rather broad term used for products (carrier- or liquid-based) containing living or dormant micro-organisms like bacteria, fungi, actinomycetes and algae alone or in combination, which on application help in fixing atmospheric N or solubilize/mobilize soil nutrients in addition to secretion of growth-promoting substances for enhancing crop growth. "Bio" means living, and "fertilizer" means a product that provides nutrients in usable form. Biofertilizers are also known as bioinoculants or microbial cultures. Strictly speaking, the term is a misnomer, albeit a widely used one. Unlike fertilizers, these are not used to provide nutrients present in them, except *Azolla* where used as green manure. Biofertilizers can be broadly classified into four categories:

- N-fixing biofertilizers: Rhizobium, Azotobacter, Azospirillum, Acetobacter, BGA and Azolla;
- P-solubilizing/mobilizing biofertilizers (PSB or PSM): P-solubilizing, e.g. *Bacillus*, *Pseudomonas* and *Aspergillus*, P-mobilizing, e.g. VAM;
- composting accelerators: (i) cellulolytic (*Trichoderma*), and (ii) lignolytic (*Humicola*);
- plant-growth promoting rhizobacteria: species of *Pseudomonas*.

Bioinoculant

A biological preparation containing living organisms, such as biofertilizers, used in agriculture for inoculation of seeds, soils or other plant materials. See biofertilizer.

Biological nitrogen fixation (BNF)

The process involving the conversion of nitrogen gas (N_2) into ammonia through a biological process (in contrast to industrial N fixation). Same as biological dinitrogen fixation. Many micro-organisms, such as *Rhizobium*, *Azotobacter* and BGA utilize molecular N_2 through the help of nitrogenase enzyme and reduce it to NH_3:

$$N_2 + 6H^+ + 6e^- \longrightarrow 2NH_3$$

It is a major source of fixed N for plant life on the earth. Estimates of global terrestrial BNF range from 100 to 290 million tonnes of N per year, of which 40–48 million tonnes is estimated to be biologically fixed in agricultural crops and fields. Mo and Co are considered to play a particularly important role in BNF.

Blue green algae (BGA)

Photosynthetic, N-fixing algae, also known as cyanobacteria. These are unicellular and aerobic organisms. Their role in paddy-fields was reported by P.K. Dey of India in 1939. More than 100 species of BGA are known to fix N. Commonly occurring BGA are *Nostoc*, *Anabaena*, *Aulosira*, *Tolypothrix*, and

Calothrix. These are used as biofertilizer for wetland rice (paddy) and can provide 25–30 kg N/ha. They also secrete hormones such as IAA and GA and improve soil structure by producing polysaccharides, which help in the binding of soil particles resulting in better soil aggregation. Also used as a soil conditioner and to prevent soil erosion through mat formation.

Borax

Sodium tetraborate compound. Formula: $Na_2B_4O_7.10H_2O$. Contains 10.5 percent B. An important B fertilizer for soil or foliar application. Standard specifications of borax based on Indian experience are:

- content of boron (as B), percent by weight, minimum: 10.5;
- matter insoluble in water, percent by weight, maximum: 1.0;
- pH: 9.0–9.5;
- lead (as Pb), percent by weight, maximum: 0.003.

Bulk density

Mass per unit bulk volume (including pores) of soil or particle that has been dried to constant weight at 105 °C. The bulk density of different biofertilizer carriers is: peat (1.02 g/cm^3), lignite (1.08 g/cm^3), and charcoal (0.43 g/cm^3). The bulk density of ammonium sulphate is 720–1 040 kg/m^3.

Bulk fertilizer

Commercial fertilizer in a non-packed form.

Cadmium (Cd)

A toxic heavy metal. Atomic weight: 112.4. Usual content in soils is 0.4 ppm. Can enter finished fertilizers through PR, which is an important raw material, and other sources. Potentially toxic to plants and animals. Of great concern to human health, Cd is associated with crippling condition known as *Itai-itai* (Japanese). PRs can contain a wide range of Cd content. See phosphate rock.

Caking

Refers to the change of fertilizer powder or granules into hard lumps. This is usually a consequence of extended storage under pressure in a humid environment. It is a sign of deterioration in physical quality. Use of anti-caking agents can help to minimize caking.

Calcium ammonium nitrate (CAN)

A mixture of ammonium nitrate and finely pulverized limestone or dolomite, granulated together. It contains 21–26 percent N, half in the form of ammoniacal-N and half in the form of nitrate-N. Its use does not make the soil acid by virtue of the Ca in it. Standard specifications of CAN based on Indian experience are:

- moisture, percent by weight, maximum: 1.0;
- total ammoniacal- and nitrate-N, percent by weight, minimum: 25.0;

- ammoniacal-N, percent by weight, minimum: 12.5;
- calcium nitrate, percent by weight, maximum: 0.5;
- particle size: not less than 80 percent of the material shall pass through 4-mm IS sieve and be retained on 1-mm IS sieve. Not more than 10 percent of the material shall be below 1-mm IS sieve.

Cation exchange capacity (CEC)
The capacity of a soil or any other substance with negatively charged exchange complex to hold cations in exchangeable form is referred as the CEC. It is a measure of the net negative charge of a soil. Expressed in me/100 g of soil (old term) or Cmol/kg (new term). The CEC depends on the type and proportion of organic matter and clay minerals present in the soil. Clay soils have a higher CEC than sandy soils.

Citric-acid-soluble P_2O_5
That part of the total P_2O_5 particularly in basic slag and bone meal that is insoluble in water but soluble in 2-percent citric acid solution and considered to be plant available.

Clay
A group of hydrated aluminium silicates of microcrystalline structure. A common constituent of soils. Smallest size particles of mineral matter in the soil, usually less than 0.002 mm in diameter. Clays play a major role in determining soil texture, soil structure, water retention, CEC and nutrient dynamics. Examples: kaolinite, illite and montmorillonite.

Coated fertilizer
A fertilizer whose granules are covered with a thin layer of a different material in order to improve its behaviour and/or modify the characteristics of the fertilizer. Commonly done to improve the physical condition of a fertilizer or reduce the rate of release of nutrients in the soil after application.

Complex fertilizer
A fertilizer that contains two or more major nutrients (N, phosphate and potash) made by a chemical reaction between the nutrient-containing raw materials. Same as multinutrient fertilizer. Examples: NP complex 23–23–0, and NPK complex 12–32–16.

Compost
An organic manure or fertilizer produced as a result of aerobic, anaerobic or partially aerobic decomposition of a wide variety of crop, animal, human and industrial wastes. Conveniently categorized as rural or urban (town) compost according to the type and location of wastes used for composting. Compost

prepared with the aid of earthworms is referred to as vermicompost. Typical nutrient content of rural compost is 0.5 percent N, 0.2 percent P_2O_5 and 0.5 percent K_2O, while that of urban compost is 1.5 percent N, 1.0 percent P_2O_5 and 1.5 percent K_2O. On average, compost also contains 10 ppm Zn, 6 ppm B and 12 ppm Mn. Nutrient status of a compost depends largely on the nutrient content of the wastes composted.

Compound fertilizer

A fertilizer having a declarable content of at least two of the nutrients N, P and K, obtained chemically (as in complex fertilizers), by mixing (as in fertilizer mixtures/bulk blends), or both.

Copper sulphate

Most common Cu fertilizer. Formula: $CuSO_4.5H_2O$ (24 percent Cu). It comes in particle sizes varying from fine powder to granular. A less hydrated form, $CuSO_4.H_2O$, contains 35 percent Cu. Standard specifications of $CuSO_4.5H_2O$ based on Indian experience are:

- copper (as Cu), percent by weight, minimum: 24.0;
- sulphur (as S), percent by weight, minimum: 12.0;
- matter insoluble in water, percent by weight, maximum: 1.0;
- soluble iron and aluminium compounds (expressed as Fe), percent by weight, maximum: 0.5;
- lead (as Pb), percent by weight, maximum: 0.003;
- pH: not less than 3.0.

Critical level (CL)

That level of concentration of a nutrient in the plant or available nutrient in the soil that is likely to result in 90 percent of the maximum yield. Where the CL is determined correctly, the probability of crop response to applied nutrient is high at below the CL and low above the CL. Same as critical limit. Used as a diagnostic tool in decision-making for nutrient application.

Critical relative humidity (CRH)

The relative humidity (usually stated at 30 °C) at which a material (fertilizer) starts absorbing moisture from the air. CRH in case of micronutrient fertilizers has not received much attention. The lower the CRH of a fertilizer, the more hygroscopic it is. Such materials need special care during storage. Some values of CRH at 30 °C are:

- urea: 75.2;
- ammonium sulphate: 79.2;
- MOP: 84.0;
- sulphate of potash: 96.3;
- DAP: 82.5.

Cyanobacteria

BGA are known also as cyanobacteria as they are procaryotic-like bacteria and their cells contain phycocyanine (blue) and green pigment. They are divided into four groups:

- unicellular, reproduced by binary fission or budding (e.g. *Gleocapsa*);
- unicellular, reproduced by multiple fission (e.g. *Chloroecidiopsis*);
- filamentous, non-heterocystous (e.g. *Plectonema*);
- filamentous, heterocystous (e.g. *Nostoc*).

Deficiency

Refers to inadequacy. In soils and plants, the state of inadequate supply or low availability of an essential nutrient for optimal plant growth. In quantitative terms, the nutrient status is below the critical level. This can be corrected by external nutrient application through fertilizers and manures. Deficiency symptoms refer to visible signs of the deficiency of a nutrient element in a growing plant or its produce, usually visible to the naked eye. Some common descriptors of nutrient deficiency symptoms in growing plants are:

- bronzing: development of bronze/copper colour on the tissue;
- chlorosis: loss of chlorophyll, resulting in loss of green colour, paleness, appearance of yellow tissue;
- decline: onset of general weakness as indicated by loss of vigour, poor growth and low productivity;
- dieback: collapse of the growing tip, affecting the youngest leaves;
- firing: burning of tissue accompanied with dark brown or reddish-brown colour;
- lesion: a localized wound of the tissue accompanied by loss of normal colour;
- necrosis: death of tissue;
- scorching: burning of the tissue accompanied by light brown coloration (this can also result from faulty spraying, salt injury, etc.).

Dicalcium phosphate

A product containing not less than 34 percent P_2O_5 in citrate-soluble form, which is considered available to plants. Formula: $CaHPO_4$.

Dolomite

An Mg-containing natural limestone mineral used for liming acid soils that also need Mg application. Formula: $CaMg(CO_3)_2$. Contains 40–45 percent CaO and 5–20 percent MgO. An important soil amendment.

Dung

The semi-solid excreta of large animals (excluding humans). Used as a manure, soil conditioner, biogas plant input and as domestic fuel. Dung is the main ingredient of FYM.

Equivalent acidity

Refers to parts by weight of calcium carbonate (as $CaCO_3$) required to neutralize the acidity resulting from the use of 100 parts by weight of an acid-forming fertilizer. The equivalent acidity of some common fertilizers is:

- anhydrous ammonia: 148;
- ammonium chloride: 128;
- ammonium sulphate: 110;
- ammonium nitrate sulphate: 93;
- urea: 84;
- DAP: 74;
- MAP: 65;
- ammonium nitrate: 63.

Equivalent basicity

The number of parts by weight of calcium carbonate (as $CaCO_3$) that corresponds in acid neutralizing capacity of 100 parts by weight of the fertilizer. In other words, it shows the neutralizing capacity, expressed as kilograms of $CaCO_3$ per 100 kg of the fertilizer. The equivalent basicity of some common fertilizers is:

- calcium nitrate: 21;
- dicalcium phosphate: 25;
- sodium nitrate: 29.

Farmyard manure (FYM)

Bulky organic manure resulting from naturally decomposed mixture of dung and urine of farm animals along with the litter (bedding material). Average, well-rotted FYM contains 0.5–1.0 percent N, 0.15–0.20 percent P_2O_5 and 0.5–0.6 percent K_2O. Desired C:N ratio in FYM should not exceed 15–20:1. In addition to NPK, it may contain about 1 500 ppm Fe, 7 ppm Mn, 5 ppm B, 20 ppm Mo, 10 ppm Co, 2 800 ppm Al, 12 ppm Cr and up to 120 ppm Pb. Often fully or partially air-dry dung is used as FYM. See bulky organic manure.

Ferrous sulphate

A common Fe fertilizer. Formula: $FeSO_4.7H_2O$. Contains 19 percent Fe and 11 percent S. Same as iron sulphate. Standard specifications of ferrous sulphate ($FeSO_4.7H_2O$) based on Indian experience are:

- ferrous iron (as Fe), percent by weight, minimum: 19.0;
- sulphur (as S), percent by weight, minimum: 10.5;
- free acid (as H_2SO_4), percent by weight, maximum: 1.0;
- ferric iron (as Fe), percent by weight, maximum: 0.5;
- matter insoluble in water, percent by weight, maximum: 1.0;
- pH: not less than 3.5;
- lead (as Pb), percent by weight, maximum: 0.003.

Fertigation

The practice of applying fertilizers together with irrigation water and not in a separate operation. More often advocated for use with drip irrigation systems than with conventional flood irrigation. In principle, all required nutrients including micronutrients can be applied through fertigation. Products available for drip irrigation should be highly water soluble and include products containing major nutrients, micronutrient salts as well as chelates of EDTA and EDDHA. Similar to chemigation.

Fertilization

The practice of applying fertilizers for plant nutrition. The fertilizers can be applied through soil, irrigation water or sprayed on plant leaves. Same as fertilizer application.

Fertilizer

A mined, refined or manufactured product containing one or more essential plant nutrients in available or potentially available forms and in commercially valuable amounts without carrying any harmful substance above permissible limits. Although organic fertilizers are also being prepared and used, they are not yet covered by the term fertilizers, largely due to tradition. Same as mineral or inorganic fertilizer. Examples: urea, SSP, zinc sulphate, borax, and copper sulphate.

Fertilizer grade

An expression used in extension and the fertilizer trade referring to the legal guarantee of the available plant nutrients expressed as a percentage by weight in a fertilizer, e.g. a 12–32–16 grade of fertilizer indicates 12 percent N, 32 percent P_2O_5 and 16 percent K_2O in that complex fertilizer.

Fertilizer mixture

A mixture prepared by physically mixing two or more finished fertilizers so as to contain two or more out of N, P and K plus any other nutrients. Mixture can be powdery or granulated. Examples: multimicronutrient mixtures, NPK mixtures, and bulk blends.

Fertilizer placement

A method of fertilizer application in which the fertilizer is placed at a specific point or zone on or below the soil surface. It minimizes soil–fertilizer contact and creates higher nutrient concentration near the point of placement than in the general field. Examples: placement in holes around tea bushes, deep placement of USGs between rice hills, and drilling of phosphatic fertilizer below the seed.

Fertilizer quality

Chemical and physical state of a finished fertilizer as specified in the accepted quality standards of a country. For example, in India, fertilizer quality should be as per the Fertilizer Control Order (FCO). Quality can be acceptable (good) or substandard (non-standard), in which case it deviates from the stated parameters. Fertilizer loses its quality when it is non-standard and/or adulterated. Fertilizer quality control refers to totality of all legislation, enforcement, testing and monitoring activities aimed at ensuring its quality as laid down in quality standards.

Filler

Any material mixed with fertilizers during production for purposes other than addition of plant nutrients so as to give anti-caking properties and for adjusting their weight to bring the percentage of nutrients so as to maintain grade composition. Must not contain any harmful or toxic substance. Examples: sand, lime, dolomite, silica, and sawdust.

Fortified fertilizer

A fertilizer to which another compound has been deliberately added in order to enhance its nutrient value. Several common fertilizers can be fortified with compounds of nutrients, such as S, B and Mo. An additional advantage of fortification is that small amounts of micronutrients needed can be applied uniformly over a field with ease. Examples: SSP fortified with B (boronated SSP), urea fortified with Zn (zincated urea), and NP/NPK complexes fortified with B or Zn.

Fused calcium and magnesium phosphate

A product derived from the fusion of PR with about 30 percent of magnesium oxide as such or as a mineral silicate. Typical fused calcium phosphate contains 27 percent P_2O_5 and 19 percent Ca while fused magnesium phosphate contains 8 percent Mg and 10 percent P_2O_5. Most of the phosphate is in citrate-soluble (available) form, although very little is water soluble. These products must be finely ground in order to be effective sources of phosphate for plants as their availability is related directly to their specific surface, which in turn is inversely proportional to their particle size.

Granular fertilizer

Solid material formed into particles of a predetermined mean size.

Granulation

Techniques using a process such as agglomeration, accretion or crushing to make a granular fertilizer.

Green manure

Refers to fresh green plant matter (usually of legumes and often specifically grown for this purpose in the main field) that is ploughed in or turned into the soil to serve as manure. Several legume plants are used as green manure crops.

These are an important source of organic matter and plant nutrients, specially N. A key component (where feasible) of integrated plant nutrition systems (IPNS). Green manure can either be grown *in situ* and incorporated or grown elsewhere and brought in for incorporation in the field to be manured. Not all plants can be used as a green manure in practical farming. Green manures may be: plants of grain legumes such as pigeon pea, green gram, and cowpea; perennial woody multipurpose legumes such as *Leucaena leucocephala* (*subabul*), *Gliricidia sepium*, *Cassia siamea*; and non-grain legumes, such as *Crotalaria, Sesbania, Centrosema, Stylosanthes* and *Desmodium*. As green manures add whatever they have absorbed from the soil, they also promote the recycling of soil nutrients from lower depths to the topsoil. The most desirable characteristics in selecting a green manure crop are: (i) local adaptability of the plant; (ii) fast growth and production of a large amount of green matter (biomass) per unit area per unit time; (iii) tolerance to soil and environmental stresses such as acidity, alkalinity and drought; (iv) resistance to pests; and (v) easy to decompose, requiring minimum gap between incorporation and planting the main crop.

Ground phosphate rock

Material obtained by grinding naturally occurring PR to a fineness meeting relevant specifications or accepted custom, generally for direct application to soils.

Growth medium

Any material such as soil and peat used as a support for plant roots that has a capacity for water retention and that may contain added or naturally occurring nutrients. Also a medium in which micro-organisms are grown such as during biofertilizer production.

Guano

Group of organic manures derived from animal excreta, usually of small animals and includes materials such as bat, Peruvian and fish guano. General N content of guano can be 0.4–9.0 percent and total P_2O_5 can be 12–26 percent. Found and used in certain areas only.

Gypsum

The naturally occurring mineral calcium sulphate. Formula: $CaSO_4.2H_2O$ (containing 18.6 percent S and 23 percent Ca). Agricultural grade gypsum is usually of 70-percent purity containing 13–15 percent S and 16–19 percent Ca. Its solubility in water is 2.5 g/litre. It is an important source of both Ca and S for plants and is commonly used as an amendment for reclaiming alkali soils.

Heavy metal

Elements with a high atomic weight and specific gravity of more than 5 (density greater than 500 kg/m^3). These include plant nutrients as well as potential pollutant/toxic metals to plants and animals (Pb, Cd, etc). Some P fertilizers may contain heavy metals that originate from the PR. Most metal micronutrients (Fe, Mo, Mn, Ni, Cu and Zn) are also heavy metals. Thus, not all heavy metals are toxic, especially where present within permissible limits. The toxicity of a metal depends on its concentration in relation to plant needs and tolerance. At excessive concentrations, even micronutrients can become toxic.

High-analysis fertilizer

An arbitrary term for a fertilizer containing more than 25 percent of one or more of the three major plant nutrients, namely, N, P (as P_2O_5) and K (as K_2O). Examples: urea, DAP, NPK complexes, polyphosphates, and elemental S products.

Hoof and horn meal

An organic manure obtained from the processing, drying and grinding of animal hooves and horns. Usually contains 13–15 percent N and 0.3–1.5 percent P_2O_5.

Humus

The highly decomposed fraction of SOM having little resemblance to the matter from which it has been derived. It is characterized as an amorphous, dark coloured, nearly odourless, stable material of high molecular weight. It is the major food reservoir of soil microbes as it contains organic C and N needed for their development. Humic material has a very high CEC (200–500 Cmol/kg soil). It improves the buffering and WHC of soil. The process of formation of humus is called humification.

$$\text{Organic residues} \xrightarrow{\text{microbial decomposition}} CO_2 + H_2O + \text{humus} + \text{nutrients}$$

Kieserite

Trade name for magnesium sulphate monohydrate. Formula: $MgSO_4.H_2O$ (16 percent Mg). Sparingly soluble in cold water but readily soluble in hot water. Its bulk density is 1.4 g/cm^3 and its angle of repose is 34 °. Used as fertilizer for soil or foliar application to provide Mg as well as S.

Liquid fertilizers

Fertilizers in liquid finished form. Examples: urea ammonium nitrate solutions, polyphosphates, thiosulphates, suspensions, and special formulations for fertigation. Same as fluid fertilizers. Several liquid fertilizers can contain micronutrients, which can be in solution, in chelated (sequestered) form or in suspended form using suspension agents such as special type of clay, usually 2 percent attapulgite.

Liquid manure

Liquid resulting from animal urine and litter juices or from a dung heap that can be used as an organic manure.

Liming material

Product containing one or both of the elements Ca and Mg, generally in the form of an oxide, hydroxide or carbonate, principally intended to maintain or raise the pH of soil.

Low-analysis fertilizer

An arbitrary term for a mineral fertilizer containing less than 25–30 percent (N), P (as P_2O_5) and K (as K_2O). Same as "dilute fertilizer". Term falling into disuse for its restrictive nature and non-recognition of other useful nutrients such as S in them. Examples: ammonium sulphate and SSP.

Luxury consumption

Absorption of a nutrient by a plant well in excess of the quantities required. Common in case of N, K and Cl but can also occur in Zn. A waste from the farmer's viewpoint as the excess nutrient absorbed does not lead to extra yield. Reduces the physiological NUE although increased crop recovery of added nutrients.

Macronutrients

Essential plant nutrients that are required by plants in relatively large amounts (as compared with micronutrients). Include: N, P, K S, Ca and Mg, as also C, H and O (non-mineral nutrients).

Magnesium sulphate

A common Mg fertilizer. In anhydrous form, $MgSO_4$ contains 20 percent Mg. In hydrated form, $MgSO_4.7H_2O$ (Epsom salt), it contains 10 percent Mg. It is readily soluble in water, has a bulk density of 1 g/cm^3 and an angle of repose of 33°. It can be used for soil application and for foliar application. See also Kieserite. Standard specifications of magnesium sulphate ($MgSO_4.7H_2O$) based on Indian experience are:

- free flowing – crystalline form;
- magnesium (as Mg), percent by weight, minimum: 9.6;
- sulphur (as S), percent by weight, minimum: 12.0;
- matter insoluble in water, percent by weight, maximum: 1.0;
- lead (as Pb), percent by weight, maximum: 0.003;
- pH (5-percent solution): 5.0–8.

Manganese sulphate

A common Mn fertilizer. Formula: $MnSO_4.H_2O$. Contains 30.5 percent Mn.

Manure

Term used traditionally for all types of plant nutrient sources including organic manures and fertilizers but now increasingly restricted to animal-dung-based bulky organic manures, composts, oilcakes, bone meal and other animal meals. See FYM and compost.

Micronutrients

Group name for essential plant nutrients B, Cl, Cu, Fe, Mn, Mo, Ni and Zn. These are required by plants in much smaller amounts than macronutrients but are equally essential. Also known as "minor elements". The glossary of the Soil Science Society of America defines them as nutrients found in concentrations of less than 100 ppm (0.01 percent) in plants and includes nine elements in the list, the above-listed elements and Co.

Mineral fertilizer

See fertilizer.

Multimicronutrient fertilizer

A fertilizer containing several micronutrients. Can be solid or liquid. Usually a physical mixture.

Municipal solid waste (MSW)

A mixture of domestic, small-scale industrial and demolition solid wastes generated within a community. About 80 percent of MSW is combustible and 82 percent of combustibles are of biological origin, hence, usable as raw material for composting.

Muriate of potash (MOP)

Same as potassium chloride. Derived from muriatic acid, the earlier name for hydrochloric acid.

Mycorrhiza

The term "mycorrhizae" (plural) means root fungus (from the Greek myces = fungus; rhiza = root). Symbiotic fungi that form a mutually beneficial association with plant roots. Mycorrhizae are of three types: (i) ectotrophic; (ii) endotrophic; and (iii) ectendotrophic. In ectomycorrhizae, a distinct fungal sheath develops on the root. In endomycorrhizae, fungal hyphae penetrate root cells. Relationship between mycorrhizae and plant roots is useful in improving the capability of plants for soil exploration and nutrient uptake. Mycorrhizae have special structures known as vesicles and arbuscules. The arbuscules help in the transfer of nutrients from the fungus to the root system, and the vesicles, which are "saclike" structures, store P as phospholipids. The survival and performance of VAM fungi is affected by the host plant, soil fertility, cropping practices, and biological and environmental factors. Maximum root colonization and sporulation occurs in low-fertility soils.

Neem cake

Residue left after extracting oil from neem seeds. A non-edible oilcake. Contains 5 percent N, 1 percent P_2O_5 and 1.5 percent K_2O. Used as an organic manure and also for coating urea, which helps to reduce the rate of nitrification and to protect applied N against losses.

Nitrate of soda

Chiefly the sodium and potassium salt of nitric acid containing not less than 15 percent nitrate-N and 10 percent potash (as K_2O).

Nitrophosphates

Products obtained by treatment of PR with nitric acid alone or in admixture with sulphuric or phosphoric acid, with or without subsequent treatment with ammonia. Their N is partly in ammoniacal and partly in nitrate form. Usually only a part of their P (30–85 percent) is water soluble, the remainder being citrate soluble. Also referred to as nitric phosphates or ammonium nitrate phosphates (ANP). Example: nitrophosphate grade 23–23–0. Typical internationally accepted technical specifications of this fertilizer specify a maximum moisture content of 1 percent by weight. Standard specifications of nitrophosphate (23–23–0) based on Indian experience are:

- moisture, percent by weight, maximum: 1.5;
- total N, percent by weight, minimum: 23.0;
- N in ammoniacal form, percent by weight, minimum: 11.5;
- N in nitrate form, percent by weight, maximum: 11.5;
- neutral ammonium citrate soluble phosphate (as P_2O_5), percent by weight, minimum: 23.0;
- water-soluble phosphate as P_2O_5, percent by weight, minimum: 18.5;
- calcium nitrate, percent by weight, maximum: 1.0;
- particle size: not less than 90 percent of the material shall pass through 4-mm IS sieve and be retained on 1-mm IS sieve. Not more than 5 percent of the material shall be below 1-mm IS sieve.

Non-acid-forming fertilizer

A fertilizer not capable of increasing the acidity or reducing the alkalinity of the soil. Example: calcium ammonium nitrate.

Oilcake

The residue left after oil has been extracted from an oilseed. Non-edible oilcakes can be used as manure, and edible oilcakes are used primarily as cattle feed. Example: groundnut cake. Having almost similar content of organic C but variable levels of N, P and K, oilcakes mineralize easily when added to soil. The C:N ratios in them are highly favourable for quick decomposition. Notwithstanding the alternative use of edible oilcakes as animal feed, both types of materials have been extensively used as organic fertilizers, either alone or in combination with mineral fertilizers.

Organic fertilizer

A fertilizer prepared from one or more processed materials of a biological nature (plant/animal) and/or unprocessed mineral materials (lime, PR, etc.) that have been altered through controlled microbial decomposition into a homogenous product with sufficient plant nutrients to be of value as a fertilizer. Usually contains a minimum of 5 percent nutrients (N + P_2O_5 + K_2O). Synonymous with organic manures and various types of composts but with greater degree of product standardization. Important carriers of all nutrients. Primary external sources of nutrients in organic farming. See compost.

Organic manure

A manure derived principally from substances of plant origin but sometimes also containing solid and liquid animal wastes. Partially humified and mineralized under the action of soil microflora, the organic manure acts primarily on the physical and biophysical components of soil fertility. A very broad term, it covers manures made from cattle dung, excreta of other animals, other animal wastes, rural and urban wastes, crop residues, and green manures. Concentrated organic manures, such as oilcakes, slaughterhouse wastes, fishmeal, guano and poultry manures are comparatively rich in NPK. The beneficial effects of organic manure go beyond the supply of nutrients – which in many instances is relatively small – by the enhancement of soil structure, water storage, CEC and biological activity. Interchangeable with organic fertilizers. Examples: compost and FYM. See also see compost, and organic fertilizer.

Peat

A dark brown or black plant residue produced by the partial decomposition and disintegration of mosses, sedges, trees and other plants. Commonly used as mixing material because of its water-retaining properties. Accepted as the best available carrier of biofertilizers. Indian peat contains 54 percent organic C, compared with 65 percent in Australian peat and 86 percent in American peat. Average composition of Indian peat is 54.2 percent C, 5.7 percent H and 1.5 percent N. It has a WHC of 149 percent, a bulk density of 2.18 g/cm^3, and a total surface area 647 m^2/g. Used in the preparation of organic fertilizers.

Phosphate-solubilizing micro-organisms (PSM)

Bacteria, fungi and actinomycetes that can solubilize insoluble forms of P. P-solubilizing bacteria (PSB) include *Bacillus megatherium* var. *phosphaticum, Bacillus polymyxa, Bacillus subtilis, Pseudomonas striata, Agrobacterium* sp., and *Acetobacter diazotrophicus*. P-solubilizing fungi (PSF) include *Aspergillus awamori, Penicillium digitatum, Penicillium bilaji*, and yeast (*Saccharomyces* sp.). P-solubilizing actinomycetes (PSA) include *Streptomyces* sp., and *Nocardia* sp. Generally, PSM secrete organic acids that dissolve insoluble phosphate. These microbes help in the solubilization of P from PR and other sparingly soluble forms of soil P by decreasing their particles size, reducing it to nearly amorphous forms. See also biofertilizer.

Phosphocompost

P-enriched compost. A type of enriched compost or fortified organic manure. It can be prepared through composting in which wastes are composted along with 12.5 or 25 percent suitable PR for 3–4 months. Preparation of one type of phosphocompost includes: crop waste 60 percent, animal dung 15 percent, FYM 2 percent, soil 2 percent, PR 15 percent, iron pyrites 5 percent, and urea 1 percent. Using an example from India, the following materials are needed to produce 1 000 tonnes phosphocompost on dry basis:

- 800 tonnes organic refuse, crop residues, leaves, grasses, weeds, etc.;
- 100 tonnes cattle dung or biogas slurry;
- 100 tonnes soil;
- 50 tonnes well-decomposed FYM/compost/ sewage sludge
- 265 tonnes suitable PR.

Their mixture is allowed to decompose in pits for three months. The contents are mixed together after 10, 20 and 45 days. Phosphocompost is ready in about three months. It contains 6–8 percent P_2O_5. During composting, about 50 percent of the insoluble P of the PR is converted into citrate-soluble P. This also provides a potential avenue for the gainful utilization of low-grade PR.

Potassium chloride (KCl)

Most common K fertilizer, contains 58–62 percent K_2O and about 48 percent Cl. Readily water soluble. Critical relative humidity of 84 percent at 30 °C. It has a higher salt index than potassium sulphate. Commercially called MOP. Typical internationally accepted technical specifications of particle size state that 95 percent of the material shall pass through 1.7-mm IS sieve and be retained on 0.25-mm IS sieve. Standard specifications of potassium chloride/MOP based on Indian experience are:

- moisture, percent by weight, maximum: 0.5;
- water-soluble potash (as K_2O), percent by weight, minimum: 60.0;
- sodium as NaCl, percent by weight (on dry basis), maximum: 3.5;
- particle size: minimum 65 percent of the material shall pass through 1.7-mm IS sieve and be retained on 0.25-mm IS sieve.

Potassium magnesium sulphate

A fertilizer providing K, Mg and S (22 percent K_2O, 11 percent Mg or 17 percent MgO and 22 percent S) all in plant-available form. Formula: $K_2SO_4.2MgSO_4$. It is a neutral salt as regards its effect on soil pH and contains less than 1.5 percent chloride. It should not be mixed with urea or CAN. Standard specifications of potassium magnesium sulphate based on Indian experience are:

- moisture, percent by weight, maximum: 0.5;
- potash content (as K_2O), percent by weight, minimum: 22.0;
- magnesium (as MgO), percent by weight, minimum: 18.0;
- sulphur (as S), percent by weight, minimum: 20.0;
- total chloride (as Cl), percent by weight (on dry basis), maximum: 2.5;

- sodium (as NaCl), percent by weight (on dry basis), maximum: 2.0.

Potassium sulphate (SOP)

An important source of K (50 percent K_2O) and S (18 percent), both in readily plant-available form. Formula: K_2SO_4. Particularly suitable for crops that are sensitive to chloride in place of potassium chloride. Very low salt index (46.1) compared with 116.3 in the case of MOP on material basis. It also stores well under damp conditions. SOP should not be mixed with CAN or urea. Typical internationally accepted technical specifications of SOP include maximum moisture content of 1 percent by weight and a maximum Na content as NaCl of 1.0 percent by weight. In addition, particle size specifications are that 90 percent of the material shall pass through 4-mm IS sieve and be retained on 1-mm IS sieve. Furthermore, not more than 5 percent material shall be below 1 mm in size. Standard specifications of potassium sulphate (SOP) based on Indian experience are:

- moisture, percent by weight, maximum: 1.5;
- potash (as K_2O), percent by weight, minimum: 50.0;
- sulphur (as S), percent by weight, minimum: 17.5;
- total chlorides (as Cl), percent by weight (on dry basis), maximum: 2.5;
- sodium (as NaCl), percent by weight (on dry basis), maximum: 2.0;

Precision farming

A farming system that uses GPS technology involving satellites and sensors on the ground and intensive information management tools to understand variations in resource conditions within fields. This information is used to apply fertilizers and other inputs more precisely and to predict crop yields more accurately.

Press mud

A by-product of sugar factories. Residue obtained by filtration of the precipitated impurities that settle out in the process of clarification of the mixed juice from sugar cane. Forms a cake of variable moisture content. The material has 55–75 percent moisture, is soft and spongy, light weight, amorphous and dark brown, and it can readily absorb moisture when dry. Depending on the process used in the sugar factory, it can be either sulphitation press mud or carbonation press mud. It contains 1.2 percent N, 2.1–2.4 percent P_2O_5, 2.0 percent K_2O, 238–288 ppm Zn and 112–132 ppm Cu. Material from factories using sulphitation process is a good source of S. Press mud from sugar factories using the carbonation process can find use as a liming material. Used as manure, as a soil amendment and as potential carrier of biofertilizer. Also known as filter cake, filter press cake, filter muck, mill mud, filter mud and filter press mud.

Prill

Spherical particle obtained by solidification of falling droplets of fertilizer during manufacture. Example: prilled urea.

Rhizobium biofertilizer

An artificially prepared *Rhizobium* culture used for seed dressing of legumes before sowing. A specific *Rhizobium* culture for a specific legume crop which has high ability for infection, nodulation, N_2 fixation and for which antibiotic resistance is needed. First commercial *Rhizobium* biofertilizer was produced as "Nitragin" in the United States of America in 1895.

Seaweeds

These are red, brown or green algae living in or by the sea. Agar agar is the product of red algae (*Rhodophyceae*). Seaweeds like *Ascophyllum nodosum*, *Laminaria digitata* and *Fucus serratus* contain gibberellin, auxins, cytokinin, etc. and are used as liquid organic fertilizer with or without fortification with minerals in many countries. Their role is more of a plant growth stimulant rather than of a nutrient supplier.

Sewage sludge

End product of the fermentation (aerobic or anaerobic) of sewage. Semi-solid product and a potential organic manure. Its general composition is 1.1–2.3 percent N, 0.8–2.1 percent P_2O_5 and 0.5–1.7 percent K_2O. It also contains Na, Ca, S, several micronutrients, toxic heavy metals, and Al. Usually, the concentration of most of these is higher in anaerobic than in aerobic sewage sludge.

Slow-release fertilizer

A fertilizer that is not readily soluble but releases its nutrients slowly over a period of time. Usually, some N fertilizers and micronutrient frits are slow release. Examples: isobutylidene diurea (IBDU), oxamide, and crotonylidene diurea (CDU). Similar to controlled-release fertilizers.

Slurry

Semi-liquid effluent from livestock sheds, consisting of urine and faeces, possibly diluted with water. Can be used as a fertilizer and as an ingredient during composting.

Soil amendment

A substance added to a poor soil to improve its fertility and more particularly its physico-chemical condition by alleviating excessive acidity, alkalinity, salinity, compactness, etc. Crop residues and bulky organic manures can be used as amendments to add nutrients and improve soil physical properties. An amendment usually incorporates plant nutrients. However, several soil amendments have a profound effect on the availability of P, Ca, Mg and micronutrients because of their effect on soil pH. Examples: lime for neutralizing excess soil acidity, and gypsum for reducing excess of alkalinity/sodicity.

Soil fertility

The component of soil productivity that deals with its available nutrient status, its ability to provide nutrients out of its own reserves for crop production and reactions with external nutrient additions. Its assessment is useful for deciding fertilizer application rates, which is the main function of soil testing laboratories. Fertilizers are needed where soil fertility is low and inadequate to support desired level of plant production. Aim of fertilizer application is to increase soil fertility. See also soil test.

Soil test

A rapid but reproducible measurement (usually chemical) made on a soil sample to assess its fertility status for a particular nutrient. Fertilizer recommendations when made for a specific field on the basis of soil tests are more balanced and more profitable than blanket/general recommendations. The higher the soil test value, the lower the fertilizer requirement and vice versa. A soil test has to be calibrated against crop response, which should result in a significant correlation before the soil test can be used for making fertilizer recommendations. Examples: Bray and Kurtz P_1 test for available P, DTPA – extractable test for Zn, and hot water extraction for available B. See also soil fertility.

Solution fertilizer

Liquid fertilizer free of solid particles. See also liquid fertilizers.

Straight fertilizer

A traditional term referring to fertilizers that contain (and are used for) one major nutrient (traditionally N, P or K) as opposed to multinutrient fertilizers. For secondary nutrients, products containing elemental S, magnesium sulphate, calcium oxide, etc. In micronutrients, borax, Zn or Fe chelates and sulphate salts of micronutrients are straight fertilizers, although the phrase is not often used for micronutrient carriers. Not a straightforward term because many "straight fertilizers" also contain other essential plant nutrients, such as S.

Sulphate of potash (SOP)

See potassium sulphate.

Sulphur bentonite

An elemental S product in which 10–15 percent bentonite clay is included during manufacturing for ease in granulation, pastille formation, handling and application. Materials with a range of particle size, hence, decomposition rates are variable. Agronomic efficiency not very different from that of elemental S.

Superphosphate

Class of fertilizers obtained by reacting PR with sulphuric acid or with phosphoric acid. Common types are single superphosphate (SSP) containing 16 percent P_2O_5

and 11–12 percent S, and triple superphosphate (TSP) containing 46 percent P_2O_5 and little S.

Suspension fertilizer
A two-phase fertilizer in which solid particles are maintained in suspension in the aqueous phase. A type of liquid fertilizer. Addition of a clay attapulgite facilitates keeping its constituents in suspension form. See also liquid fertilizers.

Toxicity
Adverse reaction of plants caused by certain constituents in the soil or water that are taken up by the plants and accumulated to high concentration. This results in plant damage, reduced yields or even death of plants. The degree of plant damage depends on the element, its uptake, concentration in the plant tissue and the sensitivity of the crop.

Triple superphosphate (TSP)
A fertilizer obtained by treating PR with phosphoric acid and containing about 46 percent P_2O_5, mainly in water-soluble form. Unlike SSP, it contains little S.

Urban compost
Compost prepared from urban and industrial wastes, city garbage, sewage sludge, etc. Its typical composition is 1.5–2.0 percent N, 1.0 percent P_2O_5 and 1.5 percent K_2O. Commercially prepared urban compost has been reported to contain 1 percent Fe, about 375 ppm Cu, 705 ppm Zn, 740 ppm Mn and small amounts of other micronutrients. Also termed town compost (as opposed to rural compost). See also compost.

Urea
A white, crystalline, non-protein organic N compound made synthetically from ammonia and CO_2. First synthesized by Wholer in 1928. Formula: $CO(NH_2)_2$. This non-electrolyte compound contains readily water-soluble 46 percent N, all in amide (NH_2) form. Most concentrated solid N fertilizer. Produced as prills or granules of varying sizes. It is hydrolysed in the soil by the enzyme urease to furnish ammonium and then nitrate ions. Used as solid N fertilizer for soils, for foliar application and as an ingredient of NP/NPK complexes. Leaves behind an acidic effect in soils. Sometimes fortified with Zn and Fe.

$$\text{Urea} \xrightarrow{\text{urease}} \text{Ammonium carbamate} \longrightarrow \text{Ammonia}$$

Vermicompost (also wormicompost)
An important type of compost and organic fertilizer that contains earthworm cocoons, excreta, beneficial micro-organisms, actinomycetes, plant nutrients, organic matter, enzymes, hormones, etc. An organic fertilizer produced by

earthworms and containing on an average 0.6 percent N, 1.5 percent P_2O_5 and 0.4 percent K_2O. In addition to NPK, it is also a source of micronutrients, having an average of 22 ppm Fe, 13 ppm Zn, 19 ppm Mn and 6 ppm Cu. A product of variable composition. Vermicomposting is an appropriate technique for the disposal of non-toxic solid and liquid organic wastes. It helps in cost-effective and efficient recycling of animal wastes (poultry, horse, piggery excreta and cattle dung), agricultural residues and industrial wastes using low energy. It improves soil health, and, thus, productivity.

Zinc sulphate

Common Zn-containing fertilizer. Produced as $ZnSO_4.7H_2O$ (21 percent Zn) or $ZnSO_4.H_2O$ (33 percent Zn). Used for soil or foliar application. Also provides S. Standard specifications of zinc sulphate heptahydrate based on Indian experience are:

- zinc (as Zn), percent by weight, minimum: 21.0;
- sulphur (as S), percent by weight, minimum: 10.0;
- cadmium (as Cd), percent by weight, maximum: 0.0025;
- arsenic (as As), percent by weight, maximum: 0.01;
- lead (as Pb), percent by weight, maximum: 0.003;
- copper (as Cu), percent by weight, maximum: 0.1;
- magnesium (as Mg), percent by weight, maximum: 0.5;
- matter insoluble in water, percent by weight, maximum: 1.0;
- pH: not less than 4.0.

For a more detailed glossary

FAO–FDCO integrated nutrient management – a glossary of terms by Tandon and Roy (2004) (also available at http://www.fao.org.landandwater/agll/ipns/index_en.jsp).

Bibliography

lloway, B.J. 2004. *Zinc in soils and crop nutrition*. Brussels, International Zinc Association. 128 pp.

Ange, A.L. 1993. Environmental risks of fertiliser use, over use or misuse. *Proc. regional FADINAP seminar on fertilisation and the environment*, pp. 49–112. Bangkok.

Arnon, D.I. & Stout, P.R. 1939. The essentiality of certain elements in minute quantity for plants with special reference to copper. *Plant Physiol.*, 14: 371–375.

Aulakh, M.S. 1985. Content and uptake of nutrients by pulses and oilseed crops. *Ind. J. Ecol.*, 12(2): 238–242.

Bar-Yosef, B. 1999. Advances in fertigation. *Adv. Agron.*, 65: 1– 77.

Beaufils, E.R. 1971. Physiological diagnosis as a guide for improving maize production based on principles developed for rubber trees. *Fert. Sve. S. Afr. J.*, 1: 1–30.

Bhumbla, D.R. 1974. Soil amendments for efficient fertiliser use. *Fert. News*, 19(7): 21–26.

Blair, G.J. 1979. *Sulphur in the tropics*. Tech. Bull. IFDC-T 12. Muscle Shoals, Alabama, USA, Sulphur Institute and International Fertiliser Development Centre.

Bolt, G.H. & Bruggenwert, M.G.M., eds. 1976. *Soil chemistry. Part A. Basic elements*. Amsterdam, The Netherlands, Elsevier. 281 pp.

Borlaug, N.E. 1993. Quoted in N.E. Borlaug & C.R. Dowswell, eds. *Fertilizer: to nourish infertile soil that feeds a fertile population that crowds a fragile world*. 61st IFA Annual Conference. Paris, IFA.

Borlaug, N.E. 1997. Quoted in IFA/UNEP, ed. *Mineral fertilizer use and the environment*. Paris, IFA. 51 pp.

Brady, N.C. & Weil, R.R. 1996. *The nature and properties of soils*. 11th edition. London, Prentice Hall International.

Brar, B.S. & Pasricha, N.S. 1998. Long-term studies on integrated use of organic and inorganic fertilizers in maize-wheat-cowpea cropping system on alluvial soil in Punjab. *In* A. Swarup, ed. *Proceedings of national workshop on long-term soil fertility management through integrated plant nutrient supply*, pp. 154–168. Bhopal, India, Indian Institute of Soil Science.

Brinton, W.F. 2000. *Compost quality standards and guidelines*. Report to NYSAR. Woods End Research Laboratory (available at http://compost.css.cornell.edu).

Bruulsema, T.W. 2004. Understanding the science behind fertiliser recommendation. *In*: *Better crops with plant food, No. 3*, pp. 16–19. Norcross, USA, PPI.

Bruulsema, T.W., Fixen, P.E. & Snyder, C.S. 2004. Fertiliser nutrient recovery in sustainable cropping systems. *Better crops with plant food, No. 4*, pp. 15–17., Norcross, USA, PPI.

Clements, H.F. 1980. *Sugarcane crop logging and crop control*. USA, University Press. 520 pp.

Cooke, G.W. 1982. *Fertilizing for maximum yield.* 3rd edition. London, Granada. 465 pp.

Darmody, R.G. & Peck, T.R. 1993. *Soil organic matter changes through time at the University of Illinois Morrow Plots.* USA, Department of Agronomy, University of Illinois at Urbana-Champaign.

DeClerck, F. & Singer, M.J. 2003. Looking back 60 years, California soils maintain overall chemical quality. *Calif. Agric.*, 57(2): 38–41.

Dobermann, A. & Witt, C. 2004. The evolution of site-specific nutrient management in irrigated rice systems of Asia. *In* A. Dobermann, C. Witt & D. Dawe, eds. *Increasing productivity of intensive rice systems through site-specific nutrient management*, pp. 75–99. Los Banos, Philippines, International Rice Research Institute.

Driessen, P.M. & Dudal, R. 1991. *The major soils of the world.* Wageningen, The Netherlands, Agricultural University.

Eckard, R.J., Chen, D., White, R.E. & Chapman, D.F. 2003. Gaseous nitrogen loss from temperate perennial grass and clover dairy pastures in south-eastern Australia. *Austral. J. Agric. Res.*, 54: 561–570.

Evans, L.T. 2003. Agricultural intensification and sustainability. *Outlook on Agric.*, 32: 83–89.

Fairhurst, T & Hardter, R. 2003. *Oil palm. Management for large and sustainable yields.* PPI/PPIC and IPI.

Fairhurst, T. & Witt, C. 2002. *A practical guide to nutrient management.* PPI/PPIC/IRRI, Singapore.

FAO. 1980. *Soil and plant testing as a basis of fertilizer recommendations.* Soils Bulletin No. 38/2. Rome. 100 pp.

FAO. 1983. *Micronutrients,* by J.C. Katyal, & N.S. Randhawa. Fertilizer and Plant Nutrition Bulletin No. 7. Rome. 82 pp.

FAO. 1989. *Sustainable agricultural production. Implications for agricultural research*, by Technical Advisory Committee. FAO Research and Technology Paper No. 4. Rome. 131 pp.

FAO. 1991. *World soil resources. An explanatory note on the FAO World Soil Resources Map at 1:25 000 000 scale*. Rome.

FAO. 1992. *Secondary nutrients,* by G. Kemmler. AGL/MISC/20/92. Rome. 73 pp.

FAO. 1993a. *Agriculture towards 2010.* FAO Conference Report C 93/24. Rome.

FAO. 1993b. Plant nutrition systems on farm and soil fertility management, by A.L. Ange. *In: The role of IPNS in sustainable and environmentally sound agricultural development*, pp. 47–76. FAO–RAPA Publication No. 1993/12. Bangkok.

FAO. 1995. *Integrated plant nutrition system.* FAO Fertilizer and Plant Nutrition Bulletin No. 12. Rome. 426 pp.

FAO. 1996. *World Food Summit, Rome Declaration and Plan of Action.* Rome. 43 pp.

FAO. 1998. *Guide to efficient plant nutrient management.* Rome. 19 pp.

FAO. 1999. *Integrated soil management in Southern and East Africa.* AGL/MISC/23/99. Rome.

FAO. 2000a. *Agriculture: towards 2015/30.* Technical Interim Report. Rome.

FAO. 2000b. *Fertilizer requirements in 2015 and 2030.* Rome. 29 pp.

FAO. 2000c. *Simple soil, water and plant testing techniques for soil resource management*, by J.A. Adepetu, H. Nabhan & A. Osinubi. AGL/MISC/28/2000. IITA and FAO, Rome. 158 pp.

FAO. 2001a. *Soil and nutrient management in SSA in support of the soil fertility initiative*. AGL/Misc.31/01. Rome. 379 pp.

FAO. 2001b. *Soil fertility management in support of food security in SSA*. Rome. 55 pp.

FAO. 2003a. *On-farm composting methods*, by R.V. Misra, R.N. Roy & H. Hiraoka. Land and Water Discussion Paper No. 2. Rome. 35 pp.

FAO. 2003b. *Assessment of soil nutrient balance – approaches and methodologies*, by R.N. Roy, R.V. Misra, J.P. Lesschen & E.M. Smaling. Fertilizer and Plant Nutrition Bulletin No. 14. Rome. 87 pp.

FAO. 2003c. Economic and environmental impact of improved nitrogen management in Asian rice-farming systems, by R.N. Roy & R.V. Misra. *In: Sustainable rice production for food security*. Proc. 20th Session International Rice Commission, Bangkok, 23–26 July 2002. Rome.

FAO. 2004a. *FAO fertilizer yearbook*. Vol. 53 – 2003. Rome. 214 pp

FAO. 2004b. *Use of phosphate rocks for sustainable agriculture*. FAO Fertilizer and Plant Nutrition Bulletin No. 13. Rome. 148 pp.

FAO. 2005. *Current world fertilizer trends and outlook to 2009/2010*. Rome. (in press)

FAO / Fertilizer Industry Coordination Committee (FAO/FIAC). 1983. *Economic, financial and budget aspects for fertilizer use and development*. Rome.

FAO / Fertilizer Industry Coordination Committee (FAO/FIAC). 1992. *The importance of crop production*. Rome. 62 pp.

FAO / International Fertilizer Industry Association (FAO/IFA). 1999. *Fertilizer strategies*. Rome. 98 pp.

FAO / International Fertilizer Industry Association (FAO/IFA). 2000. *Fertilizers and their use – a pocket guide for extension officers*. 4th edition, Rome. 70 pp.

FAO / International Fertilizer Industry Association (FAO/IFA). 2001 *Global estimates of gaseous emissions from agricultural land*. Rome. 85 pp.

FAO / International Food Policy Research Institute (FAO/IFPRI). 1998. *Proc. IFPRI/ FAO workshop on plant nutrient management, food security and sustainable agriculture: the future through 2020*. Rome. 243 pp.

FAO / Regional Asia Pacific (FAO/RAPA). 1992. *Environmental issues in land and water development*. RAPA Publication No. 1992/8. Bangkok.

FAO / United Nations Environment Programme (FAO/UNEP). 1999. *The future of our land: facing the challenge*. Rome, 71 pp.

FAOSTAT. FAO statistical database online (available at http://faostat.fao.org).

Fertiliser Association of India (FAI). 2004. *Fertiliser statistics for 2003-04*. New Delhi.

Fertilizer Advisory, Development and Information Network for Asia and the Pacific (FADINAP). 1993. *Proceedings of the regional FADINAP seminar on fertilisation and the environment*. Bangkok, ESCAP/FAO/UNIDO. 2 pp.

Finck, A. 1982. *Fertilizers and fertilization. Introduction and practical guide to crop fertilization*. Weinheim, Germany, Verlag. Chimie.

Finck, A. 1992. *Fertilizers and fertilization. Introduction and practical guide to crop fertilization.* 2nd edition (German). Weinheim, Germany, Verlag. Chimie.

Finck, A. 2001. Fertilizers and their effective use. *In: World fertilizer use manual.* IFA (available at http://www.fertilizer.org).

Finck, A. 2006. Soil nutrient management for plant growth. *In* B. Warkentin, ed. *Footprints in the soil.* IUSS Monograph. The Netherlands, Elsevier. (in press)

Goswami, N.N. 1976. Management of fertiliser phosphorus on cropping system basis. *Fert. News*, 21(9): 56–59, 63.

Graham, R.D., Welch, R.M. & Bouis, H.E. 2001. Addressing micronutrient malnutrition through enhancing the nutritional quality of staple foods: principles, perspectives and knowledge gaps. *Adv. Agron.*, 70 : 77–142.

Grewal, J.S. & Sharma, R.C. 1978. Advances in soil fertility and fertilizer use. *Fert. News*, 23(9): 38–45.

Grewal, J.S. & Sharma, R.C. 1993. Potato based cropping systems can be profitable. *Ind. Farm.*, 42(90): 11.

Gros, A. 1967. *Guide pratique de la fertilisation*. Paris, La Maison Rustique.

Gupta, R.K & Abrol, I.P. 1990. Salt–affected soils – their reclamation and management for crop production. *Adv. Soil Sci.*, 12: 233–275.

Gupta, V.K. 1995. Zinc research and agricultural production. *In* H.L.S. Tandon, ed. *Micronutrient research and agricultural production*, pp. 132–164. New Delhi, Fertiliser Development and Consultation Organisation.

Howeler, R.H. 1978. The mineral nutrition and fertilization of cassava. *In: Cassava production course*, pp. 247–292. Cali, Colombia, CIAT.

Hunsigi, G. 1993. Fertiliser management in sugarcane. *In* H.L.S. Tandon, ed. *Fertiliser management in commercial crops*, pp. 1–25. New Delhi, Fertiliser Development and Consultation Organisation.

International Fertilizer Industry Association (IFA). 1986. *The fertilizer industry – the key to world food supplies.* Paris. 100 pp.

International Fertilizer Industry Association (IFA). 1992. *World fertilizer use manual.* Paris. 632 pp.

International Fertilizer Industry Association / European Fertilizer Manufacturers Association (IFA/EFMA). 1998. *Best agricultural practices to optimize fertiliser use in Europe.* Paris. 3 pp.

International Fertilizer Industry Association / United Nations Environment Programme (IFA/UNEP). 1998. *The fertiliser industry, world food supplies and the environment.* 66 pp.

International Food Policy Research Institute (IFPRI). 1995a. *Biophysical limits to global food production (2020 Vision).* Washington, DC. 2 pp.

International Food Policy Research Institute (IFPRI). 1995b. *The right to food: widely acknowledged and poorly protected (2020 Vision).* Washington, DC. 2 pp.

International Food Policy Research Institute (IFPRI). 1997. *Is world population growth slowing? (2020 Vision).* Washington, DC. 2 pp.

International Food Policy Research Institute (IFPRI). 1999. *World food prospects: critical issues for the early 21st century (2020 Vision).* Washington, DC. 32 pp.

International Food Policy Research Institute (IFPRI). 2002. *Achieving sustainable food security for all by 2020.* International Food Policy Research Institute, Washington, DC. 14 pp.

Imas, P. 2004. New insights into chloride's role on intensive crop production. *New AG Int.*, June. 60.

IMPHOS/ICARDA. 2000. *Plant nutrient management under pressurized irrigation systems in the Mediterranean region.* Morocco. 400 pp.

International Federation of Organic Agriculture Movements (IFOAM). 1998. *Basic standards for organic production and processing.* Tholey–Theley, Germany.

Johnston, A.E. 1997. The values of long-term field experiments in agricultural, ecological and environmental research. *Adv. Agron.*, 59: 291–293.

Joshi, Y.C. 1980. Sodium/potassium index of wheat seedlings in relation to sodicity tolerance. *In: Proc. Indo-Hungarian symposium on salt-affected soils*, pp. 457–460. Karnal, India.

Juwarkar, A.S., Shende, A., Thawale, P.R., Satyanarayanan, S., Deshbratar, P.B., Bal, A.S. & Juwarkar, A. 1992. Biological and industrial wastes as sources of plant nutrients. *In* H.L.S. Tandon, ed. *Fertilisers, organic manures, recyclable wastes and biofertilisers*, pp. 72–90. New Delhi, Fertiliser Development and Consultation Organisation.

Kaarstad, O. 1997. Fertilizer's significance for cereal production and cereal yields from 1950 to 1995. *In* J.J. Mortvedt, ed. *International symposium on fertilization and the environment*, pp. 56-64. Haifa, Israel, Technion-Israel Institute of Technology.

Kant, S. & Kafkafi, U. 2002. Potassium and abiotic stresses in plants. *In* N.S. Pasricha & S.K. Bansal, eds. *Potassium for sustainable crop production*, pp. 233–251. Gurgaon, India, Potash Research Institute of India and International Potash Institute.

Kanwar, J.S. 1983. *Groundnut nutrition and fertilizer responses in India.* New Delhi, ICAR. 135 pp.

Karki, A.B., Gautam, K.M. & Karki, A. 1994. Biogas installation from elephant dung at Machan Wildlife Resort, Chitwan, Nepal. *Biogas News.*, 45.

Katyal, J.C. & Deb, D.L. 1982. Nutrient transformations in soils-micronutrients. *In: Review of soils research in India. Part I*, pp. 146–159. New Delhi, Indian Society of Soil Science.

Laegreid, M., Bockman, O.C. & Kaarstad, O. 1999. *Agriculture, fertilizers and the environment.* Cambridge, UK, CABI Publishing, University Press. 294 pp.

Mandal, A.K. & Pal, H. 1993. Fertiliser management in jute. *In* H.L.S. Tandon, ed. *Fertiliser management in commercial crops*, pp. 47–65. New Delhi, Fertiliser Development and Consultation Organisation.

Meelu, O.P. 1976. Fertiliser and crop management technology for rainfed areas of Punjab. *Fert. News*, 21(9): 34–38.

Messick, D.L. 2003. *Sulphur fertilisers: a global perspective.* TSI-FAI-IFA Workshop on Sulphur in Balanced Fertilisation, New Delhi.

Messick, D.L., de Brey, C. & Fan, M.X. 2002. *Sources of sulphur, their processing and use in fertiliser manufacture.* Proceedings No. 502. York, UK, The International Fertiliser Society. 24 pp.

Mosier, A.R., Syers, J.K. & Freney, J.R. 2004. Nitrogen fertilizer: an essential component of increased food, feed, and fibre production. *In* A.R. Mosier, K. Syers & J.R. Freney, eds. *Agriculture and the nitrogen cycle.* Washington, DC, Island Press. 291 pp.

Motsara, M.R., Bhattacharayya, P. & Srivastava, B. 1995. *Biofertiliser technology, marketing and usage*. New Delhi, Fertiliser Development and Consultation Organisation. 184 pp.

Nambiar, K.K.M. 1994. *Soil fertility and crop productivity under long-term fertiliser use in India*. New Delhi, Indian Council of Agric. Research. 144 pp.

Neue, H.E. & Mamaril, C.P. 1985. Zinc, sulphur and other micronutrients in wetland soils. *In: Wetland soil, characterization, classification and utilization*, pp. 307–319. Los Baños, Philippines, IRRI.

Oenema, O. & Heinen, M. 1999. Uncertainties in nutrient budgets due to biases and errors. *In* E.M.A. Smaling, O. Oenema & L.O. Fresco, eds. *Nutrient disequilibria in agroecosystems – concepts and case studies.* Wallingford, UK, CABI Publishing.

Palaniappan, S.P. 1992. Green manuring: nutrient potentials and management. *In* H.L.S. Tandon, ed. *Fertilisers, organic manures, recyclable wastes and biofertilisers*, pp. 52–71. New Delhi, Fertiliser Development and Consultation Organisation.

Pandey, R.K. 1991. *A primer on organic-based rice farming*. Manila, IRRI. 201 pp.

Pathak, H., Singh, U.K., Patra, A.K. & Kalra, N. 2004. Fertiliser use efficiency to improve environmental quality. *Fert. News*, 49(4): 95–106.

Peck, T.R. & Melsted, S.W. 1967. Field sampling for soil testing. *In: Soil testing and plant analysis. Soil testing. Part I*, pp. 25–35. Madison, USA, Soil Science Society of America.

Perrenoud, S. 1990. *Potassium and plant health*. 2nd edition. IPI Research Topic No. 3. Bern, International Potash Institute. 365 pp.

Pierce, F.J. & Nowak, P. 1999. Aspects of precision agriculture. *Adv. Agron.*, 67: 1–85.

Pillai, N.G. & Davis, T.A. 1963. Exhaust of macronutrients by coconut palm. *Ind. Coco. J.*, 16: 18–19.

Prasad, R.N. & Biswas, P.P. 2000. Soil resources of India. *In* Singh, G.B & Sharma, B.R., eds. *Fifty years of natural resource management research*, pp. 13–30. New Delhi, Indian Council of Agricultural Research.

Rana, R.S. 1986. Breeding crop varieties for salt-affected soils. *In: Approaches for incorporating drought and salinity resistance in crop plants*, pp. 24–55. New Delhi, Oxford and IBH.

Roy, R.N. 1992. Integrated plant nutrition systems-an overview. *In* H.L.S. Tandon, ed. *Fertilisers, organic manures, recyclable wastes and biofertilisers*, pp. 1–11. New Delhi, Fertiliser Development and Consultation Organisation.

Roy, R.N., Misra, R.V. & Montanez, A. 2002. Decreasing reliance on mineral nitrogen – yet more food. *Ambio*, 31(2): 177–183.

Sarkar, A.K. 2000. *Long term effects of fertilisers, manure and amendment on crop production and soil fertility*. Technical Bulletin No. 2/2000. Ranchi, India, Birsa Agricultural University. 57 pp.

Schroeder, D. 1984. *Soils – facts and concepts.* Bern, International Potash Institute. 140 pp.

Sekhon, G.S. & Velayutham, M. 2002. Soil fertility evaluation. *In: Fundamentals of soil science*, pp. 419–432. New Delhi, Indian Society of Soil Science.

Sharma, P.D. & Biswas, P.P. 2004. IPNS packages for dominant cropping systems in different agro climatic regions of the country. *Fert. News*, 49(10): 43–47.

Sheldon, R.P. 1987. Industrial minerals, with emphasis on phosphate rock. *In* D.J. McLaren & B.J. Skinner, eds. *Resources and world development*, pp. 347–361. New York, USA, John Wiley & Son Ltd.

Shorrocks, V.M. 1984. *Boron deficiency, its prevention and cure.* London, Borax Holding Ltd. 43 pp.

Shorrocks, V.M. & Alloway, B.J. 1988. *Copper in plant, animal and human nutrition.* Herts, UK. Copper Development Association. 84 pp.

Singh, B. 1976. Punjab Agricultural University, Ludhiana, India. (PhD thesis)

Singh, B. 1996. Environmental aspects of nitrogen use in agriculture. *In* H.L.S. Tandon, ed. *Nitrogen research and crop production*, pp. 159–170. New Delhi, Fertiliser Development and Consultation Organisation.

Smaling, E.M.A. 1993. Soil nutrient depletion in Sub-Saharan Africa. *In* H. van Reulen & W.H. Prins, eds. *The role of plant nutrients for sustainable food crop production in Sub-Saharan Africa.* Leidschendan, The Netherlands.

Smil, V. 1999. Nitrogen in crop production: an account of global flows. *Biogeochem. Cyc.*, 13: 647–662.

Stewart, M. 2002. Balanced fertilisation and the environment. *Crop nutrients and the environment* (available at www.ppi--far.org).

Subba Rao, A., Rupa, T.R. & Srivastava, S. 2001. Assessing potassium availability in Indian soils. *In* N.S. Pasricha & S.K. Bansal, eds. *Potassium in Indian agriculture*, pp. 125–157. Gurgaon, India, Potash Research Institute of India and International Potash Institute.

Swarup, A. 2000. Long term fertiliser experiments. *In* G.B. Singh & B.R. Sharma, eds. *Fifty years of natural resource management research*, pp. 221–230. Indian Council of Agricultural Research, New Delhi.

Tanaka, A. 1975. *Extension Bulletin 57.* Taipei, Taiwan Province of China, Food and Fertilizer Technology Centre.

Tandon, H.L.S. 1989. *Secondary and micronutrient recommendations for soils and crops.* New Delhi, Fertiliser Development and Consultation Organisation. 104 pp.

Tandon, H.L.S. 1992. Fertilisers and their integration with organics and biofertilisers. *In* H.L.S. Tandon, ed. *Fertilisers, organic manures, recyclable wastes and biofertilisers*, pp. 12–35. New Delhi, Fertiliser Development and Consultation Organisation.

Tandon, H.L.S. 1999. *Micronutrients in soils, crops and fertilisers.* 2nd edition. New Delhi, Fertiliser Development and Consultation Organisation. 177 pp.

Tandon, H.L.S. 2004. *Fertilisers in Indian agriculture – from 20th to 21st century.* New Delhi, Fertiliser Development and Consultation Organisation. 239 pp.

Tandon, H.L.S. & Ranganathan, V. 1988. Fertiliser management in plantation crops. *In* H.L.S. Tandon, ed. *Fertiliser management in plantation crops*, pp. 26–60. New Delhi, Fertiliser Development and Consultation Organisation.

Tisdale, S.L., Nelson, W.L. & Beaton, J.D. 1985. *Soil fertility and fertilisers.* 4th Edition. New York, USA, Macmillan Publishing Company. 754 pp.

Toynbee. A. 1965. *Four lectures given by Professor Arnold Toynbee in U.A.R.* UAR, Cairo, Public Relations Dept. 116 pp.

Tyagi, N.K. 2000. Management of salt-affected soils. *In* G.B. Singh & B.R. Sharma, eds. *Fifty years of natural resource management research*, pp. 363–401. New Delhi, Indian Council of Agricultural Research.

University of Illinois. 1994. *Illinois agronomy handbook 1995-1996.* Urbana-Champaign, USA. 205 pp.

Van Kauwenbergh, S.J. 1997. *Cadmium and other minor elements in world resources of phosphate rock.* Proc. No. 400. London, The Fertiliser Society.

Vlek, P.L.G. & Vielhauer, K. 1994. Nutrient management strategies in stressed environments. *In* S.M. Virmani, J.C. Katyal, H. Eswaran & I.P. Abrol, eds. *Stressed ecosystems and sustainable agriculture*, pp. 203–229. New Delhi, Oxford and IBH.

von Uexkull, H.R. & Mutert, E. 1993. Principles of balanced fertilisation. *In: Proc. regional seminar on fertilisation and the environment*, pp. 17–31. Bangkok, FADINAP.

Webber, M.D. & Singh, S.S. 1995. Contamination of agricultural soils. *In* D.F. Acton & L.J. Gregorich, eds. *The health of our soils – toward sustainable agriculture in Canada*, pp. 87-96 pp.

Yadav, B.R. & Khera, M.S. 1993. Analysis of irrigation waters. *In* H.L.S. Tandon, ed. *Methods of analysis of soils, plants, waters and fertilisers*, pp. 83–116. New Delhi, Fertiliser Development and Consultation Organisation.

Yadav, R.L., Prasad, K., Dwivedi, B.S., Tomar, R.K. & Singh, A.K. 2000. Cropping systems. *In* G.B. Singh & B.R. Sharma, eds. *Fifty years of natural resource management research*, pp. 411–446. New Delhi, Indian Council of Agricultural Research.

Yoshida, S. 1981. *Fundamentals of rice crop science.* Los Banos, Philippines, IRRI. 269 pp.

Zublina, J.P. 1991. *North Carolina Cooperative extension service publication AG-439-16* (updated 1997, available at http://www.soil.ncsu.edu).

Units and conversion factors

UNITS

1 metre (m) = 100 centimetres (cm) (1.0936 yards)
1 kilometre (km) = 1 000 m (kilo = thousand)
1 litre = 1 000 cubic centimetres (cc) or 1 000 millilitres (ml)
1 milligram (mg) = 1 000 micrograms (μg)
1 gram (g) = 1 000 mg
1 kilogram (kg) = 1 000 g (2.20 pounds (lb))
1 quintal = 100 kg (0.1 tonne)
1 tonne = 1 000 kg
1 hectare (ha) = 10 000 m^2 (2.471 acres)
1 percent = 1 part in 100 parts (1% = 10 000 ppm)
ppm = mg/kg or mg/litre or μg/g
1 bushel wheat (USA) = 27.215 kg
1 bushel maize (USA) = 25.410 kg

Conversion from non-SI unit to SI units

Non-SI unit	Multiply by *	To obtain SI unit
Length		
Inch	2.54	centimetres, cm (100 cm = 1 m)
Foot	0.304	metre, m
Yard	0.9144	metre, m
Statute mile	1.6093	kilometre, km
Area		
Acre	0.405	hectare, ha (10 000 m^2 = 1 ha)
Square foot	9.29×10^{-2}	square metre, m^2
Volume		
Bushel	35.24	litre
Cubic foot	2.83×10^{-2}	cubic metre, m^3
Cubic inch	2.83×10^{-2}	cubic metre, m^3
Gallon (USA)	3.78	litre
Mass		
Ounce (avdp.)	28.4	gram
Pound	0.454	kilogram, kg (10^3 g)
Hundredweight	50.8023	kilogram, kg
Long ton	1.1065	tonne
Short ton	0.90781	tonne
Yield and rate		
Bushel per acre wheat (60 lb)	67.19	kilograms per hectare, kg/ha
Bushel per acre maize (56 lb)	62.71	kilograms per hectare, kg/ha
Bushel per acre barley (48 lb)	53.75	kilograms per hectare, kg/ha
Gallon per acre (USA)	9.35	litres per hectare, litres/ha
Pounds/acre	1.121	kilograms per hectare, kg/ha

* To convert from SI to non-SI units, divide by the factor given.

Conversion from non-SI unit to SI units (Continued)

Non-SI unit	Multiply by *	To obtain SI unit
Pressure		
Atmosphere	0.101	megaPascal, MPa (10^6 Pa)
Bar	0.1	megaPascal, MPa
Temperature		
Degrees Fahrenheit (°F - 32)	0.556	degrees, °C
Energy		
British thermal unit (BTU)	1.05×10^3	joule, J
Calorie	4.19	joule, J

* To convert from SI to non-SI units, divide by the factor given.

Other conversion factors (nutrients)

From	To	Multiply by	From	To	Multiply by
N	Protein	6.25			
P	P_2O_5	2.29	P_2O_5	P	0.436
K	K_2O	1.20	K_2O	K	0.83
Ca	CaO	1.40	CaO	Ca	0.715
Mg	MgO	1.66	MgO	Mg	0.603
S	SO_4	3.0	SO_4	S	0.33
S	SO_3	2.5	SO_3	S	0.44